Solar Fuels

Scrivener Publishing
100 Cummings Center, Suite 541J
Beverly, MA 01915-6106

Advances in Solar Cell Materials and Storage

Series Editor: Nurdan Demirci Sankir and Mehmet Sankir

Scope: Because the use of solar energy as a primary source of energy will exponentially increase for the foreseeable future, this series on *Advances in Solar Cell Materials and Storage* will focus on new and novel solar cell materials and their application for storage. The scope of the series deals with the solution-based manufacturing methods, nanomaterials, organic solar cells, flexible solar cells, batteries and supercapacitors for solar energy storage, and solar cells for space.

Publishers at Scrivener
Martin Scrivener (martin@scrivenerpublishing.com)
Phillip Carmical (pcarmical@scrivenerpublishing.com)

Solar Fuels

Edited by
Nurdan Demirci Sankir
Department of Materials Science and Engineering, TOBB University of Economics and Technology, Ankara, Turkey

and

Mehmet Sankir
Department of Materials Science and Engineering, TOBB University of Economics and Technology, Ankara, Turkey

This edition first published 2023 by John Wiley & Sons, Inc., 111 River Street, Hoboken, NJ 07030, USA and Scrivener Publishing LLC, 100 Cummings Center, Suite 541J, Beverly, MA 01915, USA
© 2023 Scrivener Publishing LLC

For more information about Scrivener publications please visit www.scrivenerpublishing.com.

All rights reserved. No part of this publication may be reproduced, stored in a retrieval system, or transmitted, in any form or by any means, electronic, mechanical, photocopying, recording, or otherwise, except as permitted by law. Advice on how to obtain permission to reuse material from this title is available at http://www.wiley.com/go/permissions.

Wiley Global Headquarters
111 River Street, Hoboken, NJ 07030, USA

For details of our global editorial offices, customer services, and more information about Wiley products visit us at www.wiley.com.

Limit of Liability/Disclaimer of Warranty
While the publisher and authors have used their best efforts in preparing this work, they make no representations or warranties with respect to the accuracy or completeness of the contents of this work and specifically disclaim all warranties, including without limitation any implied warranties of merchantability or fitness for a particular purpose. No warranty may be created or extended by sales representatives, written sales materials, or promotional statements for this work. The fact that an organization, website, or product is referred to in this work as a citation and/or potential source of further information does not mean that the publisher and authors endorse the information or services the organization, website, or product may provide or recommendations it may make. This work is sold with the understanding that the publisher is not engaged in rendering professional services. The advice and strategies contained herein may not be suitable for your situation. You should consult with a specialist where appropriate. Neither the publisher nor authors shall be liable for any loss of profit or any other commercial damages, including but not limited to special, incidental, consequential, or other damages. Further, readers should be aware that websites listed in this work may have changed or disappeared between when this work was written and when it is read.

Library of Congress Cataloging-in-Publication Data

ISBN 978-1-119-75057-4

Cover image: Pixabay.Com
Cover design by Russell Richardson

Set in size of 11pt and Minion Pro by Manila Typesetting Company, Makati, Philippines

Printed in the USA

10 9 8 7 6 5 4 3 2 1

Contents

Preface — xiii

Part I: Solar Thermochemical and Concentrated Solar Approaches — 1

1 Materials Design Directions for Solar Thermochemical Water Splitting — 3
Robert B. Wexler, Ellen B. Stechel and Emily A. Carter
- 1.1 Introduction — 4
  - 1.1.1 Hydrogen via Solar Thermolysis — 7
  - 1.1.2 Hydrogen via Solar Thermochemical Cycles — 8
  - 1.1.3 Thermodynamics — 13
  - 1.1.4 Economics — 16
- 1.2 Theoretical Methods — 17
  - 1.2.1 Oxygen Vacancy Formation Energy — 18
  - 1.2.2 Standard Entropy of Oxygen Vacancy Formation — 22
  - 1.2.3 Stability — 24
  - 1.2.4 Structure — 25
  - 1.2.5 Kinetics — 26
- 1.3 The State-of-the-Art Redox-Active Metal Oxide — 26
- 1.4 Next-Generation Perovskite Redox-Active Materials — 30
- 1.5 Materials Design Directions — 33
  - 1.5.1 Enthalpy Engineering — 33
  - 1.5.2 Entropy Engineering — 37
  - 1.5.3 Stability Engineering — 41
- 1.6 Conclusions — 42
- Acknowledgments — 42
- Appendices — 43
  - Appendix A. Equilibrium Composition for Solar Thermolysis — 43
  - Appendix B. Equilibrium Composition of Ceria — 44
- References — 46

2 Solar Metal Fuels for Future Transportation   65
Youssef Berro and Marianne Balat-Pichelin

  2.1 Introduction   66
    2.1.1 Sustainable Strategies to Address Climate Change   66
    2.1.2 Circular Economy   66
    2.1.3 Sustainable Solar Recycling of Metal Fuels   68
  2.2 Direct Combustion of Solar Metal Fuels   69
    2.2.1 Stabilized Metal-Fuel Flame   70
    2.2.2 Combustion Engineering   71
    2.2.3 Designing Metal-Fueled Engines   72
  2.3 Regeneration of Metal Fuels Through the Solar Reduction of Oxides   75
    2.3.1 Thermodynamics and Kinetics of Oxides Reduction   75
    2.3.2 Effect of Some Parameters on the Reduction Yield   77
      2.3.2.1 Carbon-Reducing Agent   77
      2.3.2.2 Catalysts and Additives   78
      2.3.2.3 Mechanical Milling   78
      2.3.2.4 CO Partial Pressure   79
      2.3.2.5 Carrier Gas   79
      2.3.2.6 Fast Preheating   79
      2.3.2.7 Progressive Heating   80
    2.3.3 Reverse Reoxidation of the Produced Metal Powders   80
    2.3.4 Reduction of Oxides Using Concentrated Solar Power   81
    2.3.5 Solar Carbothermal Reduction of Magnesia   83
    2.3.6 Solar Carbothermal Reduction of Alumina   86
  2.4 Conclusions   89
  Acknowledgments   90
  References   90

3 Design Optimization of a Solar Fuel Production Plant by Water Splitting With a Copper-Chlorine Cycle   97
Samane Ghandehariun, Shayan Sadeghi and Greg F. Naterer

  Nomenclature   98
  3.1 Introduction   100
  3.2 System Description   108
  3.3 Mathematical Modeling and Optimization   113
    3.3.1 Energy and Exergy Analyses   113
    3.3.2 Economic Analysis   116
    3.3.3 Multiobjective Optimization (MOO) Algorithm   120

	3.4	Results and Discussion	121
	3.5	Conclusions	130
		References	131

4 Diversifying Solar Fuels: A Comparative Study on Solar Thermochemical Hydrogen Production Versus Solar Thermochemical Energy Storage Using Co_3O_4 — 137
Atalay Calisan and Deniz Uner
- 4.1 Introduction — 137
- 4.2 Materials and Methods — 141
- 4.3 Thermodynamics of Direct Decomposition of Water — 142
- 4.4 A Critical Analysis of Two-Step Thermochemical Water Splitting Cycles Through the Red/Ox Properties of Co_3O_4 — 143
  - 4.4.1 Red/Ox Characteristics of Co_3O_4 Measured by Temperature-Programmed Analysis — 145
  - 4.4.2 The Role of Pt as a Reduction Promoter of Co_3O_4 — 147
  - 4.4.3 A Critical Analysis of the Solar Thermochemical Cycles of Water Splitting — 149
- 4.5 Cyclic Thermal Energy Storage Using Co_3O_4 — 151
  - 4.5.1 Mass and Heat Transfer Effects During Red/Ox Processes — 152
  - 4.5.2 Cyclic Thermal Energy Storage Performance of Co_3O_4 — 152
- 4.6 Conclusions — 157
- Acknowledgements — 157
- References — 157

Part II: Artificial Photosynthesis and Solar Biofuel Production — 161

5 Shedding Light on the Production of Biohydrogen from Algae — 163
Thummala Chandrasekhar and Vankara Anuprasanna
- 5.1 Introduction — 164
- 5.2 Hydrogen or Biohydrogen as Source of Energy — 165
- 5.3 Hydrogen Production From Various Resources — 167
- 5.4 Mechanism of Biological Hydrogen Production from Algae — 168
- 5.5 Production of Hydrogen from Different Algal Species — 171
  - 5.5.1 Generation of Hydrogen in *Scenedesmus obliquus* — 171
  - 5.5.2 Production of Hydrogen in *Chlorella vulgaris* — 174

		5.5.3	Generation of Hydrogen in Model Alga *Chlamydomonas reinhardtii*	175
	5.6	Concluding Remarks		177
		Acknowledgments		177
		References		177
6	Photoelectrocatalysis Enables Greener Routes to Valuable Chemicals and Solar Fuels			185
	Dipesh Shrestha, Kamal Dhakal, Tamlal Pokhrel, Achyut Adhikari, Tomas Hardwick, Bahareh Shirinfar and Nisar Ahmed			
	6.1	Introduction		186
	6.2	C–H Functionalization in Complex Organic Synthesis		189
	6.3	Examples of Photoelectrochemical-Induced C–H Activation		190
	6.4	C–C Functionalization		192
	6.5	Electrochemically Mediated Photoredox Catalysis (e-PRC)		194
	6.6	Interfacial Photoelectrochemistry (iPEC)		197
	6.7	Reagent-Free Cross Dehydrogenative Coupling		199
	6.8	Conclusion		199
		References		200

Part III: Photocatalytic CO_2 Reduction to Fuels 205

7	Graphene-Based Catalysts for Solar Fuels			207
	Zhou Zhang, Maocong Hu and Zhenhua Yao			
	7.1	Introduction		208
	7.2	Preparation of Graphene and Its Composites		209
		7.2.1	Preparation of Graphene (Oxide)	209
		7.2.2	Preparation of Graphene-Based Photocatalysts	210
			7.2.2.1 Hydrothermal/Solvothermal Method	211
			7.2.2.2 Sol-Gel Method	212
			7.2.2.3 *In Situ* Growth Method	212
	7.3	Graphene-Based Catalyst Characterization Techniques		214
		7.3.1	SEM, TEM, and HRTEM	214
		7.3.2	X-Ray Techniques: XPS, XRD, XANES, XAFS, and EXAFS	215
		7.3.3	Atomic Force Microscopy (AFM)	217
		7.3.4	Fourier Transform Infrared Spectroscopy (FTIR)	218
		7.3.5	Other Technologies	219
	7.4	Graphene-Based Catalyst Performance		220
		7.4.1	Photocatalytic CO_2 Reduction	223

		7.4.2 Hydrogen Production by Water Splitting	229
	7.5	Conclusion and Future Opportunities	235
		Acknowledgments	237
		References	237

8 Advances in Design and Scale-Up of Solar Fuel Systems — 247
Ashween Virdee and John Andresen

- 8.1 Introduction — 248
- 8.2 Strategies for Solar Photoreactor Design — 248
  - 8.2.1 Photocatalytic Systems — 249
    - 8.2.1.1 Slurry Photoreactor — 252
    - 8.2.1.2 Fixed Bed Photoreactor — 254
    - 8.2.1.3 Twin Photoreactor (Membrane Photoreactor) — 256
    - 8.2.1.4 Microreactor — 259
  - 8.2.2 Electrochemical System — 260
    - 8.2.2.1 CO_2 Electrochemical Reactors — 263
  - 8.2.3 Photoelectrochemical (PEC) Systems — 267
- 8.3 Design Considerations for Scale-Up — 272
- 8.4 Future Systems and Large Reactors — 274
- 8.5 Conclusions — 276
- References — 277

Part IV: Solar-Driven Water Splitting — 285

9 Photocatalyst Perovskite Ferroelectric Nanostructures — 287
Debashish Pal, Dipanjan Maity, Ayan Sarkar and Gobinda Gopal Khan

- 9.1 Introduction — 288
- 9.2 Ferroelectric Properties and Materials — 289
- 9.3 Fundamental of Photocatalysis and Photoelectrocatalysis — 290
  - 9.3.1 Photocatalytic Production of Hydrogen Fuel — 290
  - 9.3.2 Photoelectrocatalytic Hydrogen Production — 291
  - 9.3.3 Photocatalytic Dye/Pollutant Degradation — 292
- 9.4 Principle of Piezo/Ferroelectric Photo(electro)catalysis — 292
- 9.5 Ferroelectric Nanostructures for Photo(electro)catalysis — 294
- 9.6 Synthesis and Design of Nanostructured Ferroelectric Photo(electro)catalysts — 295
  - 9.6.1 Hydrothermal/Solvothermal Methods — 295
  - 9.6.2 Sol-Gel Methods — 300
  - 9.6.3 Wet Chemical and Solution Methods — 303

		9.6.4 Vapor Phase Deposition Methods	305
		9.6.5 Electrospinning Methods	306
	9.7	Photo(electro)catalytic Activities of Ferroelectric Nanostructures	307
		9.7.1 Photo(electro)catalytic Activities of $BiFeO_3$ Nanostructures and Thin Films	307
		9.7.2 Photo(electro)catalytic Activities of $LaFeO_3$ Nanostructures	311
		9.7.3 Photo(electro)catalytic Activities of $BaTiO_3$ Nanostructures	314
		9.7.4 Photo(electro)catalytic Activities of $SrTiO_3$ Nanostructures	317
		9.7.5 Photo(electro)catalytic Activities of $YFeO_3$ Nanostructures	319
		9.7.6 Photo(electro)catalytic Activities of $KNbO_3$ Nanostructures	319
		9.7.7 Photo(electro)catalytic Activities of $NaNbO_3$ Nanostructures	322
		9.7.8 Photo(electro)catalytic Activities of $LiNbO_3$ Nanostructures	323
		9.7.9 Photo(electro)catalytic Activities of $PbTiO_3$ Nanostructures	323
		9.7.10 Photo(electro)catalytic Activities of $ZnSnO_3$ Nanostructures	325
	9.8	Conclusion and Perspective	327
		References	329
10	Solar-Driven H_2 Production in PVE Systems		341
	Zaki N. Zahran, Yuta Tsubonouchi and Masayuki Yagi		
	10.1	Introduction	342
	10.2	Approaches for H_2 Production *via* Solar-Driven Water Splitting	343
	10.3	Principle of Designing of PVE Systems for Solar-Driven Water Splitting	348
	10.4	Development of PVE Systems for Solar-Driven Water Splitting	352
		10.4.1 PVE Systems Based on Si PV Cells	353
		10.4.2 PVE Systems Based on Group III-V Compound PV Cells	354
		10.4.3 PVE Systems Based on Chalcogenide PV Cells	356
		10.4.4 PVE Systems Based on Perovskite PV Cells	358

		10.4.5	PVE Systems Based on Organic Heterojunction PV Cells	359

	10.5	Conclusions and Future Perspective	361
		References	361

11 Impactful Role of Earth-Abundant Cocatalysts in Photocatalytic Water Splitting — 375
Yubin Chen, Xu Guo, Zhichao Ge, Ya Liu and Maochang Liu

	11.1	Introduction	376
	11.2	Categories of Cocatalysts Utilized in Photocatalytic Water Splitting	378
		11.2.1 Metal and Non-Metal Cocatalysts	379
		11.2.2 Metal Oxides and Hydroxides	380
		11.2.3 Metal Sulfides	381
		11.2.4 Metal Phosphides and Carbides	382
		11.2.5 Molecular Cocatalysts	383
	11.3	Factors Determining the Cocatalyst Activity	384
		11.3.1 Intrinsic Properties of Cocatalysts	384
		11.3.2 Interfacial Coupling of Cocatalysts With Host Semiconductors	388
	11.4	Advanced Characterization Techniques for Cocatalytic Process	393
	11.5	Conclusion	395
		Acknowledgments	396
		References	396

Index — 411

Preface

Among all other energy sources, solar power is the one with the highest capacity and greatest potential. It is humanity's great loss to not be able to use solar energy to produce all the energy we need. This is particularly true since the environmental and sociopolitical problems caused by the use of fossil fuels have negatively affected our future welfare. Therefore, with this as the motivating factor, basic science and engineering studies have been continuing at a rapid pace with the aim of eliminating existing problems to ensure a more efficient and widespread use of solar energy. The biggest disadvantage that solar energy has to face is that it is not accessible at all times of the day and year; in other words, the energy obtained from the sun must be stored.

The energy from photovoltaic systems can be stored in flow batteries or other battery systems, as well as making it storable by converting solar energy to chemical energy, which will make it more cost-effective and versatile compared to the current method. Synthetic chemical fuels obtained by using solar energy are called solar fuels. Hydrogen, methanol, methane, ammonia, carbon monoxide, and some other hydrocarbons and/or oxygenates can be produced from abundant feedstocks such as water, carbon dioxide, and nitrogen via different solar energy-based routes. These routes include solar thermolysis, artificial photosynthesis, and photocatalytic and photoelectrochemical conversion. Therefore, we organized our book to review these routes in informative chapters submitted by distinguished authors. We, as editors, wish to thank the authors for their valuable contributions. This volume covers cutting-edge technologies and materials for efficient solar fuel generation. Additionally, it highlights the research efforts in the literature and adds a valuable component to the area. In addition to the basics, this book also discusses advanced engineering details for both scientists and engineers in academia and industry.

There are four parts and eleven chapters in the book. Part I, Solar Thermochemical and Concentrated Solar Approaches, includes four chapters. Chapter 1 summarizes hydrogen generation via solar thermolysis.

This chapter focuses on the theoretical methods, the state-of-the-art redox-active metal oxides, next-generation perovskite redox-active materials, and materials design directions. Chapter 2 covers recyclable solar transport fuels. In this chapter, all the important aspects of sustainability of solar metal fuels for future long-distance transportation through combustion/reduction cycles are discussed, including direct combustion of solar metal fuels and regeneration of metal fuels through the solar reduction of oxides. Chapter 3 discusses the design and optimization of a standalone plant for hydrogen generation powered by solar energy. Fundamental advances in the copper-chlorine (Cu-Cl) high-performance thermochemical cycle, thermodynamic and economic analyses, and optimization of the system for two objective functions, including the levelized cost of producing hydrogen and solar-to-hydrogen efficiency, are explained in this chapter. Chapter 4 presents a comparative study on solar thermochemical hydrogen production versus solar heat storage using cobalt oxide (Co_3O_4). Among the topics covered are the thermodynamics of direct decomposition of water, a critical analysis of two-step thermochemical water splitting cycles through the redox properties of Co_3O_4, and cyclic thermal energy storage using Co_3O_4.

Part II, Artificial Photosynthesis and Solar Biofuel Production, includes two chapters. Chapter 5 covers the production of biohydrogen from algae. Overall, this chapter intends to summarize the developments in hydrogen production from certain algal species, which is helpful for commercial practice in the near future. Chapter 6 summarizes state-of-the-art applications of photoelectrocatalysis (PEC) in the synthesis of valuable chemicals and solar fuels. This chapter focuses on C-H functionalization in complex organic synthesis, examples of photoelectrochemical-induced C-H activation, C-C functionalization, electrochemically mediated photoredox catalysis, interfacial photoelectrochemistry, and reagent-free cross dehydrogenative coupling.

Part III, Photocatalytic CO_2 Reduction to Fuels, includes two chapters. Chapter 7 focuses on graphene-based catalysts for solar fuels. The preparation of graphene and its composites and the performance of graphene-based catalysts are covered in this chapter. Chapter 8 covers the advances in the design and scale-up of solar fuel systems. Also discussed are strategies for solar photoreactor design, including photocatalytic and electrochemical systems for carbon dioxide reduction, design considerations for scale-up, and future systems and large reactors.

Part IV, Solar-Driven Water Splitting, includes three chapters. Chapter 9 summarizes the advanced materials and systems for solar hydrogen generation. Perovskite ferroelectric nanostructures for photocatalysis and photoelectrocatalysis are also introduced in this chapter. Chapter 10 focuses on photovoltaic-electrolyzer (PVE) systems, consisting of photovoltaic (PV) cells connected by wires with electrolyzers equipped with an anode and a cathode in an electrolyte solution as one of the most promising approaches for solar-driven water splitting. Finally, Chapter 11 offers meaningful guidance to design cost-effective and highly efficient cocatalysts for photocatalytic water splitting. In this context, the basic working principle of cocatalysts and a summary of extensively studied earth-abundant cocatalysts are provided.

In conclusion, we would like to emphasize that this third volume of the *Advances in Solar Cell Materials and Storage* series provides an overall view of the new and highly promising photoactive materials and system designs for solar fuel generation. Therefore, readers from diverse fields, including chemistry, physics, materials science, engineering, and mechanical and chemical engineering, can definitely take advantage of the information presented in this book to better understand the impacts of solar fuels.

Series Editors
Nurdan Demirci Sankir PhD and Mehmet Sankir PhD
Department of Materials Science and Nanotechnology Engineering, TOBB University of Economics and Technology
February 20, 2023

Part I
SOLAR THERMOCHEMICAL AND CONCENTRATED SOLAR APPROACHES

1

Materials Design Directions for Solar Thermochemical Water Splitting

Robert B. Wexler[1], Ellen B. Stechel[2] and Emily A. Carter[1]*

[1]*Department of Mechanical and Aerospace Engineering and the Andlinger Center for Energy and the Environment, Princeton University, Princeton, NJ, United States*
[2]*ASU LightWorks® and the School of Molecular Sciences, Arizona State University, Tempe, Arizona, United States*

Abstract

Solar thermochemical water splitting (STWS) offers a renewable route to hydrogen with the potential to help decarbonize several industries, including transportation, manufacturing, mining, metals processing, and electricity generation, as well as to provide sustainable hydrogen as a chemical feedstock. STWS uses high temperatures from concentrated sunlight or other sustainable means for high-temperature heat to produce hydrogen and oxygen from steam. For example, in its simplest form of a two-step thermochemical cycle, a redox-active metal oxide is heated to ≈1700 to 2000 K, driving off molecular oxygen while producing oxygen vacancies in the material. The reduced metal oxide then cools (ideally with the extracted heat recuperated for reuse) and, in a separate step, comes into contact with steam, which reacts with oxygen vacancies to produce molecular hydrogen while recovering the original state of the metal oxide. Despite its promising use of the entire solar spectrum to split water thermochemically, the estimated cost of hydrogen produced via STWS is ≈4 to 6× the U.S. Department of Energy (DOE) Hydrogen Shot target value of $1/kg.

One contributing approach to bridging this cost gap is the design of new materials with improved thermodynamic properties to enable higher efficiencies. The state-of-the-art (SOA) redox-active metal oxide for STWS is ceria (CeO_2) because of its close to optimal, although too high, oxygen vacancy formation enthalpy and large configurational and electronic entropy of reduction. However, ceria requires high operating temperatures, and its efficiency is insufficient. Therefore, efforts to increase the efficiency of STWS cycles have focused

---

*Corresponding author: eac@princeton.edu

on further optimizing oxygen vacancy formation enthalpies and augmenting the reduction entropy via substitution or doping and materials discovery schemes. Examples of the latter include the perovskites $BaCe_{0.25}Mn_{0.75}O_3$ and $(Ca,Ce)(Ti,Mn)O_3$. These efforts and others have revealed intuitive chemical principles for the efficient and systematic design of more effective materials, such as the strong correlation between the enthalpies of crystal bond dissociation and solid-state cation reduction with the enthalpy of oxygen vacancy formation, as well as configurational entropy augmentation via the coexistence of two or more redox-active cation sublattices.

The purpose of this chapter is to prepare the reader with an up-to-date account of STWS redox-active materials, both the SOA and promising newcomers, as well as to provide chemically intuitive strategies for improving their cycle efficiencies through materials design—in conjunction with ongoing efforts in reactor engineering and gas separations—to reach the cost points for commercial viability.

Keywords: Climate change, concentrated solar technologies, hydrogen, solar thermolysis, solar thermochemical cycles, redox-active materials, off-stoichiometric, quantum mechanics simulations

1.1 Introduction

Combatting anthropogenic climate change is one of the critical scientific and engineering challenges of our time. The associated global warming (Figure 1.1a)—predominantly brought about by greenhouse-gas emissions from burning fossil fuels [1–3]—already has led to extreme weather events that threaten the safety and food/water security of life on Earth. Averting the most disastrous effects of climate change calls—at least in part—for clean fuel alternatives to avoid the CO_2 emissions from hard-to-electrify sectors, including heavy-duty vehicles with petroleum-based combustion engines. One encouraging alternative is H_2, which has a higher-energy density per unit mass than liquid hydrocarbons and can be produced using sustainable energy in the form of concentrated solar heat via thermolysis or thermochemical water splitting (Figure 1.1b) [4]. Although not reviewed here, H_2 can also be sustainably produced from water by alternative means, for example, via photoelectrochemical water splitting [5, 6] and both high- [7] and low-temperature [8] electrolysis employing renewable (or nuclear) energy. Concentrated solar technologies (CSTs) also promise to reduce the carbon dioxide footprint of fossil-fuel-derived H_2 from steam-methane reforming, hydrocarbon (fossil or biomass) gasification, solid-oxide electrolysis, and methane cracking.

Two popular solar thermal collector/receiver/reactor designs are the tower with a heliostat field and the parabolic dish (Figure 1.1c) [9]. In the increasingly adopted solar power tower plant architecture, many heliostats focus sunlight on an elevated receiver, achieving a solar concentration ratio (C)—i.e., the factor by which a collector/receiver multiplies the intensity of sunlight impinging upon the Earth's surface—of ≈1000. For parabolic dishes, a polished metal mirror lining concentrates sunlight on a focal point, where redox-active materials could be heated to high temperatures (e.g., 1700–1800 K [10]). While dishes currently are more expensive than towers, they generally lead to a higher C [11] and recently have been used in demonstration CST-based systems [10].

The theoretical maximum efficiency of solar-to-H_2 conversion using CSTs is—under the assumption of ideal optics and a perfectly insulated receiver—the product of the solar collector, receiver, and reactor (Carnot) efficiencies [12, 13]

$$\eta_{solar-to-fuel} = \eta_{collector} \eta_{receiver} \eta_{Carnot} \tag{1.1}$$

$$\eta_{receiver} = 1 - \frac{\sigma T^4}{IC} \tag{1.2}$$

$$\eta_{Carnot} = 1 - \frac{T_{sur}}{T} \tag{1.3}$$

where σ is the Stefan–Boltzmann constant; T the temperature of the receiver; I the intensity of the direct, normal-incident sunlight; and T_{sur} is the temperature of the surroundings (e.g., 298.15 K). Suppose a heliostat field with a solar tower is used instead of a parabolic dish. In that case, $\eta_{collector}$ will be less than one due to factors including the cosine effect (i.e., due to heliostats not pointing directly at the sun and the receiver simultaneously, hence, there is a reduction in the effective reflection area) [14]. One can think of the receiver efficiency ($\eta_{receiver}$) as the fraction of absorbed

6   SOLAR FUELS

sunlight that is not reradiated by the blackbody-like receiver. Increasing C can increase the T range over which $\eta_{receiver}$ is close to 100%. For example, if parabolic dishes—with C reaching 10000—can be made economical, then a nearly perfect receiver can be achieved at ≈2000 K (Figure 1.1d,

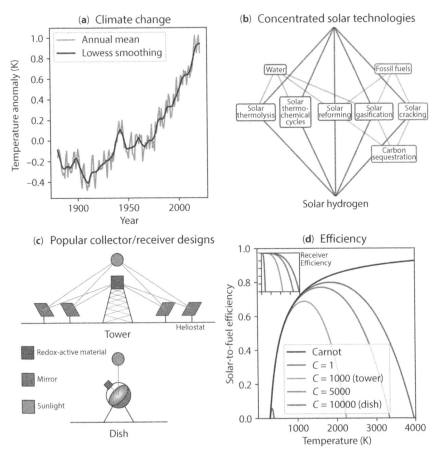

Figure 1.1 Concentrated solar hydrogen for combatting climate change. (a) Increase in global temperature since 1880. (b) Routes to solar hydrogen via concentrated solar technologies. (c) Popular collector/receiver designs for concentrated solar heat technologies. (d) Ideal solar-to-fuel efficiency ($\eta_{solar\text{-}to\text{-}fuel}$ in Equation (1.1)) and (d, inset) receiver efficiency ($\eta_{receiver}$ in Equation (1.2)—with the same ticks and tick labels as the larger panel). Note that towers can have $C > 1000$ and developing dishes with $C = 10{,}000$ is quite challenging. That said, we chose these values to indicate the effect of order-of-magnitude changes in C on the theoretical solar-to-fuel efficiency.

inset). While $\eta_{receiver}$ dominates $\eta_{solar\text{-}to\text{-}fuel}$ in the high-temperature limit, the efficiency of a Carnot engine (η_{Carnot}) governs the low-temperature regime, which decreases to zero as T approaches T_{sur} from above. Upon multiplying these three efficiencies, it becomes clear that—for a given C—there is an ideal temperature at which $\eta_{solar\text{-}to\text{-}fuel}$ is maximized (Figure 1.1d). As an example, consider a dish that provides $C = 5000$. If the receiver is heated to 1800 K, one can use ≤76% of the concentrated sunlight energy for solar-to-H_2 conversion. Here, the "less than" indicates that other loss mechanisms and engineering constraints typically produce efficiencies << 76%.

1.1.1 Hydrogen via Solar Thermolysis

Having introduced CSTs and their efficiencies for a general solar-to-H_2 process, we now consider the earliest and perhaps simplest approach to CST-based hydrogen production via solar thermolysis or *direct* solar water splitting [15]. In solar thermolysis, $H_2O(g)$ is heated to $T \geq 2500$ K, above which it can undergo the following high-temperature reactions (Figure 1.2a) [16]:

Blue line $\qquad H_2O(g) \rightleftharpoons H_2(g) + \frac{1}{2}O_2(g) \qquad (1.4)$

Orange line $\qquad H_2(g) \rightleftharpoons 2H(g) \qquad (1.5)$

Green line $\qquad \frac{1}{2}O_2(g) \rightleftharpoons O(g) \qquad (1.6)$

Red line $\qquad H_2O(g) \rightleftharpoons H(g) + OH(g) \qquad (1.7)$

At $T < 2000$ K and $p = 1$ bar, none of these reactions occur with appreciable yields, leaving $H_2O(g)$ intact (Figure 1.2b). As T reaches 2500 K, ≈4% of $H_2O(g)$ molecules split into $H_2(g)$ and $O_2(g)$ (Equation 1.4). For $T > 2500$ K, however, side reactions—such as the atomization of $H_2(g)$ (Equation 1.5) and $O_2(g)$ [Equation (1.6)], and the dissociation of $H_2O(g)$ into $H(g)$ and $OH(g)$ (Equation 1.7)—compete with the desired water-splitting reaction, leading to a maximum $H_2(g)$ mole fraction of ≈0.19 at 3400 K. In addition to its upper limit for H_2 generation, solar thermolysis is impractical [17] because it produces an explosive mixture of $H_2(g)$ and $O_2(g)$ that requires careful separation and rapid quenching to avoid recombination, which reduces efficiency. Furthermore, the T needed to produce $H_2(g)$ and not

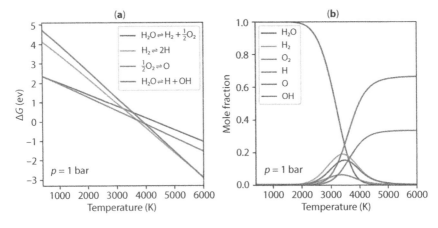

Figure 1.2 Thermodynamics of hydrogen production via solar thermolysis. (a) Gibbs free energy change (ΔG) of high-temperature reactions at $p = 1$ bar. (b) Equilibrium mole fractions at $p = 1$ bar (see Appendix A. Equilibrium Composition for Solar Thermolysis).

H(g) or OH(g)—i.e., ≈2500 K—leads to the thermal failure of the ceramics used for H_2(g) and O_2(g) separation, thus motivating—in the absence of solutions for these issues—another route to solar H_2, namely solar thermochemical water splitting (STWS) [18–26].

1.1.2 Hydrogen via Solar Thermochemical Cycles

To split water at lower temperatures and preclude the formation of undesired gas-phase molecules, one can employ thermochemical cycles, the simplest of which—and the primary subject of this book chapter—is a two-step cycle [27–34] (Figure 1.3a) with redox-active, metal-oxide materials (Figure 1.3b). In such a cycle, a metal oxide (MO_x, where x is the number of moles of O per cation) first is heated, using CSTs, to temperatures typically exceeding 1500 K and most often close to 1800 K, at which point it is reduced to a more O-poor stoichiometry ($MO_{x-\delta}$), i.e.,

$$\frac{1}{\delta}MO_x(s) \rightleftharpoons \frac{1}{\delta}MO_{x-\delta} + \frac{1}{2}O_2(g) \tag{1.8}$$

where δ is the off-stoichiometry; note that we have purposefully omitted the phase of the reduced metal oxide for reasons to be explained momentarily. Generally speaking, one would reduce at the highest temperatures within engineering and economic constraints to ensure maximal reduction (as increasing the temperature makes ΔG more negative and therefore increases δ) and fast kinetics. In the second step, the reduced metal oxide cools to a temperature where reoxidation is possible when exposed to $H_2O(g)$, which leads to water splitting and regeneration of the original metal oxide, i.e.,

$$\frac{1}{\delta}MO_{x-\delta} + H_2O(g) \rightleftharpoons \frac{1}{\delta}MO_x(s) + H_2(g) \quad (1.9)$$

Generally, $MO_{x-\delta}$ will not reoxidize to the fully stoichiometric form MO_x but will cycle between two forms of the metal-oxide stoichiometry—both partially reduced—where the difference between the two off-stoichiometries is one of the performance metrics. The reoxidation is further limited if there is a small amount of hydrogen in the gas stream, which might be expected if one separates, in the gas phase, the hydrogen from the reoxidation product stream and recycles any unconverted steam.

Unlike thermal reduction (Equation 1.8), whose ideal operating temperature is bounded only from above by the thermal stability of the material and durability of the reactor, one would perform water splitting (Equation 1.9) at temperatures high enough for fast kinetics but low enough for a good ΔG of reoxidation. This compromise often requires water splitting to be done around 1000 K or higher. Another consideration is recuperation of heat between the high temperature and low temperature steps. The larger the temperature difference, the greater the engineering challenge to limit the losses.

Until now, we have neither specified the phase of $MO_{x-\delta}$ nor the extent of reduction δ. Two-step metal-oxide thermochemical cycles are based on either volatile or non-volatile metal oxides. Volatile refers to a metal oxide for which a solid-to-gas phase transition accompanies thermal reduction. One of the most widely studied volatile cycles is ZnO(s)/Zn(g) [35–37]:

$$ZnO(s) \rightleftharpoons Zn(g) + \frac{1}{2}O_2(g) \quad (1.10)$$

$$Zn(s) + H_2O(g) \rightleftharpoons ZnO(s) + H_2(g) \quad (1.11)$$

In the thermal reduction step [Equation (1.10)], which one must carry out at temperatures above 2000 K, ZnO(s) volatilizes to Zn(g) and $O_2(g)$. While the ZnO(s)/Zn(g) cycle offers favorable efficiencies even in the absence of heat recovery (energy conversion efficiency ≈ 45% and maximum exergy efficiency ≈ 29%), its issues are similar to those faced in solar thermolysis in that the high temperatures required for significant reduction put a considerable thermal strain on the receiver/reactor [17, 38]. After thermal reduction, one generally quenches quickly to avoid the back reaction before separating Zn(s) from O_2(g). Alternatively, electrothermal gas-phase separation has been considered [39, 40]. Water splitting [Equation (1.11)], on the other hand, typically takes place at $T \leq 900$ K, revealing another difficulty for ZnO(s)/Zn(g): the need for a giant temperature swing (≥ 1100 K). Other redox couples for volatile, two-step STWS have been considered, such as post-transition-metal oxides in the SnO_2(s)/SnO(g) cycle [41–43]; however, those with greater attention currently are solid phase, a.k.a. non-volatile, redox-active materials.

Within non-volatile, redox-active metal oxides, the two main categories are stoichiometric (line compounds) and off-stoichiometric. First, we consider stoichiometric metal oxides, where stoichiometric refers to materials for which reduction and reoxidation produce pure, solid-phase, metal-containing compounds obeying full stoichiometry constraints on composition. One can further subdivide stoichiometric metal oxides into single-component and multi-component compositions. Examples best illustrate the difference between these two types of stoichiometric oxides. The prototypical single-component materials are metal-doped ferrites [44–54], whose thermal reduction and water splitting reactions are

$$(M_xFe_{1-x})_3O_4(s) \rightleftharpoons 3xMO(s) + 3(1-x)FeO(s) + \frac{1}{2}O_2(g) \quad (1.12)$$

$$3xMO(s) + 3(1-x)FeO(s) + H_2O(g) \rightleftharpoons (M_xFe_{1-x})_3O_4(s) + H_2(g) \quad (1.13)$$

where the metal (M) dopant or substituent can be Fe (in which case Fe is not a dopant and the phase is magnetite) [55–59], Zn [60], Ni [60, 61], Co [60, 62] (as well as a complete replacement of Fe with Co [63]), Mn [61], and others. Ferrites with other metals substituted in the spinel or inverse spinel structure can be tuned to provide nearly optimal reduction Gibbs free energetics and reduction temperatures lower than 2000 K [64, 65]. However, both their reduction and water-splitting kinetics are slow because O^{2-} is close-packed in both oxide structures, Fe_3O_4 and FeO(s). Therefore, it does not react beyond the surface [66]. Additionally, powdered Fe oxides sinter, rendering them uncyclable [56, 62, 67, 68]. To enhance cyclability, one can use yttria-stabilized zirconia as an inert support that incorporates active Fe ions into its crystal lattice, forming a solid solution, thus alleviating the sintering or melting of iron oxides at the working temperatures of 1200 to 1700 K [55, 69]. Note that, for ferrite cycles, a single metal oxide reduces and reoxidizes, hence the terminology "single component."

Alternatively, multi-component cycles involve the redox of more than one metal oxide component. An excellent example of this case is the cycle based on the mineral hercynite $FeAl_2O_4(s)$ [70–77]:

$$CoFe_2O_4(s) + 3Al_2O_3(s) \rightleftharpoons CoAl_2O_4(s) + 2FeAl_2O_4(s) + \frac{1}{2}O_2(g)$$

(1.14)

$$CoAl_2O_4(s) + 2FeAl_2O_4(s) + H_2O(g) \rightleftharpoons CoFe_2O_4(s) + 3Al_2O_3(s) + H_2(g)$$

(1.15)

During thermal reduction, $CoFe_2O_4(s)$—a metal-substituted ferrite—reacts with three moles of $Al_2O_3(s)$, producing $CoAl_2O_4(s)$—a pigment known as cobalt blue—along with two moles of hercynite and a half mole of $O_2(g)$. These intermediate products then split water at lower temperatures, restoring the original solids in their starting stoichiometric coefficients and generating $H_2(g)$. Both steps have two metal-oxide components in the reactants and products, so the hercynite cycle is multi-component. However, like the ferrites, this cycle suffers from poor kinetics, which is unsurprising considering one of the components is cobalt ferrite $CoFe_2O_4(s)$. Other studied multicomponent cycles include—but are not limited to—those based on the metal sulfate/oxide [e.g., $MnSO_4(s)/MnO(s)$ [78]] and metal dioxide/pyrochlore [i.e., $CeO_2(s)+MO_2(s)/Ce_2M_2O_7(s)$ where M can be, e.g., Ti [79], Si [79], or Sn [80]] redox couples. Ultimately, kinetic limitations are a hallmark of stoichiometric materials because their STWS cycles require the nucleation and growth of bulk phases. A promising path to promote faster kinetics is to use off-stoichiometric metal oxides, which tend to be mixed ionic-electronic conductors (MIECs) that form and fill oxygen vacancies (V_Os) during thermal reduction and water splitting, respectively, instead of undergoing major bulk structural phase transitions. As off-stoichiometric metal oxides, particularly MIECs because of their superior ion diffusion kinetics, currently are the subject of intense research for STWS applications and are the redox-active materials of choice for pilot plants, we focus on them here. Below we emphasize developing intuition that explains observed physicochemical phenomena, in order to determine materials design criteria that can lead to tailoring materials for more optimal thermochemical cycles.

Before we dive into the details of off-stoichiometric metal oxides for STWS, we would be remiss if we did not mention the utility of multi-step cycles. We will first describe the Cu-Cl [81] cycle (Figure 1.3c). In the hydrolysis step, $Cu(II)Cl_2(s)$ is heated to ≈ 673 K in the presence of $H_2O(g)$,

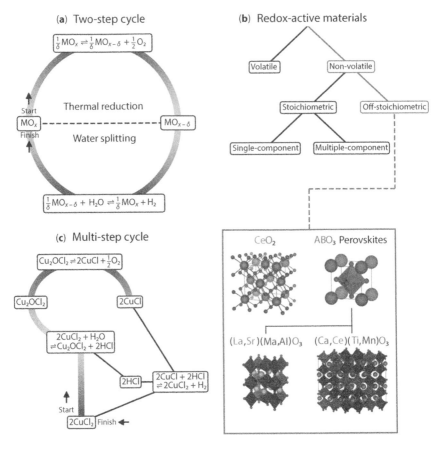

Figure 1.3 Hydrogen production via solar thermochemical cycles. (a) Schematic of a two-step cycle for a metal oxide (MO_x) that becomes off-stoichiometric ($MO_{x-\delta}$, where δ is the off-stoichiometry) upon thermal reduction (where the color of the circle denotes relative temperature). (b) Types of redox-active materials typically employed for two-step STWS, where our focus is on nonvolatile materials that become off-stoichiometric upon thermal reduction, such as CeO_2 and ABO_3 perovskites and their alloys. (c) Schematic of a multistep cycle, specifically, here, the copper chloride hybrid cycle, which involves hydrolysis (blue), thermal reduction (red), and electrolysis (black) steps at different temperatures.

forming melanothallite $Cu(II)_2OCl_2(s)$ and $HCl(g)$. Then, in the thermal reduction step, $Cu(II)_2OCl_2(s)$ is solar heated to ≈773 K using CSTs, which leads to its reductive decomposition into $Cu(I)Cl(s)$ and $O_2(g)$. One can liken this step to the thermal reduction step in the two-step STWS cycles described above. Last is the electrolysis step, where $HCl(g)$ from hydrolysis and $Cu(I)Cl(s)$ from thermal reduction react at ambient temperatures and under the application of an oxidizing electrode potential, producing $H_2(g)$ and regenerating $Cu(II)Cl_2$ for subsequent cycles. While the Cu-Cl cycle enables efficient heat recycling and offers lower operating temperatures than two-step cycles, solids handling between steps and corrosive chemical components—in addition to the difficulties associated with engineering a multi-step engine with compounding inefficiencies—challenge the practical application of the multi-step Cu-Cl cycle, as well as others, including the hybrid sulfur [$H_2SO_4(aq)/SO_2(g)$] [82–84] and sulfur-iodine cycles [85–87]. As a final remark, we acknowledge that the preceding discussion represents a limited survey of cycles and redox-active materials. There have been >300 cycles screened [19, 88]. However, since several seminal articles offer a more comprehensive overview [22, 25, 33], we truncate here our consideration of either stoichiometric or line compounds (volatile and non-volatile materials) or multi-step or hybrid cycles.

1.1.3 Thermodynamics

One of the most critical constraints for two-step solar-thermochemical cycles [14, 89–96] with off-stoichiometric metal oxides is the thermodynamic spontaneity of the thermal reduction and water splitting reactions [97, 98]. For reversible/equilibrium thermal reduction [Equation (1.8)], the Gibbs free energy change for an infinitesimal change in the off-stoichiometry $d\delta$ in a counter-current reactor [99–101] is

$$\Delta G_{red} = \frac{1}{d\delta}\left(G^o_{MO_{x-\delta-d\delta}} - G^o_{MO_{x-\delta}}\right) + \frac{1}{2}G^o_{O_2} + \frac{1}{2}RT_{red}\ln\left(\frac{p^{red}_{O_2}}{p^o}\right)$$
$$= \Delta G^o_v + \frac{1}{2}RT_{red}\ln\left(\frac{p^{red}_{O_2}}{p^o}\right) = 0 \qquad (1.16)$$

where G (G^o) is the (standard) Gibbs free energy, R is the universal gas constant, T_{red} is the thermal reduction temperature, $p^{red}_{O_2}$ is the inlet partial pressure of $O_2(g)$ for the thermal reduction reaction at T_{red}, $p°$ is the reference pressure (1 bar), and

$$\Delta G_v^o = \frac{1}{d\delta}\left(G_{MO_{x-\delta-d\delta}}^o - G_{MO_{x-\delta}}^o\right) + \frac{1}{2}G_{O_2}^o = \Delta H_v^o - T_{red}\Delta S_v^o \quad (1.17)$$

ΔG_v^o, ΔH_v^o, and ΔS_v^o are the standard V_O formation Gibbs free energy, enthalpy, and entropy at an oxygen partial pressure of p, respectively, which depend on δ (for simplicity, we omit this dependence). Figure 1.4a shows combinations of ΔH_v^o and T_{red} that satisfy $\Delta G_{red} = 0$ for $p_{O_2}^{red} = 1 \times 10^{-4}$ bar. One can read this graph as follows. For example, if a redox-active material has a T-independent $\Delta S_v^o = 15\ k_B$ and $\Delta H_v = 3$ eV, $T_{red} \approx 1800$ K or above is required for thermal reduction. In other words, the intersection of an isoenthalpic (constant-ΔH_v^o) line with a vertical line passing through ΔS_v^o gives the minimum reduction temperature.

For water splitting (Equation 1.9), the Gibbs free energy change is

$$\Delta G_{ox} = \frac{1}{d\delta}\left(G_{MO_{x-\delta}}^o - G_{MO_{x-\delta-d\delta}}^o\right) + G_{H_2}^o - G_{H_2O}^o + RT_{ox}\ln\left(\frac{p_{H_2}^{ox}}{p_{H_2O}^{ox}}\right)$$
$$(1.18)$$

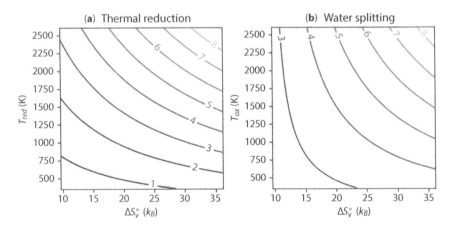

Figure 1.4 Thermodynamics of hydrogen production via a two-step solar-thermochemical cycle at (a) $p_{O_2}^{red} = 1 \times 10^{-4}$ bar for thermal reduction and (b) $\theta = 0.1$ for water splitting. The different curves are isoenthalpic lines, and the numbers (colors) are (correspond to) ΔH_v^o in eV. Spontaneous reaction may occur for T_{red} above the isoenthalpic lines for a given ΔH_v^o.

where T_{ox} is the reoxidation temperature, and $p_{H_2}^{ox}$ and $p_{H_2O}^{ox}$ are the outlet partial pressures of $H_2(g)$ and $H_2O(g)$, respectively, for the water-splitting reaction at T_{ox}. To obtain ΔG_{ox} in terms of ΔG_v^o, we consider the Gibbs free energy change of gas-phase water splitting [Equation (1.4)], i.e.,

$$\Delta G_{ws}^o = G_{H_2}^o + \frac{1}{2}G_{O_2}^o - G_{H_2O}^o = -RT_{ox}\ln K_{ws} \quad (1.19)$$

Here, K_{ws} is the equilibrium constant of water splitting (abbreviated ws). Substituting this result into Equation (1.18) and using the definition of ΔG_v^o [Equation (1.17)], one arrives at the following equality for reversible/equilibrium ΔG_{ox}:

$$\Delta G_{ox} = -\Delta G_v^o - RT_{ox}\ln\left(\frac{1-\theta}{\theta}K_{ws}\right) = 0 \quad (1.20)$$

Note that we have replaced $p_{H_2}^{ox}/p_{H_2O}^{ox}$ with $\theta/(1-\theta)$, where θ is the outlet H_2/H_2O conversion ratio. For example, if we begin with one mole of $H_2O(g)$ and the reaction proceeds 10 %—i.e., with $\theta = 0.1$—then we end with $1 - \theta = 0.9$ or 0.9 moles of $H_2O(g)$ and 0.1 moles of $H_2(g)$. For off-stoichiometric metal oxides, $\theta = 0.1$ is a reasonable target value for the water-splitting step [98]. To explore the thermodynamics of water splitting graphically, Figure 1.4b shows the relationship between T_{ox} (vertical axis), ΔS_v^o (horizontal axis), and ΔH_v^o (colors and numbers). The curvature of the $\Delta H_v^o = 3$ eV isoenthalpic line indicates that T_{ox} depends very sensitively on ΔS_v^o, with values ranging from ≈600 K to ≈1800 K for $\Delta S_v^o = 17$ k_B and 12 k_B, respectively. Therefore, modulating T_{red} and T_{ox} demands careful control of the material-specific parameters, ΔH_v^o and ΔS_v^o. Based on the water-splitting isoenthalpic lines in Figure 1.4b, it is clear that—to ensure practical temperatures and temperature swings—ΔH_v^o must be between 3 eV and 4 eV for metal oxides with typical ΔS_v^os of 12-17 k_B. Solving Equations (1.16), (1.17), and (1.20) for ΔH_v^o gives

$$\Delta H_v^o = \frac{RT_{red}\ln\left[(1-\theta)\theta^{-1}K_{ws}\left(p^o/p_{O_2}^{red}\right)^{1/2}\right]}{1 - T_{red}/T_{ox}} \quad (1.21)$$

If one selects the target reducing conditions $T_{red} = 1800$ K and $p_{O_2}^{red} = 1\times10^{-4}$ to 1×10^{-3} bar and reoxidizing conditions $T_{ox} = 1200$ K and

16   Solar Fuels

$\theta = 0.1$, then the optimal ΔH_v° is 3.5 to 3.9 eV at $p^{\circ} = 1$ bar and $K_{ws}(T_{ox} = 1200\text{ K}) = 1.2582 \times 10^{-8}$ [16].

The amount of $H_2(g)$ produced from one mole of $MO_{x-\delta}$ and one mole of water in a cycle depends on both δ and θ [102–104], where δ is the number of moles of O that one mole of the metal oxide can extract from $H_2O(g)$ after thermal reduction and θ is the conversion yield of water. One can measure δ by thermogravimetric analysis or coulometric titration [105–109]. For the former, one measures the mass of a sample over time as the temperature and partial pressure of oxygen changes. For the latter, one uses a constant current system to quantify the partial pressure of $O_2(g)$ accurately.

Kinetics also place important constraints on the design of STWS applications [110–116]. For example, low temperatures (assuming the reoxidation reaction is exothermic) improve the thermodynamics of water splitting [33] but lead to sluggish kinetics and therefore suppress the rate of $H_2(g)$ production. High temperatures alleviate these kinetics issues but disfavor the spontaneity and conversion yield of water splitting. According to Equation (1.16), reducing $p_{O_2}^{red}$ reduces T_{red} for the same reduction extent. There are two ways to reduce $p_{O_2}^{red}$, inert gas (N_2 or Ar) sweeping [101] and vacuum pumping [117], and each has its own challenges.

1.1.4   Economics

Before describing metal-oxide design directions, it is essential to consider the economics of STWS [118, 119]. The U.S. DOE recently designated the Hydrogen Shot target to be $1/kg for clean H_2 within a decade, which—if achieved—could lead to a five-fold increase in hydrogen use and mostly from clean hydrogen. One estimate in a technoeconomic assessment [119] of a plant co-producing hydrogen and electricity with ceria as the redox-active material suggests an nth of a kind commercial scale plant might produce H_2 at a cost of $4.55/kg. In that study, component prices (e.g., ≈$22,500,000 for a single 27.74 MW tower system) contributes ≈9.63% of the cost. The $4.55/kg H_2 is more than four times the target value. For hybrid cycles, where excess heat produces electricity, opportunities exist to decrease cost by increasing solar field efficiency, increasing revenue from electricity, and reducing the financial capital recovery factor. Capitalizing on these opportunities and others mentioned in a recent technoeconomic analysis by one of the authors [119], a realistic estimate of the minimum cost achievable, in the absence of some unforeseen technological disruption, is $2.09/kg H_2.

One way to decrease the price of $H_2(g)$ from STWS is to avoid the costs associated with the temperature swing between thermal reduction and water splitting (i.e., $\Delta T = T_{red} - T_{ox} \neq 0$) via isothermal [120–122] cycles, i.e., where $\Delta T = 0$. However, a recent analysis by Bayon *et al.* indicates that such cycles require operation at $T_{red} = T_{ox} \approx 2030$ K and utilization of redox-active metal oxides with specific values of ΔH_v^o and ΔS_v^o [98]. Another strategy to reduce the cost of solar thermochemical $H_2(g)$ is to design, e.g., using quantum mechanics simulation techniques, new redox-active materials that are composed of more abundant elements and provide ideal thermochemical characteristics that, for off-stoichiometric metal oxides, include an optimal oxygen vacancy formation energy $E_v \approx \Delta H_v^o$ [for a greater reduction extent, δ, than ceria and therefore $H_2(g)$ productivity [95] per unit of redox-active material] and a tunable ΔS_v^o, which—like E_v—improves δ and $H_2(g)$ productivity but also controls ΔT and thus works to meet the challenge of solid-solid heat recuperation.

1.2 Theoretical Methods

Going forward, the main emphases of this chapter are two-fold: (1) to introduce state-of-the-art and next-generation off-stoichiometric redox-active metal oxides and to explain what has made them effective or promising and (2) to outline how one might go about designing superior off-stoichiometric metal-oxide materials for two-step STWS. For the latter, many of the strategies we will outline for the redox-active material's design draw on insights from quantum mechanics simulations. Generally speaking, this endeavor requires quantum mechanics because the off-stoichiometric metal oxides for two-step STWS typically contain both ionic and covalent bonds (where

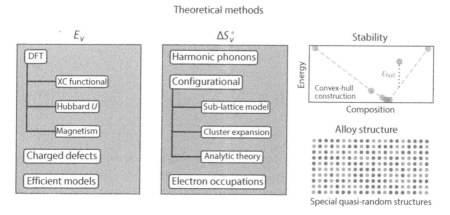

Figure 1.5 Theoretical methods for modeling redox-active metal oxides for STWS. Considerations when computing ΔG_v^o, broken down by E_v, ΔS_v^o, materials bulk stability, and structure. Each subtopic is addressed in the text.

the latter, of course, have a quantum mechanical origin), and redox-active transition metal cations (which exhibit quantum mechanical effects, such as crystal-field energy-level splitting and magnetism). Computational designers of metal oxides that undergo partial thermal reduction have four goals (Figure 1.5): (1) optimize E_v [97], (2) tune ΔS_v^o, (3) control stability (which also applies for metal oxides that undergo stoichiometric thermal reduction), and (4) construct a realistic structural model (which is especially important for alloys).

1.2.1 Oxygen Vacancy Formation Energy

The first goal necessitates an accurate method for calculating E_v. The method of choice is density functional theory (DFT), for which its foundational theorems prove that one can express the total energy of a quantum mechanical system in its ground state simply as a functional of the electron density (n), a function of only three spatial coordinates. However, to obtain accurate electron kinetic energies, its usual implementation introduces one orbital for each electron, raising the complexity to be a function of $3N$ coordinates where N is the number of electrons and with an algorithmic scaling of typically $\sim N^3$. By contrast, conventional algorithms for more exact many-body-wavefunction methods for directly solving the Schrödinger equation scale typically as N^5-N^7. DFT approximations provide an accuracy-efficiency compromise for routinely computing total energy and enthalpy changes for chemical reactions and materials of up to a few hundred atoms at 0 K and for all species in their electronic ground states. There are four terms in the DFT energy functional,

$$E[n] = \int d\mathbf{r}\, n(\mathbf{r}) V_n(\mathbf{r}) - \sum_i \int d\mathbf{r}\, \phi_i^*(\mathbf{r}) \frac{\nabla^2}{2} \phi_i(\mathbf{r}) + \frac{1}{2} \iint d\mathbf{r}\, d\mathbf{r}' \frac{n(\mathbf{r}) n(\mathbf{r}')}{|\mathbf{r}-\mathbf{r}'|} + E_{xc}[n] \qquad (1.22)$$

where $n(\mathbf{r})$ is the electron density, V_n is the nuclear potential, i is the electron index, ϕ are the one-electron wavefunctions (required because the exact kinetic energy density functional is unknown), and ∇ is the Laplacian. The first three terms deal with the classical electrostatic attraction between electrons and nuclei, the quantum mechanical kinetic energy of the electrons, and the classical electrostatic repulsion between electrons, respectively. However, the exact form of the final term, which describes electron exchange and correlation (or XC), is unknown, so approximations are necessary and always must be validated for the systems of interest.

A logical starting point for developing XC functionals [123–125] is to assume that $E_{xc} = E_x + E_c$ is that of the homogeneous electron gas, i.e., homogeneously distributed electrons in a box. This local density approximation (LDA) is helpful because an exact analytic expression and numerical solution exist for E_x and E_c, respectively. However, the actual electron density has curvature, especially for materials with localized chemical bonds, high angular-momentum electrons, or defects. As a result, one should include information about the higher-order gradients of the electron density in constructing the XC functional. Perhaps the most widely used XC functional is based on the generalized gradient approximation (GGA) of Perdew, Burke, and Ernzerhof (commonly abbreviated as PBE [126]), which includes information about the electron density and its first derivative. Compared to the LDA, the PBE GGA dramatically improves predictions of the energetics of bond breaking and formation, which involves mostly localized electrons and therefore large gradients of the electron density at the reaction site. More recently, Sun et al. showed that inclusion of information about the second derivative of the electron density in the so-called strongly constrained and appropriately normed (SCAN) meta-GGA satisfies all 17 known XC constraints and provides remarkable accuracy for many solids [127].

DFT within the PBE GGA or SCAN meta-GGA, however, suffers from self-interaction errors (SIEs) not completely eliminated by the approximate XC functional that are introduced by the interaction of each electron with the entire electron density (including its own density) in the Coulomb energy functional. SIEs can lead to spurious delocalization of electrons, which is especially problematic for open-shell and redox-active transition metal compounds, whose d electrons can be localized and spin-polarized. To at least partly ameliorate the SIE, one can apply a Hubbard U correction to the DFT total energy, e.g., using the rotationally invariant approach introduced by Dudarev et al. [128], i.e.,

$$E_U[\{n_{mm}^{l\sigma}\}] = \sum_{l,\sigma} \frac{U^l}{2} \mathrm{Tr}[\mathbf{n}^{l\sigma}(1-\mathbf{n}^{l\sigma})] \qquad (1.23)$$

where l and σ are the angular momentum and spin quantum numbers, n_{mm} is a diagonal element of the on-site occupancy matrix $\mathbf{n}$, U_l is the effective on-site Coulomb parameter, and Tr is the matrix trace of the quantity in brackets. Equation (1.23) can be understood as a total energy correction that drives the on-site occupancy matrix in the direction of idempotency, i.e., it penalizes noninteger electron occupation numbers. To correct the considerable error in the calculated redox reaction energies of many transition-metal oxides, which arises from the SIE in the PBE GGA, Wang *et al.* calibrated PBE+U values for oxides containing the following transition metals: Co, Cr, Fe, Mn, Mo, Ni, V, and W [129]. Recently, one of the authors and her coworkers showed that the SCAN+U [130, 131] framework more accurately reproduces the ground-state structure, lattice parameters, magnetic moments, and electronic properties of transition-metal oxides. We summarize the optimized U values for PBE+U and SCAN+U in Table 1.1. Since SCAN theoretically is a more accurate XC functional than PBE, i.e., it includes a more accurate description of electron exchange, it is not surprising that SCAN

Table 1.1 Hubbard U values for XC+U calculations fit to relevant oxidation energies (unless otherwise noted). For example, the U value for SCAN+U calculations of Ce oxides was fit to reproduce the experimental enthalpy of the following reaction: $4CeO_2(s) \rightleftharpoons 2Ce_2O_3(s) + O_2(g)$. References are enclosed in brackets.

Element	PBE+U (eV)	SCAN+U (eV)
Sc	3.00 [138]	0 [131]
Ti	3.00 [138]	2.5 [131]
V	3.25 [129]	1 [131]
Cr	3.7 [129]	0 [131]
Mn	3.9 [129]	2.7 [130]
Fe	5.3 [129]	3.1 [130]
Co	3.32 [129]	3 [131]
Ni	6.2 [129]	2.5 [131]
Cu	3.6 (UHF-derived) [139]	0 [131]
Mo	4.38 [129]	n/a
W	6.2 [129]	n/a
Ce	2-3 [140, 141]	2 [130]

requires a lower U correction [132]. Transition-metal oxides also frequently exhibit magnetic degrees-of-freedom such as long-range magnetic order (ferromagnetic, antiferromagnetic, ferrimagnetic, and nonmagnetic) and local spin state (low, intermediate, and high), where the latter is especially relevant for compounds containing transition-metal cations, which can have varying d-electron counts and crystal-field splittings.

Armed with Hubbard-U-corrected DFT, one can compute E_v, which can be written as

$$E_v = E_{defect}^{supercell} - E_{bulk}^{supercell} + \frac{1}{2}E_{O_2} + q(E_F - \epsilon_{VBM}) + E_{corr} \quad (1.24)$$

where $E_{defect}^{supercell}$ and $E_{bulk}^{supercell}$ are the total energy—calculated using one's preferred flavor of DFT XC and corrections—of a supercell with (defect) and without (bulk) the oxygen vacancy, E_{O_2} is the total energy of an $O_2(g)$ molecule, q is the charge of the defect (e.g., $q = 2$ for the removal of O^{2-}), $E_F - \epsilon_{VBM}$ is the Fermi energy relative to that of the valence band maximum (VBM), and E_{corr} fixes finite-size effects deriving from the use of periodic DFT and the supercell approach [133, 134]. The latter two terms only appear in the case of charged defects. A variety of correction procedures exist for the calculation of charged defects under periodic boundary conditions. In the state-of-the-art correction schemes proposed by Freysoldt, Neugebauer, and Van de Walle (FNV) [135], and Kumagai and Oba (KO) [136], they express E_{corr} as

$$E_{corr} = -E_{lat} + q\Delta\phi \quad (1.25)$$

where E_{lat} includes the interaction between the defect-induced charge density, the host material, and the neutralizing jellium (uniform compensating charge) background, and $\Delta\phi$ is a term that aligns the electrostatic potential of the defective and pristine materials with one produced by a model defect-induced charge density in an area of the material distant from the defect. The FNV procedure employs the plane-averaged electrostatic potential to calculate $\Delta\phi$, whereas the KO method uses atomic-site potentials. See ref [137] for a thorough overview of these and other correction schemes. The supercell approach can make the DFT calculations expensive. Additionally, brute-force DFT calculations do not necessarily explain why some materials favor and others disfavor V_O formation. Therefore, there is a need for efficient methods of calculating E_v—especially for metal oxides with disordered sublattices—using, e.g., phenomenological model building and machine learning; we will survey such methods in the subsection Enthalpy Engineering in the section Materials Design Directions.

1.2.2 Standard Entropy of Oxygen Vacancy Formation

The second goal calls for an approach to compute the various contributions to ΔS_v: the translational, rotational, and vibrational entropy of gas-phase O_2 (ΔS_g, where "g" means "gas"); the phonon entropy change of solid-phase $MO_{x-\delta}$ (ΔS_p, where "p" means "phonon"); the configurational entropy change upon cation sublattice reduction (ΔS_c, where "c" means "configurational"); and the electronic entropy change (ΔS_e, where "e" means "electronic"). Note that for the rest of the chapter we drop the ° from ΔS_v° for simplicity but it and its contributions remain standard entropy changes. Given the complexity and cost associated with calculating ΔS_g from first principles, one typically takes its measured value (e.g., ΔS_g = 15.9 k_B per ½O_2 at 1800 K and 1 bar) from standard databases, such as NIST-JANAF [142]. For lower temperatures (i.e., usually <1000 K), accurate approaches for calculating ΔS_p include the frozen phonon method and density functional perturbation theory [143], invoking the harmonic approximation [144]. Calculating ΔS_p for higher temperatures requires the use of techniques that capture phonon anharmonicities, such as molecular dynamics (MD) simulations based on fluctuation-dissipation theory [145–148].

Calculating ΔS_c can be approached in a few different ways. The first scheme uses ideal-solution-based models to describe the entropy associated with the disordered reduction of cation sublattices by neutral-V_O donated electrons. Per mole of cation sublattice, the ideal solution-phase configurational entropy is given by

$$\Delta S_c^{ideal} = -nR \sum_i x_i \ln x_i \qquad (1.26)$$

where n is the number of moles, i is the component index, and x is the mole fraction. In the first-principles-based sublattice formalism recently developed by Sai Gautam et al. [145–148], the excess entropy not captured by the ideal solution contributions is expressed in terms of binary interaction parameters (L) as within the compound energy formalism [150, 151]. For a system of two sublattices (e.g., one cation and one anion) and two components on each sublattice (e.g., oxidized metal M_{ox} and reduced metal M_{red} on the cation sublattice, and O and V_O on the anion sublattice), one can write the excess entropy as

$$\Delta S^{excess} = -\Delta G^{excess}/T$$
$$= y^c_{M_{ox}} y^c_{M_{red}} y^a_O L_{M_{ox},M_{red}:O}$$
$$+ y^c_{M_{ox}} y^c_{M_{red}} y^a_{V_O} L_{M_{ox},M_{red}:V_O} \quad (1.27)$$
$$+ y^c_{M_{ox}} y^a_O y^a_{V_O} L_{M_{ox}:O,V_O} + y^c_{M_{red}} y^a_O y^a_{V_O} L_{M_{red}:O,V_O}$$

where y^z_X is the site fraction of the species X on the z sublattice [c = cation; a = anion in Eq. (1.27)].

While precedence for such sublattice models accurately describing phase behavior does exist, configurational entropy does not always exhibit ideal behavior in multicomponent, multi-sublattice systems. One can directly evaluate deviations from ideal behavior by converting the grand canonical output of cluster expansion-based Monte Carlo simulations into canonical quantities [152];

$$\langle S \rangle = \frac{1}{T}\left(\langle E \rangle - \Phi - \sum_i \mu_i \langle n_i \rangle \right) \quad (1.28)$$

where $\langle E \rangle$ is the thermodynamically averaged energy, Φ is the grand potential obtained by thermodynamic integration along a fixed T or fixed chemical potential (μ) path, and n_i is the number of species i exchanged with the μ reservoir. The main challenge with such an approach is parameterizing an accurate cluster-expansion Hamiltonian, which can sometimes require hundreds of first-principles quantum calculations. Additionally, it does not explicitly consider the effect of lattice expansion and vibrations on configurational entropy [153]. Note that while analytic approaches do exist for the calculation of ΔS_v, e.g., the statistical-thermodynamic theory of Ling (which applies to high concentrations of point defects) [154], we will not discuss them here because their derivations are involved.

Finally, we must establish an approach for computing ΔS_e, i.e., the electronic entropy changes upon changing the filling of a particular electron shell (e.g., 4f). In the opposing crystal potential method of Zhou and Aberg [155], one obtains the constraining Lagrange multipliers that act as a cancellation potential against the crystal field and lead to a spherical d-electron distribution. Table 1.2 lists the magnitudes of these different kinds of entropy contributions.

Table 1.2 Magnitudes of the different kinds of entropy contributions. Exp and RA means experiment and redox-active, respectively.

Contribution	ΔS [k_B per ½O_2(g) produced]	Reference
O_2(g)	15.9	[16] (exp, T = 1800 K)
Phonon	2.5 (for $CeO_{1.97}$)	[144] (theory, T > 298.15 K)
Configurational	Ideal* 9.7 (for $CeO_{1.97}$) 10.1 (for $ABO_{2.97}$, one RA cation) 11.5 (for $ABO_{2.97}$, two RA cations, one e⁻/cation) Cluster-expansion-based Monte Carlo 5.9 ($\Delta S_p + \Delta S_c$, for $CeO_{1.97}$)	[152] (theory, T = 1480 K)
Electronic	4.5 (Ce^{4+} Ce^{3+}, in $CeO_{1.97}$)	[156] (theory, T = 1800 K)
Total	26.1 (for $CeO_{1.97}$)	[157] (exp, independent of T)

1.2.3 Stability

Of equal importance to a material's thermochemical properties (E_v and ΔS_v) is its stability with respect to decomposition into other compounds with the same summed stoichiometry. One can calculate the stability of a material by computing the Gibbs free energies of the redox-active material and all relevant secondary phases and then using the convex hull construction to calculate the phase diagram [158, 159]. For an isothermal, isobaric, closed system, the appropriate thermodynamic potential is G, which one can express as follows

$$G(T, P, \{N_i\}) = E(T, P, \{N_i\}) + PV(T, P, \{N_i\}) - TS(T, P, \{N_i\})$$
$$\approx E(T, P, \{N_i\}) - TS(T, P, \{N_i\}) \quad (1.29)$$

where E is the system's internal energy, T is the temperature, S is the entropy, P is the pressure, V is the volume, and N_i is the number of atoms of species i. Note that if one performs static DFT calculations, then T = 0 K. Therefore, one must take additional steps to evaluate the stability of the redox-active materials at $T \neq 0$ K. These steps generally involve including the relevant sources of entropy for a particular phase, which for solids mostly are due to vibrations and for gases are due to vibrations, rotations, and translations.

As mentioned above, for temperatures not too much higher than 298.15 K, the harmonic approximation provides an accurate and efficient means to compute vibrational entropies. At the temperatures relevant for thermal reduction (e.g., near 2000 K), phonon anharmonicities must be considered, which requires computation of, e.g., phonon spectra from DFT-MD simulations, which can be quite expensive. Phase transitions, such as melting, are also relevant at thermal reduction temperatures. However, currently, the only approach to accurately determine the melting point involves—in the absence of comprehensively validated classical interatomic potentials—costly DFT-MD simulations of solid-liquid coexistence.

1.2.4 Structure

One often would like to replace partially one of the metal elements in a metal oxide with another metal element to tune E_v, ΔS_v, and the stability. Such random alloys are hard to model quantum mechanically because the system dimensions that lend themselves to DFT calculations are often too small to accommodate a structure that one may regard as random. In other words, the periodicity imposed by boundary conditions introduces spurious correlations that make the modeled system deviate from the solid solution. One sometimes can alleviate this problem using so-called special quasi-random structures (SQSs) [160–162]. Finding the SQS generally amounts to minimizing an objective function that quantifies the difference between the current structure's site occupations and that of the random alloy. One such objective function (Q) is

$$Q = \sum_{\alpha \in A} \left| \Gamma_\alpha - \Gamma_\alpha^{target} \right| - \omega \quad (1.30)$$

where α is a cluster (e.g., pairs and triples of atoms within a prescribed distance cutoff), A is the list of all considered cluster definitions, and Γ is a cluster vector, whose elements are the average correlation (e.g., the product of pseudo-spin site identifiers) for each α. One compares the cluster vector of the SQS (Γ) with that of the random alloy (Γ^{target}), where the sum of their element-wise absolute deviations describes how much they differ. The second term in Equation (1.30) controls the importance of the distribution of the cluster vector deviations, where ω is the radius of the largest pair cluster, such that all clusters with the same or smaller radii have $\Gamma_\alpha - \Gamma_\alpha^{target} = 0$. SQS cells are the best possible approximations to random

1.2.5 Kinetics

Finally, we note that kinetics can play an essential role in the efficiency of STWS cycles. From transition state theory, the critical kinetic parameter during thermal reduction is the activation energy of V_O diffusion ($E_{a,diff}$). For the water-splitting step, $E_{a,diff}$ again affects the kinetics though not as significantly as the activation energy of $H_2O(g)$ dissociation ($E_{a,diss}$). That said, some metal oxides dissociatively adsorb $H_2O(g)$ with no barrier [163–165], i.e., $E_{a,diss} = 0$. The activation energy typically is computed using a transition-state search algorithm, such as the nudged elastic band (NEB) method [166], which finds saddle points and minimum energy paths between predetermined reactants and products. The method works by optimizing several intermediate interpolated structures (images) along the reaction path. Each image finds the lowest energy possible while maintaining equal spacing to neighboring images. This constrained optimization is achieved by adding spring forces along the band between images and projecting out the forces' component due to the potential perpendicular to the band. In the original implementation of the NEB method, the highest energy image will not always be at a saddle point. The climbing image (CI) modification drives this image up to the nearest saddle point by removing its spring forces along the band [167, 168]. In this way, the image tries to maximize its energy along the band and minimize in all other directions. Other methods for finding the transition state include the modified single iteration synchronous-transit approach of Trottier *et al.* [169] and the modified CI-NEB approach of Caspersen and Carter [170], which, respectively, expedite the transition state search in solid-state reactions and extend the CI-NEB approach to solid-solid phase transitions that involve changes in the cell shape and volume. Once E_as are obtained, one can use them, along with pre-exponential factor estimates, in the construction of microkinetic models that determine the steady-state or time-evolving reaction yields as a function of, e.g., temperature, pressure, and the chemical potentials of the species in the system [171].

1.3 The State-of-the-Art Redox-Active Metal Oxide

To date, the most widely implemented off-stoichiometric metal oxide for STWS is $CeO_2(s)$ (ceria) [157, 172–185]. $CeO_2(s)$ is fluorite-structured and crystallizes in the cubic $Fm\overline{3}m$ space group. Ce^{4+} bonds to eight equivalent O^{2-} atoms in a body-centered cubic geometry. Each O^{2-} bonds to four identical Ce^{4+} atoms to form a mixture of edge and corner-sharing OCe_4 tetrahedra. A favorable property of ceria is its ability to exchange oxygen via storing and releasing oxygen reversibly [186], i.e.,

$$\frac{1}{\delta}CeO_2(s) \rightleftharpoons \frac{1}{\delta}CeO_{2-\delta}(s) + \frac{1}{2}O_2(g) \qquad (1.31)$$

Experimental findings show that ΔH_v varies with δ as

$$\Delta H_v(\delta) = b + m \log \delta = 4.09 - 0.33 \log \delta \qquad (1.32)$$

where the intercept and slope are in eV [187]. Using a statistical thermodynamics model [187] of dilute defect clusters ($Ce^{3+}V_OCe^{3+}$), Bulfin et al. derived the following equation of state for the equilibrium composition off-stoichiometric ceria:

$$\left(\frac{\delta}{\delta_m - \delta}\right)^{1/\delta_m} = \left(\frac{p_{O_2}}{p^o}\right)^{-1/2} \exp\left(\frac{\Delta S_g + \Delta S_p}{R}\right) \exp\left(-\frac{\Delta H_v(S)}{RT}\right)$$

(1.33)

Here, δ_m is the maximum oxygen off-stoichiometry (Appendix B. Equilibrium Composition of Ceria). Note that we are using the model of Bulfin et al. [187] and not the state-of-the-art thermodynamic model of Zinkevich [150] for pedagogical purposes as the former provides a simple, intuitive, and closed-form equation of state. At constant p_{O_2}, an increase in T leads to the reduction of $CeO_{2-\delta}(s)$, i.e., an increase in δ (Figure 1.6a). At constant T, a decrease in p_{O_2} also leads to an increase in δ. While thermal reduction of ceria typically requires temperatures around 1800 K, ceria exhibits excellent thermal stability with good resistance to sintering (which slows kinetics considerably) and a high melting point of 2670 K.

The main drawback of $CeO_2(s)$ is its too high $\Delta H_v = 4.4$ eV compared to the ideal range of 3.4 to 3.9 eV. Metal doping or substitution to decrease ΔH_v has been considered extensively. Ceria dopants/substitutions generally fall into five categories based on their valence: monovalent, divalent, trivalent, and multi-valent dopants, and tetravalent substitutions (Figure 1.6b). First, we will list some examples from these categories. Two monovalent dopants studied are Li [188, 189] and K [190], where the former retains material porosity more effectively than pristine ceria [191]. Divalent dopants mostly have been sampled from the s-block alkaline earth metals—Mg [188, 192], Ca [188, 192, 193], Sr [188, 192, 193], and Ba [188]—though d-block Zn [194] also has been considered. Among the trivalent dopants studied, there are representative elements from the p block (Al [195]), d block (Sc [192, 194] and Y [192, 193, 196, 197]), and f block (La [196, 197], Sm [193, 197], Gd [193, 196, 197], and Dy [192]). While p-block elements, such as Si [79] and Sn [188], have been considered, the tetravalent d-block elements Zr [35, 110, 111, 191–193, 198–201] and Hf [191, 192] especially are promising, where the former is the most widely studied substitution. Experiments by Le Gal et al. show that the Ce^{4+} reduction yield ($y_{Ce^{3+}} \approx 2\delta$) increases linearly with Zr content (y_{Zr}) (Figure 1.6c) [196]. Therefore, Zr doping can be used to control the thermodynamics of V_O formation in ceria. Note that, even though Ti is directly above Zr in the periodic table, it softens samples at $T > 1700$ K, thereby limiting its use for STWS [191, 194]. Finally, multi-valent dopants have been examined as well, most of which are transition metals from the 3d block (V [194, 202], Cr [194, 203], Mn [178, 194], Fe [194, 204], Co [194], Ni [178, 194, 205], and Cu [178]), 4d block (Nb [79] and Rh [206]), and 5d block (Ta [175]); f-block Pr [191, 196] has been investigated too.

Now, we will develop an intuition for the relative effectiveness of Zr substitution compared to doping with other elements (Figure 1.6d) [38]. First, subvalent (<4+) doping decreases E_v [207] and increases δ [208, 209] to establish local charge neutrality. For example, the equilibrium composition of 10% Mg-doped ceria is $Ce_{0.9}Mg_{0.1}O_{1.9}$. Therefore, subvalent dopants decrease the number of reducible (i.e., fully coordinated) Ce^{4+}. Divalent transition-metal dopants also experience crystal field effects—e.g., the adoption of a square-planar coordination geometry—that increase the oxygen storage capacity of ceria [210]. Additionally, trivalent doping can affect the kinetics of V_O diffusion, where $E_{a,diff}$ correlates strongly with the ionic radius of the dopant [209]. Overall, aliovalent substitution does not improve significantly the Ce^{4+} reduction yield, $O_2(g)$ released upon thermal reduction, or STWS efficiency.

For tetravalent substitutions, a long-held assumption was that Zr and Hf decrease E_v by compensating for ceria expansion upon reduction. However, Muhich and Steinfeld recently suggested that Zr and Hf dopants increase the δ of ceria because they store energy in tensile-strained Zr- or Hf-O bonds, which is released upon O-vacancy formation [207]. Here, we provide an alternative hypothesis: Zr and Hf weaken O^{2-}-Ce^{4+} crystal bonds via bond order conservation [211]. To quantify crystal bond strength, we use the crystal bond dissociation energies (BDEs) developed by the authors [212]:

SOLAR THERMOCHEMICAL WATER SPLITTING MATERIALS DESIGN 29

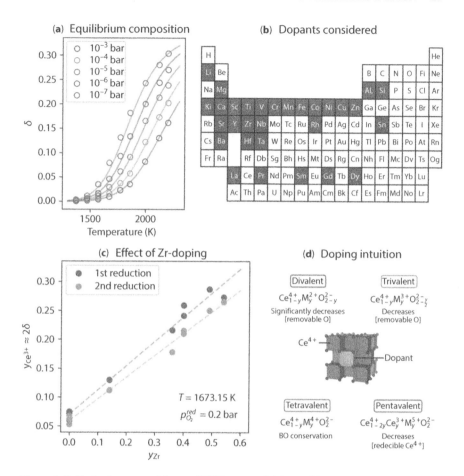

Figure 1.6 Ceria: the state-of-the-art STWS material. (a) Equilibrium composition of CeO_x as a function of temperature, showing the experimental data of Panlener et al. [157] and Dawicke and Blumenthal [213], (circles) and the kinetic model of Bulfin et al. [214] (lines, see Appendix B. Equilibrium Composition of Ceria). (b) Periodic table showing elements that have (blue) and have not (white) been considered as dopants or substitutions in ceria, based on the recent review by Bhosale et al. [174] (c) Percentage of Ce^{4+} reduced—which is $\approx 2\delta$—during the first (blue) and second (orange) cycles of thermal reduction for STWS by ceria with different Zr doping concentrations [196]. (d) Summarizing the effect of dopant valence on properties critical to STWS yield [207]. "[x]" can be read as "the concentration of x." For example, "[removable O]" means "the concentration of removable oxygen."

$$E_b[O^{2-} - M^{n+}] = \frac{-\Delta E_f[MO_{n/2}] + E_c[M] + (n/4)BDE[O_2]}{N_b[O^{2-} - M^{n+}]} \quad (1.34)$$

where n is the oxidation state of the metal (M) in the unreduced material, $\Delta E_f[MO_{n/2}]$ is the binary metal-oxide ($MO_{n/2}$) formation energy/enthalpy, $E_c[M]$ is the cohesive energy of the pure metal element, $BDE[O_2]$ is the bond dissociation energy of $O_2(g)$ per atom, and $N_b[O^{2-}-M^{n+}]$ is the number of $O^{2-}-M^{n+}$ crystal bonds per $MO_{n/2}$ formula unit. For Ce^{4+}, Zr^{4+}, and Hf^{4+}, the experimental E_b is 2.56 eV, 3.27 eV, and 3.32 eV. Based on these crystal BDEs, O^{2-} forms stronger crystal bonds with Zr^{4+} and Hf^{4+} than Ce^{4+}. Therefore, Zr- and Hf-doping weaken $O^{2-}-Ce^{4+}$ crystal bonds on average, decreasing E_v and increasing δ. Despite the enthalpy control offered by Zr and Hf substitutions, the reoxidation thermodynamics of substituted ceria are not as favorable as for pure ceria [112, 149], which highlights a fundamental thermodynamic constraint in the design of metal oxides for thermochemical redox cycles: if reduction is made easier, then reoxidation necessarily is made harder [27, 110].

1.4 Next-Generation Perovskite Redox-Active Materials

Ceria remains the redox-active metal oxide of choice for STWS analysis, lab scale, and demonstration scale, and its undoped form provides the best thermodynamics. Therefore, ways to improve ceria properties are limited primarily to nonchemical changes, e.g., morphology engineering. Researchers more recently have explored more flexible materials classes like metal-oxide perovskites with the goal of designing superior materials. ABO_3 perovskites commonly crystallize in six lattice systems: cubic (e.g., $Pm\bar{3}m$ $SrTiO_3$), hexagonal (e.g., $P6_3/mmc$ $SrMnO_3$), rhombohedral (e.g., $R\bar{3}c$ $LaCrO_3$), tetragonal (e.g., P4mm $BaTiO_3$), orthorhombic (e.g., Pnma $CaTiO_3$), and monoclinic (e.g., $P2_1/b$ $CeVO_3$). The A- and B-site cations usually bond to O^{2-} in 12- and 6-coordinate geometry, respectively. The latter typically forms corner-sharing BO_6 octahedra but also can adopt edge- and face-sharing octahedra. Octahedra can take on tilt angles and patterns and exhibit Jahn-Teller distortion, thus demonstrating the diverse design degrees-of-freedom in the metal-oxide perovskite materials class.

Like ceria, redox-active ABO_3 perovskites can store and release oxygen reversibly

$$\frac{1}{\delta}ABO_3(s) \rightleftharpoons \frac{1}{\delta}ABO_{3-\delta}(s) + \frac{1}{2}O_2(g) \qquad (1.35)$$

where $ABO_{3-\delta}$ is an off-stoichiometric, metal-oxide perovskite. Perovskites mostly fall under three categories: La-based perovskites, alkaline-earth-based perovskites, and layered, Ruddlesden-Popper perovskites. For La-based perovskites, La^{3+} occupies the A-site and a 3+ cation from either the p-block (e.g., Al^{3+} or Ga^{3+}) or d-block (e.g., Mn^{3+} or Fe^{3+}) resides on the B-site. The most widely investigated La-based perovskite composition is Sr-doped $LaMnO_3$ (LSM) [105, 215–217], which yields an estimated $\eta_{solar-to-fuel}$ lower (16%) than ceria (22%) at 1800 K but higher (13%) than ceria (7%) at 1600 K. The use of Al as a dopant in Sr-doped manganate aluminates $(Sr,La)(Mn,Al)O_3$ (Figure 1.7a) [218–220] enhances thermal reduction at 1623 K and has been shown to be stable for at least 80 cycles. Other A-site (Ca [221–223]) and B-site (Fe [216, 224] and Co [216]) substituents have been considered; however, LSM achieves the largest $H_2(g)$ production capacity. Recently, Chen et al. reported that Sr- and Co-doped $LaGaO_3$ produces more O_2 (at $T_{red} = 1623$ K and $p_{O_2}^{red} = 5 \times 10^{-6}$ bar) and H_2 (at $T_{ox} = 1073$ K and $\theta < 0.01$) per mass of redox-active material than LSM and ceria [225]. Reports of $H_2(g)$ *production per mass of redox-active material*, however, underscores one of the key problems the STWS field is trying to remedy: an apples-to-apples comparison between different perovskites (and ceria) currently is not possible from the literature reports. For example, lighter elements will automatically look better when higher productivities are reported in moles per redox-active material mass. That said, in the absence of apples-to-apples comparisons, the following lessons about Al-doped LSM can be learned: (1) Sr^{2+}-doping on the A-site produces Mn^{4+}, which is very reducible and therefore can be used to tune E_v, and (2) Al^{3+}-doping increases cycling stability because $LaAlO_3$ has a high melting temperature $T_m = 2350$ K.

For alkaline-earth-based perovskites, Ca^{2+}, Sr^{2+}, or Ba^{2+} reside on the A-site, and 4+ cations from the d-block occupy the B-site. Until recently, the most promising alkaline-earth-based perovskite oxide for STWS was $BaCe_{0.25}Mn_{0.75}O_3$ (BCM, Figure 1.7b) [226], which exhibits faster reoxidation kinetics than $Sr_{0.4}La_{0.6}Mn_{0.6}Al_{0.4}O_3$ and yields 2.5× more $H_2(g)$ per atom than ceria when reduced at 1623 K. Note that these conditions correspond to lower $\eta_{solar-to-fuel}$ than at the target $T_{red} = 1723$ K from the second law of thermodynamics; the ideal scenario is to design a redox-active, metal-oxide perovskite that outperforms ceria at ≈ 1800 K. That said, BCM shows that one can use compositional engineering (i.e., with Ce^{4+} and Mn^{4+} cations on the B-site) to tune the thermodynamics (i.e., E_v) of metal-oxide perovskites. Other ABO_3 perovskites—like $SrTi_{0.5}Mn_{0.5}O_3$ (STM) [227]—have been proposed; however, STM provides an E_v that is too low (≈2-2.5 eV vs. the ideal range of 3.4-3.9 eV) for maximally efficient water splitting. Recently, Sai Gautam et al. evaluated the E_v and thermodynamic stability of $Ca_{1-x}Ce_xMO_3$ perovskites, where x = {0, 0.5, 1} and M = {Sc, Ti, V, Cr, Mn, Fe, Co, and Ni} (Figure 1.7c) [228]. $Ca_{0.5}Ce_{0.5}MnO_3$ (CCM) was identified to be a promising candidate, based on its favorable predicted E_v = 3.65-3.96 eV (range for symmetry-distinct V_Os), which is similar to BCM in that Mn^{4+} is redox-active but is dissimilar in that in CCM, Ce^{4+} is redox-active and on the A-site. Interestingly, Sai Gautam et al. pointed out that the reduction of both the A- and B-sites should give additional configurational degrees of freedom to increase the ΔS_v of CCM, rather than if only one cation is redox-active. While CCM is thermodynamically (meta)stable, with an E_{hull} = 39 meV/atom, experimentally it fails to cycle. We recently postulated and validated that Ti-doping increases its stability, enables cycling, and does not degrade its attractive thermodynamic properties (E_v and ΔS_v) for STWS [229]. While the field of metal-oxide perovskites for STWS is fairly young—i.e., about one decade's worth of research—the early returns have yielded some promising candidates (BCM and CCTM), and vast regions of materials design space remain unexplored, such as postquaternary compositions [early examples include $(Ba,Sr)(Co,Fe)O_3$ [216], Ce-doped $(Ba,Sr)MnO_3$ [230], and $(Y,Ca,Sr)MnO_3$ [231]] and layered structures [e.g., $(Ce,Sr)MnO_4$] [232, 233].

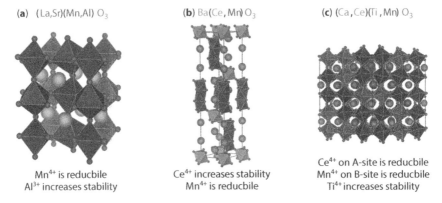

(a) $(La,Sr)(Mn,Al)O_3$
Mn^{4+} is reducbile
Al^{3+} increases stability

(b) $Ba(Ce,Mn)O_3$
Ce^{4+} increases stability
Mn^{4+} is reducbile

(c) $(Ca,Ce)(Ti,Mn)O_3$
Ce^{4+} on A-site is reducbile
Mn^{4+} on B-site is reducbile
Ti^{4+} increases stability

Figure 1.7 Promising perovskite metal oxides for STWS and the beneficial properties of the cations that comprise them.

1.5 Materials Design Directions

At this point, it is helpful to summarize what we know about the design of off-stoichiometric metal oxides for STWS. First, we know that the competing thermodynamics of thermal reduction and water splitting dictates a compromise (not too high and not too low) $\Delta H_v \approx E_v$ in the range of 3.4 to 3.9 eV, whereas ceria offers $E_v \approx 4.4$ eV [98]. Second, we know that configurational entropy is the most tunable contribution to ΔS_v, where cerium offers both ion/defect-disorder and electronic contributions. Third, we understand that the metal oxide must be the most stable compound at its composition and melt at temperatures well over 2000 K, where ceria does not form secondary phases upon thermal reduction and melts at 2670 K. Finally, we know that STWS kinetics should be fast; however, kinetics are usually fast for V_O diffusion and water splitting in off-stoichiometric/MIEC metal oxides compared to the kinetics of stoichiometric (line compounds) redox-active materials that undergo phase changes upon reduction and reoxidation [179]. We will not consider kinetics further here, but it certainly is important to consider in terms of the cycle times (longer cycle times will have a large impact on the economics) once satisfying the thermodynamic constraints.

1.5.1 Enthalpy Engineering

To design off-stoichiometric metal oxides with greater promise than ceria, one must identify materials with a lower $\Delta H_v \approx E_v$, between 3.4 and 3.9 eV. There are two main approaches: (1) high-throughput computational screening of E_v [108, 234–236] and (2) the development of electronic structure and thermodynamic descriptors for E_v. For example, Emery et al. took the first approach, calculating E_v using spin-polarized PBE+U for more than 11,000 ternary metal-oxide perovskites in cubic and distorted crystal systems containing s-, p-, d-, and f-block metals on the A- and B-sites [236]. While that work has yet to identify any suitable (meeting multiple criteria) redox-active materials for STWS, it provides a valuable data set for future analysis of E_v and stability trends in perovskites.

In contrast, the second approach offers physical intuition for E_v that one can use to minimize both the number and cost of calculations required for materials discovery. One of the first electronic descriptors proposed for E_v was partial charges [120, 237]. In 2014, Michalsky et al. showed that E_v correlates positively ($R^2 \approx 0.63$) with the partial charge on the metal atom for a diverse collection of metal-oxide surfaces [120]. The logic here would

34 Solar Fuels

be that the more positive the partial charge on the metal atom, the stronger its electrostatic attraction to O^{2-} would be and, therefore, the higher its E_v should be. One year later, Ezbiri et al. found an even stronger correlation between E_v and the partial charge on the oxygen atom for several ABO_3 perovskites [237]. Here, one can apply the same logic as for the metal partial charges, mutatis mutandis.

That same year, Deml et al. published a phenomenological model for the E_v of 45 main-group and transition-metal oxides, covering a range of compositions and crystal structures. Their model approximates E_v as

$$E_v = 0.72\left[0.60\left(E_{Op} + \frac{3}{4}E_g + 2.60\langle\Delta\chi\rangle\right) + |\Delta H_f|\right] - 2.07 \quad (1.36)$$

E_{Op} is the energy difference between the valence band maximum and the O 2p band center, E_g is the PBE+U bandgap, $\langle\Delta\chi\rangle$ is the average Pauling electronegativity difference between O and its nearest metal neighbors, and ΔH_f is the formation enthalpy of the metal oxide [238]. This model introduces two new electronic descriptors—E_{Op} and E_g. E_{Op} relates to the oxygen partial charge because the greater the energy difference between the valence band maximum and the center of the O 2p band, the greater the occupation of the O 2p band and the more negative the partial charge on O; for this reason, E_v correlates positively with E_{Op}. In contrast, E_g includes the effect of electron (de)localization, where smaller and larger E_g correspond to greater delocalization and localization, respectively. When a neutral V_O forms, the departing oxygen leaves behind two electrons that reduce the lattice. If E_g is small, those electrons donate to metallic-like, delocalized bands that manifest in spatial delocalization over multiple ions, which stabilizes the V_O, corresponding to lower E_vs [239]. Conversely, if E_g is large, then those donated electrons localize on neighboring cations in high-energy conduction-band states, which destabilizes V_O. Therefore, larger E_gs correspond to higher E_vs, thereby explaining the positive correlation between E_v and E_g for both large and small band gap materials.

The model of Deml et al. also introduces two thermodynamic descriptors, $\langle\Delta\chi\rangle$ and ΔH_f. Whereas $\langle\Delta\chi\rangle$ ostensibly could correlate with BDEs of neutral metal-oxygen diatomic molecules [240], ΔH_f captures—albeit indirectly—the metal-oxygen bond strength in crystals and the effect of bulk stability on E_v [120, 241]. The orange data in Figure 1.8a shows the predictive capability of this model, as the PBE+U-calculated and model-predicted E_v are in excellent agreement with only two outliers. While it is unclear why the $\langle\Delta\chi\rangle$ and ΔH_f proportions are $0.72 \cdot 0.60 \cdot 2.60 = 1.87$ [products of the coefficients for $\langle\Delta\chi\rangle$ in Equation (1.36)] and 0.72, respectively, Deml et al. suggest that the combination of E_{Op} and E_g estimates the energy to donate V_O-generated electrons from the O 2p band to defect states in the gap. Recently, we developed crystal features analogous to gas-phase BDEs and standard reduction potentials, namely crystal BDEs [E_b, Equation (1.34)] and crystal reduction potentials [212], which we define as

$$V_r[M^{n+} \to M^{m+}] = -E_r[M^{n+} \to M^{m+}]/(n-m)F \quad (1.37)$$

where n and m are the oxidation states of the oxidized and reduced metals in the ground-state polymorphs of their binary metal-oxide crystals $MO_{n/2}(s)$ and $MO_{m/2}(s)$, respectively, F is the Faraday constant, and E_r is the (free) energy change of $MO_{n/2}(s)$ reduction to $MO_{m/2}(s)$ and $(n-m)/4$ $O_2(g)$. Subsequently, we constructed a thermodynamic model in the spirit of Hess' Law and Born-Haber cycles using E_b, V_r, E_g, and the energy above the convex hull E_{hull}. Our model for Deml et al.'s data (where we computed V_r and E_g from PBE+U data [242, 243] on their set of crystals) is

$$E_v = -1.2 \max_{NN} V_r + 0.3 E_g + 1.87 \text{ eV} \quad (1.38)$$

where we choose the maximum V_r value among the nearest-neighbor (NN) cations of a specific V_O, emphasizing the essential role of V_r in controlling E_v, seeing that it is one of only two features needed to attain excellent agreement between the PBE+U-calculated and model-predicted E_v (blue data in Figure 1.8a). Our results also show that the presence of V_r in the model eliminates the two outliers. Additionally, $-c_r = 1.2$ has a physical interpretation as the number of electrons donated by a specific O^{2-} to its most reducible, nearest-cation neighbors upon V_O formation in a polar-covalent metal oxide.

We also built a state-of-the-art database of 341 SCAN+U E_vs in ternary metal-oxide perovskites for an assortment of A-site (Ca, Sr, Ba, La, and Ce) and B-site (Ti, V, Cr, Mn, Fe, Co, and Ni) cations, crystal systems (cubic, tetragonal, orthorhombic, hexagonal, rhombohedral, and monoclinic), and diverse electronic structures, from insulators to metals. Our model for room-temperature-stable ($E_{hull} \leq 298.15\ k_B$) ABO_3 perovskite structures is

$$E_v = 0.1 \sum_{NN} E_b - 1.5 \max_{NN} V_r + 0.4 E_g - 55.8 E_{hull} + 0.4 \text{ eV} \quad (1.39)$$

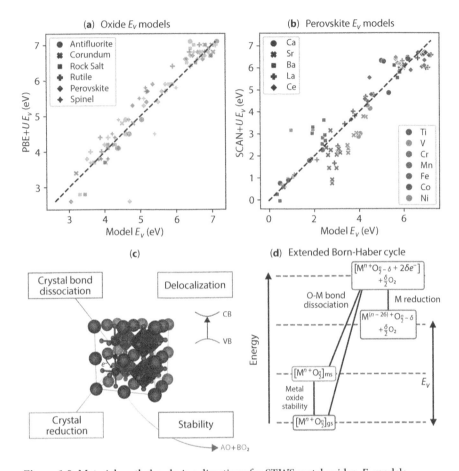

Figure 1.8 Materials enthalpy design directions for STWS metal oxides. E_v models for (a) six classes of oxides and (b) perovskite oxides. In panel (a), blue and orange correspond to the models in refs [212] and [238]. The former is based on crystal bond dissociation energies, crystal reduction potentials, band gaps, and energies above the convex hull. The latter is based on formation enthalpies, O 2p band centers, band gaps, and Pauling electronegativities averaged over the nearest neighbors of the V_O. Panel (b) shows the performance of the blue model in panel (a). Dashed diagonal lines in panels (a) and (b) correspond to lines of perfect agreement of DFT+U and the model. An annotated structure that summarizes the contributions to E_v is shown in panel (c). Panel (d) depicts V_O formation as an extended Born–Haber cycle with three steps: metal-oxide destabilization (if the metal oxide is not in its ground-state polymorph), O–M crystal bond dissociation, and solid-state M reduction.

where the sum in the E_b term is over the NNs to the V_O. Figure 1.8b shows that Equation (1.39) accurately reproduces SCAN+U E_vs for a diverse collection of metal-oxide perovskite compositions, with four intuitive terms that describe different energy contributions to E_v (Figure 1.8c). E_b is the energy to break O-M crystal bonds; interestingly, its coefficient of 0.1 suggests that V_O formation decreases the local bond order by only 10%. V_r is the energy to reduce the V_O's neighboring cations, where its coefficient suggests that 1.5 of the V_O-generated electrons localize on neighboring cations. We interpret E_g as the energy associated with the donated electrons' (de)localization. The sum of the V_r and E_g coefficients is 1.9, indicating that, of the two (1.9 ≈ 2) electrons left behind by a neutral V_O, on average 75% localize on neighboring cations while the other 25% delocalize. Finally, E_{hull} is the energy associated with metastability, and its significant coefficient is the result of small E_{hull} values (≤0.025 eV/atom) for room-temperature-stable ABO_3 perovskites. One can cast this intuition into a familiar form for a generic metal oxide: a modified Born-Haber cycle (Figure 1.8d). First, the energy increases by the metastability of the metal oxide, followed by an energy increase associated with O-M crystal bond dissociation. Depending on the cations' reducibility, the final step—cation reduction—can lead either to a decrease or an increase in the energy. The difference between the energy of the first and last step gives E_v.

Based on these insights, one can categorize the task of enthalpy engineering into four subtasks: to control crystal bond strength, crystal reducibility, electron (de)localization, and crystal stability concurrently to acquire a $\Delta H_v \approx E_v$ between 3.4 and 3.9 eV. For materials stable at 0 K, E_{hull} = 0; therefore, enthalpy engineering only involves three degrees of freedom. Fortunately, E_b and V_r are calculable, and E_g is available from existing measurements, which should help the field—experimentalists and theorists alike—to screen for stable metal-oxide materials with values of these features that satisfy the optimality constraints for E_v.

1.5.2 Entropy Engineering

The second direction in the design of off-stoichiometric metal oxides beyond ceria is to discover materials classes with tunable ΔS_v. As mentioned above, there are four main contributions to ΔS_v. First is the entropy of $\frac{1}{2}O_2(g)$ translations, rotations, vibrations, and electronic degrees of freedom, ΔS_g (= 15.9 k_B at 1800 K and $p_{O_2}^{red}$ = 1 bar [142], where the standard pressure is chosen for convenience and the practical maximum is 0.2 bar, i.e., O_2 in air, and lower than ambient p_{O_2}s are desirable to increase the entropic driving force). However, ΔS_g cannot be engineered in such a way that it benefits one redox-active material over another [244]. Second is the vibrational entropy change, ΔS_p ("p" means phonons), which—e.g., for the thermal reduction step of STWS—can be written in terms of the change in vibrational entropy with a change in the oxygen off-stoichiometry as

$$\Delta S_p = \frac{\partial S_p[MO_{x-\delta}]}{\partial \delta} \quad (1.40)$$

Here, MO_x can be either an off-stoichiometric or stoichiometric metal oxide. For off-stoichiometric metal oxides, ΔS_p is small (2.5 k_B [144] for $CeO_{1.97}$) because V_O formation, except at high concentrations, has a modest impact on the low-energy phonon modes that control S_p. Therefore, for ceria and metal-oxide perovskites, ΔS_p is difficult to engineer to shift its status as the smallest contribution to ΔS_v. Certainly, ΔS_p can be larger for solar thermochemical cycles with stoichiometric materials due to solid-state phase changes (with attendant crystal composition, bonding, coordination, and lattice changes). However, materials that undergo major cation rearrangements upon reduction typically present kinetics and durability problems that may prevent their practical implementation.

The third contribution to ΔS_v is the entropy associated with a change in the number of unique electronic microstates, ΔS_e [133]. To illustrate where ΔS_e comes from, consider the example of the reduction of Ce^{4+} to Ce^{3+} in ceria, where the empty 4f shell of eight-fold-coordinated Ce^{4+} becomes singly occupied (Figure 1.9a) [156, 244]. In the absence of spin-orbit coupling (SOC) and crystal-field splitting (CFS), the 4f states of Ce are degenerate, which, using Boltzmann's entropy formula, gives

$$\Delta S_e = S_e[Ce^{3+}] - S_e[Ce^{4+}] = S_e[Ce^{3+}] = -k_B \sum_{i=1}^{14} p_i \ln p_i \quad (1.41)$$

$$= k_B \ln 14 \approx 2.64 k_B$$

where $p_i = 1/14$ is the probability of the electron occupying the ith 4f state. Per V_O, $\Delta S_e = 5.28 \, k_B$ because V_O formation reduces two Ce^{4+} to Ce^{3+}. In the presence of SOC, the degenerate 4f states split into $^2F_{5/2}$ and $^2F_{7/2}$ levels, with a measured separation of ≈0.28 eV [245, 246] that leads to $\Delta S_e \approx 4.63 \, k_B$ per V_O at 1800 K. Clearly, SOC decreases ΔS_e by penalizing—via the Boltzmann distribution—microstates in which the electron occupies the higher energy levels. In the presence of both SOC and CFS, the $^2F_{5/2}$

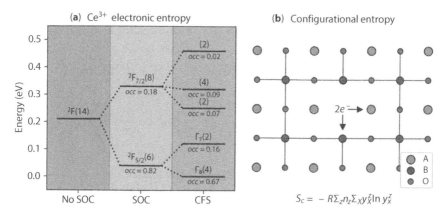

Figure 1.9 Directions for augmenting the (a) electronic and (b) configurational entropy upon V_O formation in the thermal reduction step of STWS. Panel (a) shows that, for ceria, eight-fold-coordinated Ce^{3+} in a cubic crystal field has many thermally accessible 4f states (number of microstates shown in parentheses), leading to a significant on-site electronic entropy change upon the reduction of Ce^{4+} (which has one possible configuration). SOC is spin-orbit coupling; CFS is crystal-field splitting; occ is the occupation of the state at 1800 K. Panel (b) illustrates the possibility of simultaneous reduction of both the A- and B-site cations, which increases the configuration entropy of the two reducing electrons left behind by a neutral V_O.

and $^2F_{7/2}$ levels further split into five energy levels, as shown in the green panel of Figure 1.9a, which further reduces modestly ΔS_e to $\approx 4.48\, k_B$. Since energy-level degeneracy lifting only leads to a slight decrease in ΔS_e (0.80 k_B for Ce^{4+}/Ce^{3+} at 1800 K), one can estimate its value from measurements of atomic spectra [245]. Incidentally, Naghavi et al. found that while S_e is sizable in all lanthanides, ΔS_e reaches a maximum value for Ce^{4+}/Ce^{3+} reduction. Therefore, the take-home message is that to increase ΔS_e for off-stoichiometric metal oxides other than pristine ceria, one should dope or substitute with Ce^{4+} such that Ce^{4+} is redox-active (e.g., in CCTM but not BCM), E_v is optimal, and stability against decomposition and melting is maintained.

The fourth and final contribution to ΔS_v is the configurational entropy, ΔS_c, accompanying ion and defect disorder [152]. In other words, V_O formation creates oxidized/reduced metal disorder on the redox-active cation sublattices and O/V_O disorder on the anion sublattice that both contribute entropy to ΔS_v. For simplicity in discussing the engineerability of ΔS_c, we assume that all ion/defect configurations are equally likely, i.e., we can describe the disorder as an ideal solution. For $CeO_{2-\delta}(s)$, the mathematical expressions for S_c for both the cation and anion sublattices are given by

Cation $\quad S_{c,cation} = -k_B[(1-2\delta)\ln(1-2\delta) + 2\delta\ln(2\delta)] \quad (1.42)$

Anion $\quad S_{c,anion} = -2k_B\left[\left(1-\frac{\delta}{2}\right)\ln\left(1-\frac{\delta}{2}\right) + \frac{\delta}{2}\ln\frac{\delta}{2}\right] \quad (1.43)$

where $(1-2\delta)$ is the fraction of Ce^{4+}, 2δ is the fraction of Ce^{3+}, $1-\delta/2$ is the fraction of O^{2-}, $\delta/2$ is the fraction of neutral V_O, and the "2" before k_B is the number of moles of O^{2-} per formula unit of pristine ceria. To compare against measurements, one calculates ΔS_c in the limit of an infinitesimal change in δ at the off-stoichiometry achieved, i.e., by taking the derivative of the sum of Equations (1.42) and (1.43) with respect to δ:

$$\Delta S_c(\delta) = \frac{\partial(S_{c,cation} + S_{c,anion})}{\partial \delta} = -k_B\left[2\ln\left(\frac{2\delta}{1-2\delta}\right) + \ln\left(\frac{\delta}{2-\delta}\right)\right] \quad (1.44)$$

For $\delta = 0.03$ (i.e., $CeO_{1.97}$), $\Delta S_c = 9.7\ k_B$, which is approximately two times larger than ΔS_e and four times larger than ΔS_v. V_Os in ceria have short-range order [247] and consequently, the real ΔS_c is nonideal and less than the ideal solution model [152].

Despite this reduction for real ion/defect solutions, ΔS_c still is the second-largest contribution to ΔS_v, and hence its modulation is a vital redox-active, metal-oxide engineering consideration. ΔS_c modulation is significant for $ABO_{3-\delta}$ perovskites. They commonly contain only one redox-active cation (ordinarily the B-site cation), resulting in configurational entropies per atom less than ceria for all δ/n, where n is the number of atoms in the formula unit (i.e., n = 3 for ceria and n = 5 for metal-oxide perovskites). Recently, we predicted theoretically that metal-oxide perovskites comprising two redox-active cations (e.g., Ce^{4+} and Mn^{4+}) exhibit a larger ΔS_c than those that undergo single cation reduction [228]. For $ABO_{3-\delta}(s)$ that experience simultaneous cation reduction, the mathematical expression for $S_{c,cation}$ is

$$S_{c,cation} = -2k_B[(1-\delta)\ln(1-\delta) + \delta\ln\delta] \quad (1.45)$$

For simplicity, we suppose that the two V_O-generated electrons reduce the A- and B-site cations to the same degree. In the limit of $\Delta\delta \to 0$, ΔS_c for metal-oxide perovskites with simultaneous cation reduction is derived from the derivative of the sum of Equations (1.43) and (1.45) with respect to δ:

$$\Delta S_c(\delta) = -k_B \left[2\ln\left(\frac{\delta}{1-\delta}\right) + \ln\left(\frac{\delta}{3-\delta}\right) \right] \quad (1.46)$$

For $ABO_{2.97}$ (i.e., $\delta = 0.03$), $\Delta S_c = 11.5\ k_B/5$ atoms $= 2.3\ k_B$/atom, which is $0.9\ k_B$ smaller than that of $CeO_{1.97}$ ($9.7\ k_B/3$ atoms $= 3.2\ k_B$/atom). This simple analysis indicates that metal-oxide perovskites cannot produce ΔS_c per atom greater than ceria. Therefore, while identifying oxide materials classes that offer the possibility of simultaneous cation reduction enables entropy engineering, the key design direction for non-ceria STWS metal oxides is optimizing E_v (unless multiple redox can be realized in doped ceria).

1.5.3 Stability Engineering

We will briefly mention a third materials design direction: to engineer the stability of off-stoichiometric metal oxides. Such engineering, however, requires a better atomic-scale understanding of temperature-dependent phase diagrams and melting points (T_m) of materials. At temperatures where the harmonic approximation breaks down, which are relevant in STWS, one must either include the anharmonic force constants in evaluating the vibrational free energy or perform MD simulations using the fluctuation-dissipation theorem to capture anharmonic contributions crucial in controlling phase-transition temperatures. An alternative approach to account for anharmonicities is the utilization of enhanced sampling (e.g., nested sampling), which allows for the direct calculation of the partition function (and hence thermodynamic properties) [248]. All three of these approaches are computationally challenging if one desires a quantum mechanical description of the material. Therefore, new methods, such as machine-learned potentials, are starting to be used to ease the computational burden of these calculations [249]. The modeling of material melting faces the same challenges, as—in the absence of an accurate analytic theory or phenomenology for T_m prediction—MD simulations of phase coexistence are necessary with either computationally expensive quantum mechanics techniques or more efficient machine-learned potentials.

However, the latter often require computationally demanding parameterization based on accurate quantum mechanics computations as well. That said, melting point measurements are available in the literature; therefore, instead of a materials design direction, we recommend a target for theoretical method development could be data-driven T_m prediction for multi-component solid solutions using the modern tools of data science and machine learning based on experimental data of simpler but related compounds.

1.6 Conclusions

Solar thermochemical water splitting could be a crucial component of a coordinated technological effort to mitigate the effects of climate change. The potential of thermochemical technologies based on concentrated solar radiation is exceptionally promising, given that they utilize the entire solar spectrum and can generate local temperatures that all but preclude kinetic limitations for crucial chemical reactions. Two-step solar-thermochemical water-splitting cycles typically use redox-active metal oxides that can be reversibly reduced and reoxidized over a large number of cycles. The most promising of these are off-stoichiometric/MIEC or oxygen-vacancy-forming metal oxides, such as ceria and redox-active perovskites—due to their unhindered kinetics and the latter's large composition space. This chapter introduced the fundamentals of thermochemical water splitting at high temperatures, focusing on thermodynamics and discovery of off-stoichiometric metal oxides. We also provided a high-level overview of the computational methods available to calculate the various critical quantities controlling the efficiency of solar thermochemical water splitting cycles to produce hydrogen. Having surveyed important literature on ceria and redox-active metal-oxide perovskites, we closed with a forward-looking assessment addressing what avenues researchers might follow next in the pursuit of off-stoichiometric metal-oxide materials that could lead to widespread deployment of this technology.

Acknowledgments

The authors gratefully acknowledge research support from the HydroGEN Advanced Water Splitting Materials Consortium, established as part of the Energy Materials Network under the U.S. Department of Energy, Office of Energy Efficiency and Renewable Energy, Fuel Cell Technologies Office, under Award Number DE-EE0008090. The authors also acknowledge the computational resources sponsored by the Department of Energy's Office of Energy Efficiency and Renewable Energy located at the National Renewable Energy Laboratory. The authors also thank Princeton University for computing resources. The views and opinions of the authors expressed herein do not necessarily state or reflect those of the United States Government or any agency thereof. Neither the United States Government nor any agency thereof, nor any of their employees, makes any warranty, expressed or implied, or assumes any legal liability or responsibility for the accuracy, completeness, or usefulness of any information, apparatus, product, or process disclosed, or represents that its use would not infringe privately owned rights.

Appendices

Appendix A. Equilibrium Composition for Solar Thermolysis

One can calculate the equilibrium composition of the gas-phase mixture present in solar thermolysis using the Gibbs-free-energy-minimization method, subject to system constraints, which starts with the definition of the partial molar Gibbs free energy

$$\frac{nG}{RT} = \sum_i n_i \left(\frac{\Delta G^o_{f,i}}{RT} + \ln y_i p \right) \qquad (1.47)$$

where n is the total number of moles, G is the Gibbs free energy per mole of the mixture, R is the ideal gas constant, T is the temperature in K, n_i is the number of moles of component i, y_i is the mole fraction of component i, p is the pressure, and $\Delta G^o_{f,i}$ is the Gibbs free energy of formation of component i at standard conditions. $\Delta G^o_{f,i}$ can be obtained from online databases—such as the NIST-JANAF thermochemical tables—or reference books—e.g., Barin's Thermochemical Data of Pure Substances. Here, we use the Fact-Web application, which offers user-friendly access to the FactSage pure substances database, to download the $\Delta G^o_{f,i}$ of $H_2O(g)$, $H_2(g)$, $O_2(g)$, $H(g)$, $O(g)$, and $OH(g)$ at 1 bar and temperatures ranging from 200 K to 6000 K in increments of 100 K. The objective of the method above is to find the set of n_i that minimize the Gibbs free energy of the mixture

$$\min_{n_i} \frac{nG}{RT} \qquad (1.48)$$

which, of course, is a requirement—enforced by the second law of thermodynamics—for a system at equilibrium, where the T and p are held constant. Additionally, for a closed system, we must apply an atomic balance constraint, namely

$$n_{H,in} = n_{H,out} \qquad (1.49)$$

$$n_{O,in} = n_{O,out} \qquad (1.50)$$

such that the number of moles (n) of H and O atoms in molecules fed to (*in*) and leaving (*out*) the reactor are equal.

To solve Equations (1.48)-(1.50), we used the Sequential Least-Squares Programming (SLSQP) method implemented in scipy [250] with a convergence threshold of 1×10^{-6}. We also placed boundaries on y_i such that—at equilibrium—they must acquire physically reasonable values between 0 and 1, inclusive. We reasonably assume that no $H_2O(g)$ dissociates at the lowest T. In other words, we set $n_{H_2O(g)}=1$ and $n_{H_2(g)} = n_{O_2(g)} = n_{H(g)} = n_{O(g)} = n_{OH(g)} = 0$. For $T >$ lowest T, we supplied the n_i from the previous (lower) T as an initial guess for the equilibrium composition. The Python code used to generate Figure 1.2b can be found at https://github.com/wexlergroup/stws/ as a Jupyter notebook.

Appendix B. Equilibrium Composition of Ceria

One approach to computing the equilibrium composition of $CeO_2(s)$ as a function of T and p_{O_2} is to solve a set of kinetic equations under the steady-state approximation [214]. We use this semi-empirical model and not the state-of-the-art thermodynamic model of Zinkevich [150] as the former provides a simple, intuitive, and closed-form equation of state. One can write the thermal reduction of $CeO_2(s)$ as follows

$$CeO_2(s) \rightleftharpoons CeO_{2-\delta}(s) + \frac{\delta}{2}O_2(g) \qquad (1.51)$$

where δ is the oxygen off-stoichiometry produced by oxygen vacancy (V_O) formation. The rate of change of the V_O concentration $[V_O]$ is the difference between the rates of the forward and backward reactions [Equation (1.51)],

$$\frac{d[V_O]}{dt} = k_{red}[O] - k_{ox}[V_O]p_{O_2}^n \qquad (1.52)$$

Here, k is a rate constant, *red* stands for reduction, *ox* stands for reoxidation, [O] is the removable oxygen concentration, p_{O_2} is the partial pressure of $O_2(g)$, and n is the oxygen gas power dependency that depends on the maximum δ, which we denote as δ_m. Setting $d[V_O]/dt = 0$ at equilibrium and isolating the concentration terms, we obtain

$$\frac{[V_O]}{[O]} = \frac{k_{red}}{k_{ox}} p_{O_2}^{-n} = \frac{\delta}{\delta_m - \delta} \qquad (1.53)$$

Note that we have rewritten the ratio of removed to removable O—i.e., $[V_O]/[O]$—in terms of δ and δ_m. To introduce the energetics of V_O formation in our treatment, we assume an Arrhenius relationship between a reaction's rate and its activation barrier (E_a)

$$k = k_0 e^{-\frac{E_a}{RT}} \qquad (1.54)$$

where k_0 is an empirical prefactor. Replacing k_{red} and k_{ox} with their associated Arrhenius rate expressions and labeling the difference between the E_as for reduction and reoxidation as ΔE, we can rewrite Equation (1.53) as

$$\frac{\delta}{\delta_m - \delta} = \frac{k_{red,0}}{k_{ox,0}} p_{O_2}^{-n} e^{-\frac{\Delta E}{RT}} \qquad (1.55)$$

which, upon solving for δ, yields

$$\delta = \frac{\delta_m}{\frac{k_{ox,0}}{k_{red,0}} p_{O_2}^n e^{\frac{\Delta E}{RT}} + 1} \qquad (1.56)$$

In the original derivation of this model, Bulfin et al. fit Equation (1.56) to equilibrium measurements of ceria for the triplets (δ, T, p_{O_2}) and obtained $k_{red,0}/k_{ox,0}$ = 8700 ± 800 bar^n, n = 0.218 ± 0.0013, δ_m = 0.35, and ΔE = 195.6 ± 1.2 kJ/mol [214]. Note that the V_O formation energy per ½ O_2 is $E_v = \Delta E/2n \approx 4.6$ eV. The Python code used to generate Figure 1.6a can be found at https://github.com/wexlergroup/stws/ as a Jupyter notebook.

References

1. *Climate Change 2014: Synthesis Report. Contribution of Working Groups I, II and III to the Fifth Assessment Report of the Intergovernmental Panel on Climate Change*, p. 151, IPCC Geneva, Switzerland, 2014, https://www.ipcc.ch/report/ar5/syr/.
2. *Global Climate Change Impacts in the United States: A State of Knowledge Report*, Cambridge University Press, New York, NY, 2009.
3. Oreskes, N., The scientific consensus on climate change. *Science*, 306, 1686–1686, 2004.
4. Steinfeld, A., Solar thermochemical production of hydrogen—A review. *Sol. Energy*, 78, 603–615, 2005.
5. Walter, M.G., Warren, E.L., McKone, J.R., Boettcher, S.W., Mi, Q., Santori, E.A., Lewis, N.S., Solar water splitting cells. *Chem. Rev.*, 110, 6446–6473, 2010.
6. Rajan, A.G., Martirez, J.M.P., Carter, E.A., Why do we use the materials and operating conditions we use for heterogeneous (photo) electrochemical water splitting? *ACS Catal.*, 10, 11177–11234, 2020.
7. Laguna-Bercero, M.A., Recent advances in high temperature electrolysis using solid oxide fuel cells: A review. *J. Power Sources*, 203, 4–16, 2012.
8. Ayers, K., Danilovic, N., Ouimet, R., Carmo, M., Pivovar, B., & Bornstein, M., Perspectives on low-temperature electrolysis and potential for renewable hydrogen at scale. *Annu. Rev. Chem. Biomol. Eng.*, 10, 219–239, 2019.
9. Zhao, Y., Dunn, A., Lin, J., Shi, D., *Novel Nanomaterials for Biomedical, Environmental and Energy Applications*, pp. 415–434, Elsevier, Amsterdam, Netherlands, 2019.
10. Schäppi, R., Rutz, D., Dähler, F., Muroyama, A., Haueter, P., Lilliestam, J., Patt, A., Furler, P., Steinfeld, A., Drop-in fuels from sunlight and air. *Nature*, 601, 63–68, 2022.
11. Zhou, E., Xu, K., Wang, C., *Analysis of the Cost and Value of Concentrating Solar Power in China*, National Renewable Energy Laboratory, Golden, CO, 2019, https://www.nrel.gov/docs/fy20osti/74303.pdf.
12. Fletcher, E.A., Solarthermal processing: A review. *J. Sol. Energy Eng.*, 123, 63–74, 2000.

13. Romero, M. and Steinfeld, A., Concentrating solar thermal power and thermochemical fuels. *Energy Environ. Sci.*, 5, 9234–9245, 2012.
14. Cheng, W.-H., de la Calle, A., Atwater, H.A., Stechel, E.B., Xiang, C., Hydrogen from sunlight and water: A side-by-side comparison between photoelectrochemical and solar thermochemical water-splitting. *ACS Energy Lett.*, 6, 3096–3113, 2021.
15. Fletcher, E.A. and Moen, R.L., Hydrogen-and oxygen from water: The use of solar energy in a one-step effusional process is considered. *Science*, 197, 1050–1056, 1977.
16. Bale, C.W., Bélisle, E., Chartrand, P., Decterov, S.A., Eriksson, G., Gheribi, A.E., Hack, K., Jung, I.-H., Kang, Y.-B., Melançon, J., Pelton, A.D., Petersen, S., Robelin, C., Sangster, J., Spencer, P., Van Ende, M.-A., FactSage thermochemical software and databases, 2010–2016. *Calphad*, 54, 35–53, 2016.
17. Steinfeld, A., Solar hydrogen production via a two-step water-splitting thermochemical cycle based on Zn/ZnO redox reactions. *Int. J. Hydrog. Energy*, 27, 611–619, 2002.
18. Pretzel, C.W. and Funk, J.E., *Developmental Status of Solar Thermochemical Hydrogen Production*, p. 114, Sandia Natl. Lab, Albuquerque, New Mexico, 1987.
19. Perret, R., *Solar Thermochemical Hydrogen Production Research (STCH)*, SAND2011-3622, 1219357, Sandia National Laboratories, Albuquerque, New Mexico, 2011.
20. Department of Energy (DOE), Office of Energy Efficiency and Renewable Energy (EERE), *Fuel Cell Technologies Office Annual FOA, Funding Opportunity Announcement (FOA) Number: DE-FOA-0001647*, Department of Energy, Washington, DC, 2016.
21. Rao, C.N.R. and Dey, S., Solar thermochemical splitting of water to generate hydrogen. *Proc. Natl. Acad. Sci.*, 114, 13385–13393, 2017.
22. Abanades, S., Charvin, P., Flamant, G., Neveu, P., Screening of water-splitting thermochemical cycles potentially attractive for hydrogen production by concentrated solar energy. *Energy*, 31, 2805–2822, 2006.
23. Perkins, C. and Weimer, A.W., Solar-thermal production of renewable hydrogen. *AIChE J.*, 55, 286–293, 2009.
24. Funk, J.E., Thermochemical hydrogen production: Past and present. *Int. J. Hydrog. Energy*, 26, 185–190, 2001.
25. Kodama, T. and Gokon, N., Thermochemical cycles for high-temperature solar hydrogen production. *Chem. Rev.*, 107, 4048–4077, 2007.
26. Muhich, C.L., Ehrhart, B.D., Al-Shankiti, I., Ward, B.J., Musgrave, C.B., Weimer, A.W., A review and perspective of efficient hydrogen generation via solar thermal water splitting: A review and perspective of efficient hydrogen generation. *Wiley Interdiscip. Rev. Energy Environ.*, 5, 261–287, 2016.
27. Miller, J.E., McDaniel, A.H., Allendorf, M.D., Considerations in the design of materials for solar-driven fuel production using metal-oxide thermochemical cycles. *Adv. Energy Mater.*, 4, 1300469, 2014.

28. Steinfeld, A., Kuhn, P., Reller, A., Palumbo, R., Murray, J., Tamaura, Y., Solar-processed metals as clean energy carriers and water-splitters. *Int. J. Hydrog. Energy*, 23, 767–774, 1998.
29. Zheng, Z., Liu, T., Liu, Q., Lei, J., Fang, J., A distributed energy system integrating SOFC-MGT with mid-and-low temperature solar thermochemical hydrogen fuel production. *Int. J. Hydrog. Energy*, 46, 19846–19860, 2021.
30. Budama, V.K., Johnson, N.G., McDaniel, A., Ermanoski, I., Stechel, E.B., Thermodynamic development and design of a concentrating solar thermochemical water-splitting process for co-production of hydrogen and electricity. *Int. J. Hydrog. Energy*, 43, 17574–17587, 2018.
31. Bilgen, E. and Bilgen, C., Solar hydrogen production using two-step thermochemical cycles. *Int. J. Hydrog. Energy*, 7, 637–644, 1982.
32. Kodama, T., High-temperature solar chemistry for converting solar heat to chemical fuels. *Prog. Energy Combust. Sci.*, 29, 567–597, 2003.
33. Scheffe, J.R. and Steinfeld, A., Oxygen exchange materials for solar thermochemical splitting of H_2O and CO_2: A review. *Mater. Today*, 17, 341–348, 2014.
34. Funk, J.E. and Reinstrom, R.M., Energy requirements in production of hydrogen from water. *Ind. Eng. Chem. Process Des. Dev.*, 5, 336–342, 1966.
35. Loutzenhiser, P.G., Meier, A., Steinfeld, A., Review of the two-step H_2O/CO_2-splitting solar thermochemical cycle based on Zn/ZnO redox reactions. *Materials*, 3, 4922–4938, 2010.
36. Stamatiou, A., Loutzenhiser, P.G., Steinfeld, A., Solar syngas production via H_2O/CO_2-splitting thermochemical cycles with Zn/ZnO and FeO/Fe_3O_4 redox reactions. *Chem. Mater.*, 22, 851–859, 2010.
37. Stamatiou, A., Loutzenhiser, P.G., Steinfeld, A., Syngas production from H_2O and CO_2 over Zn particles in a packed-bed reactor. *AIChE J.*, 58, 625–631, 2012.
38. Abanades, S., Metal oxides applied to thermochemical water-splitting for hydrogen production using concentrated solar energy. *ChemEngineering*, 3, 63, 2019.
39. Palumbo, R., High temperature solar electrothermal processing—III. Zinc zinc Oxide at 1200–1675K using non-consumable anode. *Energy*, 13, 319–332, 1988.
40. Fletcher, E.A., Solarthermal and solar quasi-electrolytic processing and separations: Zinc from zinc oxide as an example. *Ind. Eng. Chem. Res.*, 38, 2275–2282, 1999.
41. Abanades, S., Charvin, P., Lemont, F., Flamant, G., Novel two-step SnO_2/SnO water-splitting cycle for solar thermochemical production of hydrogen. *Int. J. Hydrog. Energy*, 33, 6021–6030, 2008.
42. Charvin, P., Abanades, S., Lemont, F., Flamant, G., Experimental study of $SnO_2/SnO/Sn$ thermochemical systems for solar production of hydrogen. *AIChE J.*, 54, 2759–2767, 2008.

43. Levêque, G., Abanades, S., Jumas, J.-C., Olivier-Fourcade, J., Characterization of two-step tin-based redox system for thermochemical fuel production from solar-driven CO_2 and H_2O splitting cycle. *Ind. Eng. Chem. Res.*, 53, 5668–5677, 2014.
44. Scheffe, J.R., McDaniel, A.H., Allendorf, M.D., Weimer, A.W., Kinetics and mechanism of solar-thermochemical H_2 production by oxidation of a cobalt ferrite-zirconia composite. *Energy Environ. Sci.*, 6, 963–973, 2013.
45. Teknetzi, I., Nessi, P., Zaspalis, V., Nalbandian, L., Ni-ferrite with structural stability for solar thermochemical H_2O/CO_2 splitting. *Int. J. Hydrog. Energy*, 42, 26231–26242, 2017.
46. Lorentzou, S., Zygogianni, A., Pagkoura, C., Karagiannakis, G., Konstandopoulos, A.G., Saeck, J.P., Breuer, S., Lange, M., Lapp, J., Fend, T., Roeb, M., Gonzalez, A.J., Delgado, A.V., Brouwer, J.P., Makkus, R.C., Kiartzis, S.J., Hydrosol-plant: Structured redox reactors for H_2 production from solar thermochemical H_2O splitting. *AIP Conf. Proc.*, 2033, 130010, 2018.
47. Bhosale, R.R., Thermodynamic analysis of Ni-ferrite based solar thermochemical H_2O splitting cycle for H_2 production. *Int. J. Hydrog. Energy*, 44, 61–71, 2019.
48. Diver, R.B., Miller, J.E., Allendorf, M.D., Siegel, N.P., Hogan, R.E., Solar thermochemical water-splitting ferrite-cycle heat engines. *J. Sol. Energy Eng.*, 130, 041001, 2008.
49. Bhosale, R.R., Shende, R.V., Puszynski, J.A., Sol-gel derived $NiFe_2O_4$ modified with ZrO_2 for hydrogen generation from solar thermochemical water-splitting reaction. *MRS Online Proc. Libr.*, 1387, 5, 2011.
50. Allen, K.M., Coker, E.N., Auyeung, N., Klausner, J.F., Cobalt ferrite in YSZ for use as reactive material in solar thermochemical water and carbon dioxide splitting, part I: Material characterization. *JOM*, 65, 1670–1681, 2013.
51. Allen, K.M., Auyeung, N., Rahmatian, N., Klausner, J.F., Coker, E.N., Cobalt ferrite in YSZ for use as reactive material in solar thermochemical water and carbon dioxide splitting, part ii: Kinetic modeling. *JOM*, 65, 1682–1693, 2013.
52. Dimitrakis, D.A., Syrigou, M., Lorentzou, S., Kostoglou, M., Konstandopoulos, A.G., On kinetic modelling for solar redox thermochemical H_2O and CO_2 splitting over $NiFe_2O_4$ for H_2, CO and syngas production. *Phys. Chem. Chem. Phys.*, 19, 26776–26786, 2017.
53. Bhosale, R.R., Mn-ferrite based solar thermochemical water splitting cycle: A thermodynamic evaluation. *Fuel*, 256, 115847, 2019.
54. Dimitrakis, D.A., Tsongidis, N., II, Konstandopoulos, A.G., Reduction enthalpy and charge distribution of substituted ferrites and doped ceria for thermochemical water and carbon dioxide splitting with DFT+U. *Phys. Chem. Chem. Phys.*, 18, 23587–95, 2016.
55. Gokon, N., Hasegawa, T., Takahashi, S., Kodama, T., Thermochemical two-step water-splitting for hydrogen production using Fe-YSZ particles and a ceramic foam device. *Energy*, 33, 1407–1416, 2008.

56. Gokon, N., Murayama, H., Umeda, J., Hatamachi, T., Kodama, T., Monoclinic zirconia-supported Fe3O4 for the two-step water-splitting thermochemical cycle at high thermal reduction temperatures of 1400–1600°C. *Int. J. Hydrog. Energy*, 34, 1208–1217, 2009.
57. Nakamura, T., Hydrogen production from water utilizing solar heat at high temperatures. *Sol. Energy*, 19, 467–475, 1977.
58. Charvin, P., Abanades, S., Flamant, G., Lemort, F., Two-step water splitting thermochemical cycle based on iron oxide redox pair for solar hydrogen production. *Energy*, 32, 1124–1133, 2007.
59. Steinfeld, A., Sanders, S., Palumbo, R., Design aspects of solar thermochemical engineering-a case study: Two-step water-splitting cycle using the Fe3O4/FeO redox system. *Sol. Energy*, 65, 43–53, 1999.
60. Allendorf, M.D., Diver, R.B., Siegel, N.P., Miller, J.E., Two-step water splitting using mixed-metal ferrites: Thermodynamic analysis and characterization of synthesized materials. *Energy Fuels*, 22, 4115–4124, 2008.
61. Tamaura, Y., Steinfeld, A., Kuhn, P., Ehrensberger, K., Production of solar hydrogen by a novel, 2-step, water-splitting thermochemical cycle. *Energy*, 20, 325–330, 1995.
62. Kodama, T., Kondoh, Y., Yamamoto, R., Andou, H., Satou, N., Thermochemical hydrogen production by a redox system of ZrO2-supported Co (II)-ferrite. *Sol. Energy*, 78, 623–631, 2005.
63. Bulfin, B., Vieten, J., Agrafiotis, C., Roeb, M., Sattler, C., Applications and limitations of two step metal oxide thermochemical redox cycles: A review. *J. Mater. Chem. A*, 5, 18951–18966, 2017.
64. Tamaura, Y., Kojima, M., Sano, T., Ueda, Y., Hasegawa, N., Tsuji, M., Thermodynamic evaluation of water splitting by a cation-excessive (Ni, Mn) ferrite. *Int. J. Hydrog. Energy*, 23, 1185–1191, 1998.
65. Tamaura, Y., Ueda, Y., Matsunami, J., Hasegawa, N., Nezuka, M., Sano, T., Tsuji, M., Solar hydrogen production by using ferrites. *Sol. Energy*, 65, 55–57, 1999.
66. Ohlhausen, J.A.T., Coker, E.N., Ambrosini, A., Miller, J.E., ToF-SIMS analysis of iron oxide particle oxidation by isotopic and multivariate analysis: ToF-SIMS of FeO$_x$ particle oxidation by isotopic and multivariate. *Surf. Interface Anal.*, 45, 320–323, 2013.
67. Kodama, T., Gokon, N., Yamamoto, R., Thermochemical two-step water splitting by ZrO2-supported NixFe3–xO4 for solar hydrogen production. *Sol. Energy*, 82, 73–79, 2008.
68. Gokon, N., Murayama, H., Nagasaki, A., Kodama, T., Thermochemical two-step water splitting cycles by monoclinic ZrO2-supported NiFe2O4 and Fe3O4 powders and ceramic foam devices. *Sol. Energy*, 83, 527–537, 2009.
69. Ishihara, H., Kaneko, H., Hasegawa, N., Tamaura, Y., Two-step water-splitting at 1273–1623K using yttria-stabilized zirconia-iron oxide solid solution via co-precipitation and solid-state reaction. *Energy*, 33, 1788–1793, 2008.

70. Scheffe, J.R., Li, J., Weimer, A.W., A spinel ferrite/hercynite water-splitting redox cycle. *Int. J. Hydrog. Energy*, 35, 3333–3340, 2010.
71. Muhich, C.L., Evanko, B.W., Weston, K.C., Lichty, P., Liang, X., Martinek, J., Musgrave, C.B., Weimer, A.W., Efficient generation of H_2 by splitting water with an isothermal redox cycle. *Science*, 341, 540, 2013.
72. Arifin, D., Aston, V.J., Liang, X., McDaniel, A.H., Weimer, A.W., $CoFe_2O_4$ on a porous Al_2O_3 nanostructure for solar thermochemical CO_2 splitting. *Energy Environ. Sci.*, 5, 9438, 2012.
73. Hoskins, A.L., Millican, S.L., Czernik, C.E., Alshankiti, I., Netter, J.C., Wendelin, T.J., Musgrave, C.B., Weimer, A.W., Continuous on-sun solar thermochemical hydrogen production via an isothermal redox cycle. *Appl. Energy*, 249, 368–376, 2019.
74. Muhich, C.L., Ehrhart, B.D., Witte, V.A., Miller, S.L., Coker, E.N., Musgrave, C.B., Weimer, A.W., Predicting the solar thermochemical water splitting ability and reaction mechanism of metal oxides: A case study of the hercynite family of water splitting cycles. *Energy Environ. Sci.*, 8, 3687–3699, 2015.
75. Trottier, R.M., Bare, Z.J.L., Millican, S.L., Musgrave, C.B., Musgrave, C.B., Musgrave, C.B., Musgrave, C.B., Predicting spinel disorder and its effect on oxygen transport kinetics in hercynite. *ACS Appl. Mater. Interfaces*, 12, 23831–23843, 2020.
76. Al-Shankiti, I.A., Bayon, A., Weimer, A.W., Reduction kinetics of hercynite redox materials for solar thermochemical water splitting. *Chem. Eng. J. Amst. Neth.*, 389, 124429, 2020.
77. Millican, S.L., Clary, J.M., Musgrave, C.B., Lany, S., Redox defect thermochemistry of $FeAl_2O_4$ hercynite in water splitting from first-principles methods. *Chem. Mater.*, 34, 519, 2022.
78. Bhosale, R.R., Al Momani, F., Rashid, S., Solar thermochemical H_2 production via $MnSO_4$/MnO water splitting cycle: Thermodynamic equilibrium and efficiency analysis. *Int. J. Hydrog. Energy*, 45, 10324–10333, 2020.
79. Charvin, P., Abanades, S., Beche, E., Lemont, F., Flamant, G., Hydrogen production from mixed cerium oxides via three-step water-splitting cycles. *Solid State Ion.*, 180, 1003–1010, 2009.
80. Ruan, C., Tan, Y., Li, L., Wang, J., Liu, X., Wang, X., A novel $CeO_{2-x}SnO_2$/$Ce_2Sn_2O_7$ pyrochlore cycle for enhanced solar thermochemical water splitting. *AIChE J.*, 63, 3450–3462, 2017.
81. Sadeghi, S. and Ghandehariun, S., Thermodynamic analysis and optimization of an integrated solar thermochemical hydrogen production system. *Int. J. Hydrog. Energy*, 45, 28426–28436, 2020.
82. O'Brien, J.A., Hinkley, J.T., Donne, S.W., Lindquist, S.-E., The electrochemical oxidation of aqueous sulfur dioxide: A critical review of work with respect to the hybrid sulfur cycle. *Electrochim. Acta*, 55, 573–591, 2010.
83. Gorensek, M.B., Hybrid sulfur cycle flowsheets for hydrogen production using high-temperature gas-cooled reactors. *Int. J. Hydrog. Energy*, 36, 12725–12741, 2011.

84. Allen, J.A., Rowe, G., Hinkley, J.T., Donne, S.W., Electrochemical aspects of the Hybrid Sulfur Cycle for large scale hydrogen production. *Int. J. Hydrog. Energy*, 39, 11376–11389, 2014.
85. Okeefe, D., Allen, C., Besenbruch, G., Brown, L., Norman, J., Sharp, R., Mccorkle, K., Preliminary results from bench-scale testing of a sulfur-iodine thermochemical water-splitting cycle☆. *Int. J. Hydrog. Energy*, 7, 381–392, 1982.
86. Kubo, S., Nakajima, H., Kasahara, S., Higashi, S., Masaki, T., Abe, H., Onuki, K., A demonstration study on a closed-cycle hydrogen production by the thermochemical water-splitting iodine–sulfur process. *Nucl. Eng. Des.*, 233, 347–354, 2004.
87. Goldstein, S., Borgard, J., Vitart, X., Upper bound and best estimate of the efficiency of the iodine sulphur cycle. *Int. J. Hydrog. Energy*, 30, 619–626, 2005.
88. Brown, L.C., Funk, J.F., Showalter, S.K., Initial screening of thermochemical water-splitting cycles for high efficiency generation of hydrogen fuels using nuclear power, US Department of Energy Technical Report, GA-A23373, 761610, 1999.
89. Bulfin, B., Lange, M., de Oliveira, L., Roeb, M., Sattler, C., Solar thermochemical hydrogen production using ceria zirconia solid solutions: Efficiency analysis. *Int. J. Hydrog. Energy*, 41, 19320–19328, 2016.
90. Ehrhart, B.D., Muhich, C.L., Al-Shankiti, I., Weimer, A.W., System efficiency for two-step metal oxide solar thermochemical hydrogen production-part 1: Thermodynamic model and impact of oxidation kinetics. *Int. J. Hydrog. Energy*, 41, 19881–19893, 2016.
91. Ehrhart, B.D., Muhich, C.L., Al-Shankiti, I., Weimer, A.W., System efficiency for two-step metal oxide solar thermochemical hydrogen production-part 2: Impact of gas heat recuperation and separation temperatures. *Int. J. Hydrog. Energy*, 41, 19894–19903, 2016.
92. Ehrhart, B.D., Muhich, C.L., Al-Shankiti, I., Weimer, A.W., System efficiency for two-step metal oxide solar thermochemical hydrogen production-part 3: Various methods for achieving low oxygen partial pressures in the reduction reaction. *Int. J. Hydrog. Energy*, 41, 19904–19914, 2016.
93. Brendelberger, S., Rosenstiel, A., Lopez-Roman, A., Prieto, C., Sattler, C., Performance analysis of operational strategies for monolithic receiver-reactor arrays in solar thermochemical hydrogen production plants. *Int. J. Hydrog. Energy*, 45, 26104–26116, 2020.
94. Budama, V.K., Brendelberger, S., Roeb, M., Sattler, C., Performance analysis and optimization of solar thermochemical water-splitting cycle with single and multiple receivers. *Energy Technol.*, 10, 2021.
95. Lundberg, M., Model calculations on some feasible two-step water splitting processes. *Int. J. Hydrog. Energy*, 10, 18, 369–376, 1993.

96. Siegel, N.P., Miller, J.E., Ermanoski, I., Diver, R.B., Stechel, E.B., Factors affecting the efficiency of solar driven metal oxide thermochemical cycles. *Ind. Eng. Chem. Res.*, 52, 3276–3286, 2013.
97. Meredig, B. and Wolverton, C., First-principles thermodynamic framework for the evaluation of thermochemical H_2O- or CO_2-splitting materials. *Phys. Rev. B*, 80, 245119, 2009.
98. Bayon, A., de la Calle, A., Stechel, E.B., Muhich, C., Operational limits of redox metal oxides performing thermochemical water splitting. *Energy Technol.*, 10, 2100222, 2022.
99. Li, S., Wheeler, V.M., Kreider, P.B., Lipiński, W., Thermodynamic analyses of fuel production via solar-driven non-stoichiometric metal oxide redox cycling. Part 1. Revisiting flow and equilibrium assumptions. *Energy Fuels*, 32, 10838–10847, 2018.
100. Li, S., Wheeler, V.M., Kreider, P.B., Bader, R., Lipiński, W., Thermodynamic analyses of fuel production via solar-driven non-stoichiometric metal oxide redox cycling. Part 2. Impact of solid–gas flow configurations and active material composition on system-level efficiency. *Energy Fuels*, 32, 10848–10863, 2018.
101. de la Calle, A., Ermanoski, I., Stechel, E.B., Towards chemical equilibrium in thermochemical water splitting. Part 1: Thermal reduction. *Int. J. Hydrog. Energy*, 47, 10474–10482, 2022.
102. Roeb, M., Neises, M., Monnerie, N., Call, F., Simon, H., Sattler, C., Schmücker, M., Pitz-Paal, R., Materials-related aspects of thermochemical water and carbon dioxide splitting: A review. *Materials*, 5, 2015–2054, 2012.
103. D'Souza, L., Thermochemical hydrogen production from water using reducible oxide materials: A critical review. *Mater. Renew. Sustain. Energy*, 2, 7, 2013.
104. Carrillo, R.J. and Scheffe, J.R., Advances and trends in redox materials for solar thermochemical fuel production. *Sol. Energy*, 156, 3–20, 2017.
105. Scheffe, J.R., Weibel, D., Steinfeld, A., Lanthanum–strontium–manganese perovskites as redox materials for solar thermochemical splitting of H_2O and CO_2. *Energy Fuels*, 27, 4250–4257, 2013.
106. Merkulov, O.V., Markov, A.A., Leonidov, I.A., Patrakeev, M.V., Kozhevnikov, V.L., Oxygen nonstoichiometry and thermodynamic quantities in solid solution $SrFe_{1-x}Sn_xO_{3-\delta}$. *J. Solid State Chem.*, 262, 121–126, 2018.
107. Bulfin, B., Call, F., Lange, M., Lübben, O., Sattler, C., Pitz-Paal, R., Shvets, I.V., Thermodynamics of CeO_2 thermochemical fuel production. *Energy Fuels*, 29, 1001–1009, 2015.
108. Vieten, J., Bulfin, B., Huck, P., Horton, M., Guban, D., Zhu, L., Lu, Y., Persson, K.A., Roeb, M., Sattler, C., Materials design of perovskite solid solutions for thermochemical applications. *Energy Environ. Sci.*, 12, 1369–1384, 2019.
109. Tagawa, H., Oxygen nonstoichiometry in perovskite-type oxide, undoped and Sr–doped $LaMnO_3$. *ECS Proc., Vol.*, 1997–40, 785–794, 1997.

110. Miller, J.E., Allendorf, M.D., Diver, R.B., Evans, L.R., Siegel, N.P., Stuecker, J.N., Metal oxide composites and structures for ultra-high temperature solar thermochemical cycles. *J. Mater. Sci.*, 43, 4714–4728, 2008.
111. Abanades, S., Legal, A., Cordier, A., Peraudeau, G., Flamant, G., & Julbe, A., Investigation of reactive cerium-based oxides for H_2 production by thermochemical two-step water-splitting. *J. Mater. Sci.*, 45, 4163–4173, 2010.
112. Bulfin, B., Call, F., Vieten, J., Roeb, M., Sattler, C., Shvets, I.V., Oxidation and reduction reaction kinetics of mixed cerium zirconium oxides. *J. Phys. Chem. C*, 120, 2027–2035, 2016.
113. Fedunik-Hofman, L., Bayon, A., Donne, S.W., Kinetics of solid-gas reactions and their application to carbonate looping systems. *Energies*, 12, 2981, 2019.
114. Irvine, J., Rupp, J.L.M., Liu, G., Xu, X., Haile, S., Qian, X., Snyder, A., Freer, R., Ekren, D., Skinner, S., Celikbilek, O., Chen, S., Tao, S., Shin, T.H., O'Hayre, R., Huang, J., Duan, C., Papac, M., Li, S., Celorrio, V., Russell, A., Hayden, B., Nolan, H., Huang, X., Wang, G., Metcalfe, I., Neagu, D., Martín, S.G., Roadmap on inorganic perovskites for energy applications. *J. Phys. Energy*, 3, 031502, 2021.
115. Davenport, T.C., Yang, C.-K., Kucharczyk, C.J., Ignatowich, M.J., Haile, S.M., Maximizing fuel production rates in isothermal solar thermochemical fuel production. *Appl. Energy*, 183, 1098–1111, 2016.
116. Ignatowich, M.J., Bork, A.H., Davenport, T.C., Rupp, J.L.M., Yang, C., Yamazaki, Y., Haile, S.M., Impact of enhanced oxide reducibility on rates of solar-driven thermochemical fuel production. *MRS Commun.*, 7, 873–878, 2017.
117. Ermanoski, I., Siegel, N.P., Stechel, E.B., A new reactor concept for efficient solar-thermochemical fuel production. *J. Sol. Energy Eng.*, 135, 031002, 2013.
118. Falter, C. and Sizmann, A., Solar thermochemical hydrogen production in the USA. *Sustainability*, 13, 7804, 2021.
119. Budama, V.K., Johnson, N.G., Ermanoski, I., Stechel, E.B., Techno-economic analysis of thermochemical water-splitting system for co-production of hydrogen and electricity. *Int. J. Hydrog. Energy*, 46, 1656–1670, 2021.
120. Michalsky, R., Botu, V., Hargus, C.M., Peterson, A.A., Steinfeld, A., Design principles for metal oxide redox materials for solar-driven isothermal fuel production. *Adv. Energy Mater.*, 5, 1401082, 2014.
121. Warren, K.J., Tran, J.T., Weimer, A.W., A thermochemical study of iron aluminate-based materials: A preferred class for isothermal water splitting. *Energy Environ. Sci.*, 15, 2022.
122. Roeb, M. and Sattler, C., Isothermal water splitting. *Science*, 341, 470–471, 2013.
123. Ceperley, D.M. and Alder, B.J., Ground state of the electron gas by a stochastic method. *Phys. Rev. Lett.*, 45, 566–569, 1980.
124. Perdew, J.P. and Zunger, A., Self-interaction correction to density-functional approximations for many-electron systems. *Phys. Rev. B*, 23, 5048–5079, 1981.

125. Perdew, J.P. and Wang, Y., Accurate and simple analytic representation of the electron-gas correlation energy. *Phys. Rev. B*, 45, 13244–13249, 1992.
126. Perdew, J.P., Burke, K., Ernzerhof, M., Generalized gradient approximation made simple. *Phys. Rev. Lett.*, 77, 3865–3868, 1996.
127. Sun, J., Ruzsinszky, A., Perdew, J.P., Strongly constrained and appropriately normed semilocal density functional. *Phys. Rev. Lett.*, 115, 036402, 2015.
128. Dudarev, S.L., Botton, G.A., Savrasov, S.Y., Humphreys, C.J., Sutton, A.P., Electron-energy-loss spectra and the structural stability of nickel oxide: An LSDA+U study. *Phys. Rev. B*, 57, 1505–1509, 1998.
129. Wang, L., Maxisch, T., Ceder, G., Oxidation energies of transition metal oxides within the GGA+U framework. *Phys. Rev. B*, 73, 195107, 2006.
130. Sai Gautam, G. and Carter, E.A., Evaluating transition metal oxides within DFT-SCAN and SCAN+U frameworks for solar thermochemical applications. *Phys. Rev. Mater.*, 2, 095401, 2018.
131. Long, O.Y., Sai Gautam, G., Carter, E.A., Evaluating optimal U for 3d transition-metal oxides within the SCAN+U framework. *Phys. Rev. Mater.*, 4, 045401, 2020.
132. Wexler, R.B., Gautam, G.S., Carter, E.A., Exchange-correlation functional challenges in modeling quaternary chalcogenides. *Phys. Rev. B*, 102, 054101, 2020.
133. Lany, S., Communication: The electronic entropy of charged defect formation and its impact on thermochemical redox cycles. *J. Chem. Phys.*, 148, 071101, 2018.
134. Pastor, E., Sachs, M., Selim, S., Durrant, J.R., Bakulin, A.A., Walsh, A., Electronic defects in metal oxide photocatalysts. *Nat. Rev. Mater.*, 7, 503, 2022.
135. Freysoldt, C., Neugebauer, J., Van de Walle, C.G., Fully Ab initio finite-size corrections for charged-defect supercell calculations. *Phys. Rev. Lett.*, 102, 016402, 2009.
136. Kumagai, Y. and Oba, F., Electrostatics-based finite-size corrections for first-principles point defect calculations. *Phys. Rev. B*, 89, 195205, 2014.
137. Walsh, A., Correcting the corrections for charged defects in crystals. *NPJ Comput. Mater.*, 7, 72, 2021.
138. Stevanović, V., Lany, S., Zhang, X., Zunger, A., Correcting density functional theory for accurate predictions of compound enthalpies of formation: Fitted elemental-phase reference energies. *Phys. Rev. B*, 85, 115104, 2012.
139. Yu, K. and Carter, E.A., Communication: Comparing ab initio methods of obtaining effective U parameters for closed-shell materials. *J. Chem. Phys.*, 140, 121105, 2014.
140. Loschen, C., Carrasco, J., Neyman, K.M., Illas, F., First-principles LDA + U and GGA + U study of cerium oxides: Dependence on the effective U parameter. *Phys. Rev. B*, 75, 035115, 2007.

141. Da Silva, J.L.F., Ganduglia-Pirovano, M.V., Sauer, J., Bayer, V., Kresse, G., Hybrid functionals applied to rare-earth oxides: The example of ceria. *Phys. Rev. B*, 75, 045121, 2007.
142. Chase Jr., M.W., NIST-JANAF thermochemical tables. *J. Phys. Chem. Ref. Data Monogr.*, 9, 1998.
143. Wu, X., Vanderbilt, D., Hamann, D.R., Systematic treatment of displacements, strains, and electric fields in density-functional perturbation theory. *Phys. Rev. B*, 72, 035105, 2005.
144. Grieshammer, S., Zacherle, T., Martin, M., Entropies of defect formation in ceria from first principles. *Phys. Chem. Chem. Phys.*, 15, 15935–15942, 2013.
145. Campañá, C. and Müser, M.H., Practical Green's function approach to the simulation of elastic semi-infinite solids. *Phys. Rev. B*, 74, 075420, 2006.
146. Kong, L.T., Bartels, G., Campañá, C., Denniston, C., Müser, M.H., Implementation of Green's function molecular dynamics: An extension to LAMMPS. *Comput. Phys. Commun.*, 180, 1004–1010, 2009.
147. Kong, L.T., Denniston, C., Müser, M.H., An improved version of the Green's function molecular dynamics method. *Comput. Phys. Commun.*, 182, 540–541, 2011.
148. Kong, L.T., Phonon dispersion measured directly from molecular dynamics simulations. *Comput. Phys. Commun.*, 182, 2201–2207, 2011.
149. Sai Gautam, G., Stechel, E.B., Carter, E.A., A first-principles-based sub-lattice formalism for predicting off-stoichiometry in materials for solar thermochemical applications: The example of ceria. *Adv. Theory Simul.*, 3, 2000112, 2020.
150. Zinkevich, M., Djurovic, D., Aldinger, F., Thermodynamic modelling of the cerium–oxygen system. *Solid State Ion.*, 177, 989–1001, 2006.
151. Hillert, M., The compound energy formalism. *J. Alloys Compd.*, 320, 161–176, 2001.
152. Gopal, C.B. and van de Walle, A., Ab initio thermodynamics of intrinsic oxygen vacancies in ceria. *Phys. Rev. B*, 86, 134117, 2012.
153. van de Walle, A. and Asta, M., Self-driven lattice-model Monte Carlo simulations of alloy thermodynamic properties and phase diagrams. *Modelling Simul. Mater. Sci. Eng.*, 10, 521, 2002.
154. Ling, S., High-concentration point-defect chemistry: Statistical-thermodynamic approach applied to nonstoichiometric cerium dioxides. *Phys. Rev. B*, 49, 864–880, 1994.
155. Zhou, F. and Åberg, D., Crystal-field calculations for transition-metal ions by application of an opposing potential. *Phys. Rev. B*, 93, 085123, 2016.
156. Naghavi, S.S., Emery, A.A., Hansen, H.A., Zhou, F., Ozolins, V., Wolverton, C., Giant onsite electronic entropy enhances the performance of ceria for water splitting. *Nat. Commun.*, 8, 285, 2017.
157. Panlener, R.J., Blumenthal, R.N., Garnier, J.E., A thermodynamic study of nonstoichiometric cerium dioxide. *J. Phys. Chem. Solids*, 36, 1213–1222, 1975.

158. Ong, S.P., Wang, L., Kang, B., Ceder, G., Li–Fe–P–O_2 phase diagram from first principles calculations. *Chem. Mater.*, 20, 1798–1807, 2008.
159. Ong, S.P., Jain, A., Hautier, G., Kang, B., Ceder, G., Thermal stabilities of delithiated olivine MPO4 (M=Fe, Mn) cathodes investigated using first principles calculations. *Electrochem. Commun.*, 12, 427–430, 2010.
160. Zunger, A., Wei, S.-H., Ferreira, L.G., Bernard, J.E., Special quasirandom structures. *Phys. Rev. Lett.*, 65, 353–356, 1990.
161. van de Walle, A., Tiwary, P., de Jong, M., Olmsted, D.L., Asta, M., Dick, A., Shin, D., Wang, Y., Chen, L.-Q., Liu, Z.-K., Efficient stochastic generation of special quasirandom structures. *Calphad*, 42, 13–18, 2013.
162. van de Walle, A., Multicomponent multisublattice alloys, nonconfigurational entropy and other additions to the alloy theoretic automated toolkit. *Calphad*, 33, 266–278, 2009.
163. Raybaud, P., Digne, M., Iftimie, R., Wellens, W., Euzen, P., Toulhoat, H., Morphology and surface properties of boehmite (γ-AlOOH): A density functional theory study. *J. Catal.*, 201, 236–246, 2001.
164. Gong, X.-Q. and Selloni, A., Reactivity of anatase TiO_2 nanoparticles: The role of the minority (001) surface. *J. Phys. Chem. B*, 109, 19560–19562, 2005.
165. Wexler, R.B. and Sohlberg, K., Role of proton hopping in surface charge transport on tin dioxide as revealed by the thermal dependence of conductance. *J. Phys. Chem. A*, 118, 12031–12040, 2014.
166. Mills, G., Jónsson, H., Schenter, G.K., Reversible work transition state theory: application to dissociative adsorption of hydrogen. *Surf. Sci.*, 324, 305–337, 1995.
167. Henkelman, G. and Jónsson, H., Improved tangent estimate in the nudged elastic band method for finding minimum energy paths and saddle points. *J. Chem. Phys.*, 113, 9978–9985, 2000.
168. Henkelman, G., Uberuaga, B.P., Jónsson, H., A climbing image nudged elastic band method for finding saddle points and minimum energy paths. *J. Chem. Phys.*, 113, 9901–9904, 2000.
169. Trottier, R.M., Millican, S.L., Musgrave, C.B., Musgrave, C.B., Musgrave, C.B., Musgrave, C.B., Modified single iteration synchronous-transit approach to bound diffusion barriers for solid-state reactions. *J. Chem. Theory Comput.*, 16, 5912–5922, 2020.
170. Caspersen, K.J. and Carter, E.A., Finding transition states for crystalline solid-solid phase transformations. *Proc. Natl. Acad. Sci.*, 102, 6738–6743, 2005.
171. Xu, R. and Wiesner, T.F., Dynamic model of a solar thermochemical water-splitting reactor with integrated energy collection and storage. *Int. J. Hydrog. Energy*, 37, 2210–2223, 2012.
172. Chueh, W.C., Falter, C., Abbott, M., Scipio, D., Furler, P., Haile, S.M., Steinfeld, A., High-flux solar-driven thermochemical dissociation of CO_2 and H_2O using nonstoichiometric ceria. *Science*, 330, 1797, 2010.

173. Arifin, D. and Weimer, A.W., Kinetics and mechanism of solar-thermochemical H_2 and CO production by oxidation of reduced CeO_2. *Sol. Energy*, 160, 178–185, 2017.
174. Bhosale, R.R., Takalkar, G., Sutar, P., Kumar, A., AlMomani, F., Khraisheh, M., A decade of ceria based solar thermochemical H_2O/CO_2 splitting cycle. *Int. J. Hydrog. Energy*, 44, 34–60, 2019.
175. Le Gal, A. and Abanades, S., Dopant incorporation in ceria for enhanced water-splitting activity during solar thermochemical hydrogen generation. *J. Phys. Chem. C.*, 116, 13516–13523, 2012.
176. Costa Oliveira, F.A., Barreiros, M.A., Haeussler, A., Caetano, A.P.F., Mouquinho, A., II, Oliveira e. Silva, P.M., Novais, R.M., Pullar, R.C., Abanades, S., High performance cork-templated ceria for solar thermochemical hydrogen production via two-step water-splitting cycles. *Sustain. Energy Fuels*, 4, 3077–3089, 2020.
177. Ganzoury, M.A., Fateen, S.-E.K., El Sheltawy, S.T., Radwan, A.M., Allam, N.K., Thermodynamic and efficiency analysis of solar thermochemical water splitting using Ce-Zr mixtures. *Sol. Energy*, 135, 154–162, 2016.
178. Kaneko, H., Miura, T., Ishihara, H., Taku, S., Yokoyama, T., Nakajima, H., Tamaura, Y., Reactive ceramics of CeO_2–MO_x (M=Mn, Fe, Ni, Cu) for H_2 generation by two-step water splitting using concentrated solar thermal energy. *Energy*, 32, 656–663, 2007.
179. Ackermann, S., Scheffe, J.R., Steinfeld, A., Diffusion of oxygen in ceria at elevated temperatures and its application to H_2O/CO_2 splitting thermochemical redox cycles. *J. Phys. Chem. C*, 118, 5216–5225, 2014.
180. Muhich, C., Hoes, M., Steinfeld, A., Mimicking tetravalent dopant behavior using paired charge compensating dopants to improve the redox performance of ceria for thermochemically splitting H_2O and CO_2. *Acta Mater.*, 144, 728–737, 2018.
181. Furler, P., Scheffe, J.R., Steinfeld, A., Syngas production by simultaneous splitting of H_2O and CO_2 via ceria redox reactions in a high-temperature solar reactor. *Energy Env. Sci.*, 5, 6098–6103, 2011.
182. Hao, Y., Yang, C.-K., Haile, S.M., High-temperature isothermal chemical cycling for solar-driven fuel production. *Phys. Chem. Chem. Phys.*, 15, 17084–17092, 2013.
183. Chueh, W.C. and Haile, S.M., Ceria as a thermochemical reaction medium for selectively generating syngas or methane from H_2O and CO_2. *ChemSusChem*, 2, 735–739, 2009.
184. Chueh, W.C. and Haile, S.M., A thermochemical study of ceria: Exploiting an old material for new modes of energy conversion and CO_2 mitigation. *Philos. Trans. R. Soc. Math. Phys. Eng. Sci.*, 368, 3269–3294, 2010.
185. Marxer, D., Furler, P., Scheffe, J., Geerlings, H., Falter, C., Batteiger, V., Sizmann, A., Steinfeld, A., Demonstration of the entire production chain to renewable kerosene via solar thermochemical splitting of H_2O and CO_2. *Energy Fuels*, 29, 3241–3250, 2015.

186. Reinhardt, K. and Winkler, H., *Ullmann's Encyclopedia of Industrial Chemistry*, Wiley-VCH Verlag GmbH & Co. KGaA (Ed.), Wiley-VCH Verlag GmbH & Co. KGaA, Hoboken, New Jersey, 2000.
187. Bulfin, B., Hoffmann, L., Oliveira, L., Knoblauch, N., Call, F., Roeb, M., Sattler, C., Schmücker, M., Statistical thermodynamics of non-stoichiometric ceria and ceria zirconia solid solutions. *Phys. Chem. Chem. Phys.*, 18, 23147–23154, 2016.
188. Takalkar, G., Bhosale, R.R., Rashid, S., AlMomani, F., Shakoor, R.A., Al Ashraf, A., Application of Li-, Mg-, Ba-, Sr-, Ca-, and Sn-doped ceria for solar-driven thermochemical conversion of carbon dioxide. *J. Mater. Sci.*, 55, 11797–11807, 2020.
189. Meng, Q.-L., Lee, C., Shigeta, S., Kaneko, H., Tamaura, Y., Solar hydrogen production using $Ce_{1-x}Li_xO_{2-\delta}$ solid solutions via a thermochemical, two-step water-splitting cycle. *J. Solid State Chem.*, 194, 343–351, 2012.
190. Portarapillo, M., Russo, D., Landi, G., Luciani, G., Di Benedetto, A., K-doped CeO_2–ZrO_2 for CO_2 thermochemical catalytic splitting. *RSC Adv.*, 11, 39420–39427, 2022.
191. Bonk, A., Maier, A.C., Schlupp, M.V.F., Burnat, D., Remhof, A., Delmelle, R., Steinfeld, A., Vogt, U.F., The effect of dopants on the redox performance, microstructure and phase formation of ceria. *J. Power Sources*, 300, 261–271, 2015.
192. Meng, Q.-L., Lee, C., Ishihara, T., Kaneko, H., Tamaura, Y., Reactivity of CeO_2-based ceramics for solar hydrogen production via a two-step water-splitting cycle with concentrated solar energy. *Int. J. Hydrog. Energy*, 36, 13435–13441, 2011.
193. Scheffe, J.R. and Steinfeld, A., Thermodynamic analysis of cerium-based oxides for solar thermochemical fuel production. *Energy Fuels*, 26, 1928–1936, 2012.
194. Liu, Z., Ma, H., Sorrell, C.C., Koshy, P., Hart, J.N., *Enhancement of Light Absorption and Oxygen Vacancy Formation in CeO2 by Transition Metal Doping: A DFT Study*, 2022, arXiv:2012.06195.
195. Nair, M.M. and Abanades, S., Tailoring hybrid nonstoichiometric ceria redox cycle for combined solar methane reforming and thermochemical conversion of H_2O/CO_2. *Energy Fuels*, 30, 6050–6058, 2016.
196. Le Gal, A., Abanades, S., Bion, N., Le Mercier, T., Harle, V., Reactivity of doped ceria-based mixed oxides for solar thermochemical hydrogen generation via two-step water-splitting cycles. *Energy Fuels*, 27, 6068–6078, 2013.
197. Call, F., Roeb, M., Schmücker, M., Sattler, C., Pitz-Paal, R., Ceria doped with zirconium and lanthanide oxides to enhance solar thermochemical production of fuels. *J. Phys. Chem. C*, 119, 6929–6938, 2015.
198. Scheffe, J.R., Jacot, R., Patzke, G.R., Steinfeld, A., Synthesis, characterization, and thermochemical redox performance of Hf^{4+}, Zr^{4+}, and Sc^{3+} doped ceria for splitting CO_2. *J. Phys. Chem. C*, 117, 24104–24114, 2013.

199. Hao, Y., Yang, C.-K., Haile, S.M., Ceria-zirconia solid solutions ($Ce_{1-x}Zr_xO_{2-\delta}$, $x \leq 0.2$) for solar thermochemical water splitting: A thermodynamic study. *Chem. Mater.*, 26, 6073–6082, 2014.
200. Rothensteiner, M., Bonk, A., Vogt, U.F., Emerich, H., van Bokhoven, J.A., Structural changes in $Ce_{0.5}Zr_{0.5}O_{2-\delta}$ under temperature-swing and isothermal solar thermochemical looping conditions determined by *in situ* Ce K and Zr K edge x-ray absorption spectroscopy. *J. Phys. Chem. C*, 120, 13931–13941, 2016.
201. Petkovich, N.D., Rudisill, S.G., Venstrom, L.J., Boman, D.B., Davidson, J.H., Stein, A., Control of heterogeneity in nanostructured $Ce_{1-x}Zr_xO_2$ binary oxides for enhanced thermal stability and water splitting activity. *J. Phys. Chem. C*, 115, 21022–21033, 2011.
202. Riaz, A., Kremer, F., Kim, T., Sattayaporn, S., Tsuzuki, T., Lipiński, W., Lowe, A., Experimental demonstration of vanadium-doped nanostructured ceria for enhanced solar thermochemical syngas production. *Nano Energy*, 81, 105639, 2021.
203. Singh, P. and Hegde, M.S., $Ce_{0.67}Cr_{0.33}O_{2.11}$: A new low-temperature O_2 evolution material and H_2 generation catalyst by thermochemical splitting of water. *Chem. Mater.*, 22, 762–768, 2010.
204. Kaneko, H., Ishihara, H., Taku, S., Naganuma, Y., Hasegawa, N., Tamaura, Y., Cerium ion redox system in $CeO_{2-x}Fe_2O_3$ solid solution at high temperatures (1,273–1,673 K) in the two-step water-splitting reaction for solar H_2 generation. *J. Mater. Sci.*, 43, 3153–3161, 2008.
205. Kaneko, H. and Tamaura, Y., Reactivity and XAFS study on $(1-x)CeO_2$–$xNiO$ ($x=0.025-0.3$) system in the two-step water-splitting reaction for solar H_2 production. *J. Phys. Chem. Solids*, 70, 1008–1014, 2009.
206. Lin, F., Rothensteiner, M., Alxneit, I., van Bokhoven, J.A., Wokaun, A., First demonstration of direct hydrocarbon fuel production from water and carbon dioxide by solar-driven thermochemical cycles using rhodium–ceria. *Energy Environ. Sci.*, 9, 2400–2409, 2016.
207. Muhich, C. and Steinfeld, A., Principles of doping ceria for the solar thermochemical redox splitting of H_2O and CO_2. *J. Mater. Chem. Mater. Energy Sustain.*, 5, 15578–15590, 2017.
208. Nolan, M., Enhanced oxygen vacancy formation in ceria (111) and (110) surfaces doped with divalent cations. *J. Mater. Chem.*, 21, 9160, 2011.
209. Nakayama, M. and Martin, M., First-principles study on defect chemistry and migration of oxide ions in ceria doped with rare-earth cations. *Phys. Chem. Chem. Phys.*, 11, 3241, 2009.
210. Scanlon, D.O., Morgan, B.J., Watson, G.W., The origin of the enhanced oxygen storage capacity of $Ce_{1-x}(Pd/Pt)_xO_2$. *Phys. Chem. Chem. Phys.*, 13, 4279, 2011.
211. Wexler, R.B., Gautam, G.S., Carter, E.A., Optimizing kesterite solar cells from Cu_2ZnSnS_4 to $Cu_2CdGe(S, Se)_4$. *J. Mater. Chem. A*, 9, 9882–9897, 2021.

212. Wexler, R.B., Gautam, G.S., Stechel, E.B., Carter, E.A., Factors governing oxygen vacancy formation in oxide perovskites. *J. Am. Chem. Soc.*, 143, 13212–13227, 2021.
213. Dawicke, J.W. and Blumenthal, R.N., Oxygen association pressure measurements on nonstoichiometric cerium dioxide. *J. Electrochem. Soc.*, 133, 904–909, 1986.
214. Bulfin, B., Lowe, A.J., Keogh, K.A., Murphy, B.E., Lübben, O., Krasnikov, S.A., Shvets, I.V., Analytical model of CeO_2 oxidation and reduction. *J. Phys. Chem. C*, 117, 24129–24137, 2013.
215. Fu, M., Xu, H., Li, X., Mechanism of oxygen vacancy assisted water-splitting of $LaMnO_3$: Inorganic perovskite prediction for fast solar thermochemical H_2 production. *Inorg. Chem. Front.*, 7, 2381–2387, 2020.
216. Demont, A., Abanades, S., Beche, E., Investigation of perovskite structures as oxygen-exchange redox materials for hydrogen production from thermochemical two-step water-splitting cycles. *J. Phys. Chem. C*, 118, 12682–12692, 2014.
217. Bakken, E., Norby, T., Stølen, S., Redox energetics of perovskite-related oxides. *J. Mater. Chem.*, 12, 317–323, 2002.
218. McDaniel, A.H., Miller, E.C., Arifin, D., Ambrosini, A., Coker, E.N., O'Hayre, R., Chueh, W.C., Tong, J., Sr-and Mn-doped $LaAlO_{3-\delta}$ for solar thermochemical H_2 and CO production. *Energy Environ. Sci.*, 6, 2424–2428, 2013.
219. McDaniel, A.H., Ambrosini, A., Coker, E.N., Miller, J.E., Chueh, W.C., O'Hayre, R., Tong, J., Nonstoichiometric perovskite oxides for solar thermochemical H_2 and CO production. *Energy Procedia*, 49, 2009–2018, 2014.
220. Deml, A.M., Stevanović, V., Holder, A.M., Sanders, M., O'Hayre, R., Musgrave, C.B., Tunable oxygen vacancy formation energetics in the complex perovskite oxide $Sr_xLa_{1-x}Mn_yAl_{1-y}O_3$. *Chem. Mater.*, 26, 6595–6602, 2014.
221. Takacs, M., Hoes, M., Caduff, M., Cooper, T., Scheffe, J.R., Steinfeld, A., Oxygen nonstoichiometry, defect equilibria, and thermodynamic characterization of $LaMnO_3$ perovskites with Ca/Sr A-site and Al B-site doping. *Acta Mater.*, 103, 700–710, 2016.
222. Cooper, T., Scheffe, J.R., Galvez, M.E., Jacot, R., Patzke, G., Steinfeld, A., Lanthanum manganite perovskites with Ca/Sr A-site and Al B-site doping as effective oxygen exchange materials for solar thermochemical fuel production. *Energy Technol.*, 3, 1130–1142, 2015.
223. Wang, L., Al-Mamun, M., Liu, P., Wang, Y., Gui Yang, H., Zhao, H., $La_{1-x}Ca_xMn_{1-y}Al_yO_3$ perovskites as efficient catalysts for two-step thermochemical water splitting in conjunction with exceptional hydrogen yields. *Chin. J. Catal.*, 38, 1079–1086, 2017.
224. Luciani, G., Landi, G., Aronne, A., Di Benedetto, A., Partial substitution of B cation in $La_{0.6}Sr_{0.4}MnO_3$ perovskites: A promising strategy to improve the redox properties useful for solar thermochemical water and carbon dioxide splitting. *Sol. Energy*, 171, 1–7, 2018.

225. Chen, Z., Jiang, Q., Cheng, F., Tong, J., Yang, M., Jiang, Z., Li, C., Sr- and Co-doped $LaGaO_{3-\delta}$ with high O_2 and H_2 yields in solar thermochemical water splitting. *J. Mater. Chem. Mater. Energy Sustain.*, 7, 6099–6112, 2019.
226. Barcellos, D.R., Sanders, M.D., Tong, J., McDaniel, A.H., O'Hayre, R.P., $BaCe_{0.25}Mn_{0.75}O_{3-\delta}$, a promising perovskite-type oxide for solar thermochemical hydrogen production. *Energy Environ. Sci.*, 11, 3256–3265, 2018.
227. Qian, X., He, J., Mastronardo, E., Baldassarri, B., Wolverton, C., Haile, S.M., Favorable redox thermodynamics of $SrTi_{0.5}Mn_{0.5}O_{3-\delta}$ in solar thermochemical water splitting. *Chem. Mater.*, 32, 9335–9346, 2020.
228. Sai Gautam, G., Stechel, E.B., Carter, E.A., Exploring Ca-Ce-M-O (M = 3d transition metal) oxide perovskites for solar thermochemical applications. *Chem. Mater.*, 32, 9964–9982, 2020.
229. Wexler, R. B., Gautam, G. S., Bell, R., Shulda, S., Strange, N. A., Trindell, J. A., Sugar, J. D., Nygren, E., Sainio, S., McDaniel, A. H., Ginley, D., Carter, E. A., Stechel, E. B., Multiple and nonlocal cation redox in Ca-Ce-Ti-Mn oxide perovskites for solar thermochemical applications, *Energy Environ. Sci.*, 2023.
230. Heo, S.J., Sanders, M., O'Hayre, R., Zakutayev, A., Double-site substitution of Ce into (Ba, Sr)MnO_3 perovskites for solar thermochemical hydrogen production. *ACS Energy Lett.*, 6, 3037–3043, 2021.
231. Dey, S., Naidu, B.S., Rao, C.N.R., $Ln_{0.5}A_{0.5}MnO_3$ (Ln = Lanthanide, A = Ca, Sr) perovskites exhibiting remarkable performance in the thermochemical generation of CO and H_2 from CO_2 and H_2O. *Chem.-Eur. J.*, 21, 7077–7081, 2015.
232. Barcellos, D.R., Coury, F.G., Emery, A., Sanders, M., Tong, J., McDaniel, A., Wolverton, C., Kaufman, M., O'Hayre, R., Phase identification of the layered perovskite $Ce_xSr_{2-x}MnO_4$ and application for solar thermochemical water splitting. *Inorg. Chem.*, 58, 12, 7705, 2019.
233. Bergeson-Keller, A.M., Sanders, M.D., O'Hayre, R.P., Reduction thermodynamics of $Sr_{1-x}Ce_xMnO_3$ and $Ce_xSr_{2-x}MnO_4$ perovskites for solar thermochemical hydrogen production. *Energy Technol.*, 10, 2100515. 2021.
234. Kubicek, M., Bork, A.H., Rupp, J.L.M., Perovskite oxides—A review on a versatile material class for solar-to-fuel conversion processes. *J. Mater. Chem. A*, 5, 11983–12000, 2017.
235. Haeussler, A., Abanades, S., Jouannaux, J., Julbe, A., Non-stoichiometric redox active perovskite materials for solar thermochemical fuel production: A review. *Catalysts*, 8, 611, 2018.
236. Emery, A.A., Saal, J.E., Kirklin, S., Hegde, V., II, Wolverton, C., High-throughput computational screening of perovskites for thermochemical water splitting applications. *Chem. Mater.*, 28, 5621–5634, 2016.
237. Ezbiri, M., Allen, K.M., Galvez, M.E., Michalsky, R., Steinfeld, A., Design principles of perovskites for thermochemical oxygen separation. *ChemSusChem*, 8, 1966–1971, 2015.

238. Deml, A.M., Holder, A.M., O'Hayre, R.P., Musgrave, C.B., Stevanović, V., Intrinsic material properties dictating oxygen vacancy formation energetics in metal oxides. *J. Phys. Chem. Lett.*, 6, 1948–1953, 2015.
239. Muñoz-García, A.B., Ritzmann, A.M., Pavone, M., Keith, J.A., Carter, E.A., Oxygen transport in perovskite-type solid oxide fuel cell materials: Insights from quantum mechanics. *Acc. Chem. Res.*, 47, 3340–3348, 2014.
240. Pavone, M., Ritzmann, A.M., Carter, E.A., Quantum-mechanics-based design principles for solid oxide fuel cell cathode materials. *Energy Environ. Sci.*, 4, 4933, 2011.
241. Zeng, Z., Calle-Vallejo, F., Mogensen, M.B., Rossmeisl, J., Generalized trends in the formation energies of perovskite oxides. *Phys. Chem. Chem. Phys.*, 15, 7526–7533, 2013.
242. Jain, A., Ong, S.P., Hautier, G., Chen, W., Richards, W.D., Dacek, S., Cholia, S., Gunter, D., Skinner, D., Ceder, G., Persson, K.A., Commentary: The materials project: A materials genome approach to accelerating materials innovation. *APL Mater.*, 1, 011002, 2013.
243. Jain, A., Hautier, G., Ong, S.P., Moore, C.J., Fischer, C.C., Persson, K.A., Ceder, G., Formation enthalpies by mixing GGA and GGA + U calculations. *Phys. Rev. B*, 84, 045115, 2011.
244. Naghavi, S.S., He, J., Wolverton, C., $CeTi_2O_6$-a promising oxide for solar thermochemical hydrogen production. *ACS Appl. Mater. Interfaces*, 12, 21521–21527, 2020.
245. Kramida, A., Ralchenko, Y., Reader, J., NIST ASD Team, *NIST Atomic Spectra Database (Version 5.9)*, National Institute of Standards and Technology, Gaithersburg, MD, 2021, https://physics.nist.gov/asd.
246. Yen, W.M., General Factors Governing the efficiency of luminescent devices. *Phys. Solid State*, 47, 1393, 2005.
247. Murgida, G.E., Ferrari, V., Ganduglia-Pirovano, M.V., Llois, A.M., Ordering of oxygen vacancies and excess charge localization in bulk ceria: A DFT + U study. *Phys. Rev. B*, 90, 115120, 2014.
248. Pártay, L.B., Csányi, G., Bernstein, N., Nested sampling for materials. *Eur. Phys. J. B*, 94, 159, 2021.
249. Deringer, V.L., Caro, M.A., Csányi, G., Machine learning interatomic potentials as emerging tools for materials science. *Adv. Mater.*, 31, 1902765, 2019.
250. Virtanen, P., Gommers, R., Oliphant, T.E., Haberland, M., Reddy, T., Cournapeau, D., Burovski, E., Peterson, P., Weckesser, W., Bright, J., van der Walt, S.J., Brett, M., Wilson, J., Millman, K.J., Mayorov, N., Nelson, A.R.J., Jones, E., Kern, R., Larson, E., Carey, C.J., Polat, İ., Feng, Y., Moore, E.W., VanderPlas, J., Laxalde, D., Perktold, J., Cimrman, R., Henriksen, I., Quintero, E.A., Harris, C.R., Archibald, A.M., Ribeiro, A.H., Pedregosa, F., van Mulbregt, P., SciPy 1.0 Contributors, SciPy 1.0: Fundamental algorithms for scientific computing in python. *Nat. Methods*, 17, 261–272, 2020.

2
Solar Metal Fuels for Future Transportation

Youssef Berro and Marianne Balat-Pichelin*

Laboratoire PROcédés, Matériaux et Energie Solaire, PROMES-CNRS UPR, Font-Romeu-Odeillo, France

---

Abstract

The two primary critical challenges facing our society nowadays, the depletion of petroleum reserves and the global warming, necessitate the substitution of transport fossil fuels with sustainable, recyclable, and cleaner fuels. The direct combustion of metal fuels is attractive due to their high-energy density and the ability to regenerate them through the reduction of their oxide products. Their regeneration is economically beneficial and environmental-friendly when performed using renewable energy sources, providing sustainability for the combustion/reduction cycles. Thus, the following chapter highlights the research advancement employed for the development of a metal-fueled combustion engine and a solar recycling reduction process for the regeneration of metal fuels. To this day, an 80% heat-to-mechanical conversion efficiency is attained when burning Mg particles in a swirled-stabilized metal-air combustor, with the ability to collect 98% of the produced submicron magnesia products. Using concentrated solar energy, a recycling yield of 68% and 96% is achieved for Al and Mg, respectively, by optimizing the operating parameters allowing to produce pure stable microsized metal powders. Further investigation is required to improve the combustion engineering of Al particles and to reduce the formation of unwanted Al-oxycarbides by-products during the carbothermal reduction of alumina.

Keywords: Carbothermal reduction, metal fuels, concentrated solar power (CSP), sustainable recycling, energy-carrying materials, combustion engineering

---

*Corresponding author: marianne.balat@promes.cnrs.fr

Nurdan Demirci Sankir and Mehmet Sankir (eds.) Solar Fuels, (65–96) © 2023 Scrivener Publishing LLC

2.1 Introduction

On one hand, energy is the key parameter for the economic growth of developing countries. However, on the other hand, energy consumption, in particular the burning of fossil fuels, is the main source of greenhouse gases emissions causing global warming. Thus, the greatest challenge facing our society is to reduce the carbon content of energy while increasing the efficiency of its use to maintain economic growth.

2.1.1 Sustainable Strategies to Address Climate Change

Since 1760, the industrial revolution was the primary reason for climatic change due to the burning of fossil fuels to provide the required energy for the industrial sector. Nowadays, community efforts focus on solving the two critical issues confronting this sector, the global warming and the depletion of fossil fuels [1]. Many global strategies and agreements were admitted to ensure clean and sustainable energy and to combat climate change, such as the United Nations Framework Convention on Climate Change (UNFCCC in 1994), the Sustainable Development Goals (SDGs in 2015) and the Paris Agreement (in 2016). Based on a System Dynamics Energy-Economy-Environment model, integrating the energy demand with the depletion of fossil fuels and alternative energy sources, Capellán-Pérez et al. suggested that transportation is the most critical sector [2]. Considering environmental aspect, transport is the top sector producing greenhouse gases [3], accounting up to 28% of the global greenhouse gases in the developed countries [4]. Considering the energy demand, transportation was the main consumer of the energy provided from petroleum sources in 2019. Indeed, around 70% of the petroleum energy was issued to provide around 91% of the transportation consumption [5]. Considering the alternative solutions, electricity dominates the consumption of the energy produced from renewable sources (56%), while the transportation sector profits only from around 12% (providing only 5% of the consumption demand) [5]. Therefore, it is critical to consider new technologies to substitute fossil fuels used in the transport sector with sustainable, recyclable and cleaner fuels.

2.1.2 Circular Economy

The European Union (EU) adopted the "Raw Material Initiative" in 2008 promoting the production and recycling of primary materials, and leading to the adoption of the "Circular Economy" strategy in 2015. In this context, the French Environment and Energy Management Agency (ADEME) defined seven pillars of the circular economy: responsible consumption; eco-design; functional economy; industrial and territorial ecology; reemploy, reuse and repair; recycling; and sustainable sourcing. In 2020, the European Commission adopted the "New Circular Economy Action Plan" as one of the main building blocks of the "European Green Deal" to reduce pressure on natural resources, create sustainable growth and jobs, halt biodiversity loss and achieve the EU 2050 climate neutrality target.

Reviewing the concept, current practices and the assessment of the circular economy in China proved that immense efforts are required to move from rhetoric to implementation. Additional to the current measures, such as pricing reforms and preferential tax policies, new measures, such as environmental taxes should be included. More importantly, technological innovations and research efforts in the field of energy savings, alternatives and recycling should be supported [6]. Figure 2.1 shows that the design of new production techniques is critical to closing the loop. A recent investigation showed that adopting renewable energy technologies in the circular economy principles and models, through eco-design and recycling, can optimize the energy transition aiming to sustainably address the economic growth and the increasing energy demand in Africa [7].

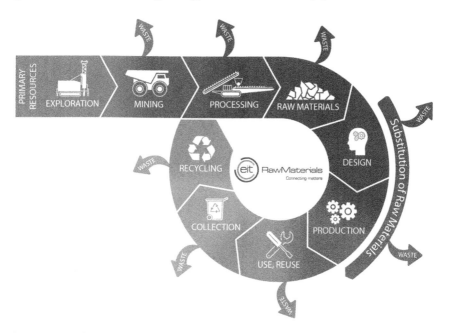

Figure 2.1 EIT RawMaterials vision for the circular economy [eitrawmaterials.eu].

2.1.3 Sustainable Solar Recycling of Metal Fuels

Historically, metal powders were used as combustibles in the aerospace and rocket research fields as during Mars missions where magnesium is used as fuel and Martian CO_2 as oxidizer [8]. They were identified as energy-carrying materials competing with hydrocarbons due to their high-energy densities [9]. However, the drawbacks related to their production through conventional processes restrict their use as transport fuels. For example, aluminum cannot be considered an alternative energy-carrying fuel when produced through the Bayer/Hall-Héroult process as the thermal and electrical efficiencies of the aluminum-based storage system did not exceed 43% and 22%, respectively [9]. Utgikar *et al.* showed that, when using conventional production processes, aluminum metal fuels are less efficient and more polluting than hydrogen and gasoline with around 11 MJ km^{-1} and 21.1 Bt CO_2e per year [10]. Therefore, the employ of new technologies for the production and recycling of metal powders using renewable energy sources becomes critical to the worth of their use as transport fuels. Figure 2.2 shows the suggested innovative process, based on the combustion/reduction cycles, allowing the use of metal fuels as transport fuels as they can be regenerated, using concentrated solar energy, from the oxide products collected in the metal-fueled burner [11].

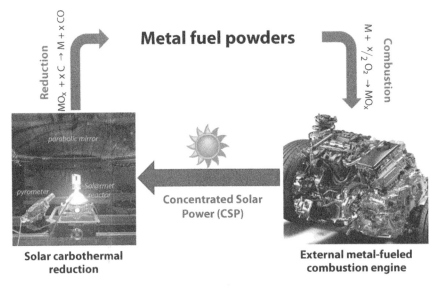

Figure 2.2 Sustainable production of metal fuels through combustion/reduction cycles [11].

2.2 Direct Combustion of Solar Metal Fuels

Metal powders, having high-energy densities, are attractive to be used as transport fuels through their direct combustion. However, this criterion is not sufficient to consider their employ as recyclable solar transport fuels. Both the combustion properties of the metal powders and the ease of their regeneration (through the reduction of their oxide products) must be considered when choosing the appropriate transport metal fuel [12]. Moreover, the environmental aspect should be considered as for beryllium due to its toxicity and carcinogenicity [13].

Considering the combustion properties of metals, Figure 2.3a shows that boron has the highest specific energy of 40 kWh l^{-1}, around four times that of diesel. Nevertheless, the slow kinetics of its combustion and the formation of sub-oxides, forming a borate glass coating on the combustor,

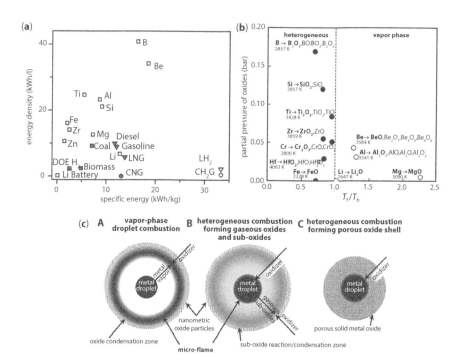

Figure 2.3 (a) Volumetric and gravimetric energy densities of various fuels including metal powders (CNG compressed natural gas, LNG liquefied natural gas, CH$_2$G compressed hydrogen gas, LH$_2$ liquid hydrogen), (b) Combustion properties of metal particles as a function of the Tf/Tb ratio (flame to boiling temperatures), (c) Combustion modes of the metal powders [12, 18].

makes its use unattractive [14]. Iron has a high-energy density but low specific energy, meaning it is too heavy to be carried in the car, as around 200 kg of iron powders are required to give a power equivalent to 34 kg of diesel, additionally to 300 kg of iron oxide produced. Furthermore, Figures 2.3b and 2.3c show that, depending on the flame temperature of the metals, only some of them (Be, Al, Mg, Li) burn in the vapor phase according to mode A, similarly to hydrocarbons. However, the oxides produced during modes A and B are nano-sized which is restrictive to their collection, making it necessary to use HEPA-type or electrostatic filters. Nevertheless, we will discuss in the following sections the ability to control the combustion modes and kinetics through combustion engineering to produce micron-sized oxides that can be collected easier using a cyclone. Thus, the choice of metal/oxide couple is independent of this restriction.

Considering the recycling of the metal fuels through the solar reduction of oxide products, silicon possess a high calorific value (30.74 MJ kg^{-1}) similar to that of Al and higher than that of Mg [9, 15]. However, the reduction temperature required for its regeneration is about 4500 K, compared to 4000 K for alumina and 3700 K for magnesia [16]. The titanium also requires high reduction temperatures for its regeneration, additionally to the formation of various oxide by-products during the reaction [17]. Thus, we considered the Al/Al$_2$O$_3$ and Mg/MgO couples as the best candidates to consider as recyclable solar fuels for transportation.

2.2.1 Stabilized Metal-Fuel Flame

The combustion modes, shown in Figure 2.3, are not only dependent on the combustion properties of the metal fuels but also on the operating parameters during their burning as the initial temperature, the pressure and the metal/oxidizer concentration ratio in the burner. For example, aluminum might burn heterogeneously, with the formation of a molten oxide layer on the aluminum droplet, if the oxygen concentration in the flow decreases below 10% as the flame-to-boiling temperature ratio (T_f/T_b) becomes lower than one [19].

Moreover, the classification of the combustion modes is assumed true only in the case of particle ignition where a micro-flame is formed around the fuel particle. However, particles can burn efficiently without the formation of a micro-flame when operating at high gas temperatures. This can be achieved in self-propagating flame fronts when burning relatively dense suspensions of metal powders in the oxidizer [20]. In this context, researchers studied the stabilization of a metal-fuel flame in a Bunsen burner by shooting micro-sized metal particles in a flow of oxidizing gas [21].

The combustion process is fast and self-sustained by the molecular diffusion of heat from the hot zone to the unburned reactants, with burning velocities dependent on the heat release and thermal diffusion rates [22]. Particles in the metal-fuel suspension can burn according to various combustion modes (A, B or C, Figure 2.4c) forming individual micro-flames in the stabilized macroscopic flame thermal wave. In this case, all the combustion modes lead to a similar stabilized laminar flame having a burning velocity and a flame thickness depending on the reactivity and combustion time of particles [23].

2.2.2 Combustion Engineering

When burning metal fuels in a stabilized flame front, oxides could form and condense on or very near the particle surface and remain attached to it at the selected flame temperature [24]. This allows producing oxides with a size range close to that of the burned metal powders (micro-sized), which is beneficial for their collection and recycling. Thus, the creation of the stabilized metal-fuel flame permits overcoming the formation of nano-sized oxides obtained during the conventional diffusion-limited combustion modes A and B (Figure 2.4c).

The reaction rate of a metal-fuel stabilized flame is proportional to the flame burning speed and inversely proportional to the flame thickness. The flame thickness is dependent on the burned metal as it is proportional to the thermal diffusivity of the metal vapors. In general, the flame thicknesses of metal flames are close to those of hydrocarbons [22]. The flame speed is proportional to the laminar burning velocity that characterizes the combustion intensity for a particular fuel-oxidizer mixture [25]. Figure 2.4 shows the dependence of the burning velocities of metal fuels from the oxidizer percentage and the burned particles size. Aluminum, magnesium and iron have burning velocities of around 20 to 30 cm s^{-1}, close to that of methane. As for methane, the burning velocity of aluminum rises when increasing the oxidizer content, however, this dependence is less important than for the methane flame [20]. The burning velocity of 15 µm Mg particles is around three times higher than that of 70 µm particles as smaller particles have higher surface-to-volume ratios, thus leading to higher reaction rates [26].

The burning velocity of both aluminum and methane improves, as the fuel concentration increases, to reach 20 and 35 m s^{-1}, respectively at a fuel/oxygen equivalence ratio equal to one. Above this ratio, the burning velocity of aluminum remains constant while that of methane drops gradually [25, 27]. This can be explained by that, contrarily to methane, adding

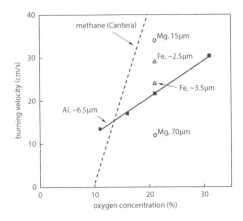

Figure 2.4 Burning velocities of suspensions of methane and some metal fuels as a function of the oxygen content in the air [18].

more metal particles does not displace air due to their much higher densities [28]. This advantage makes the stabilization of a metal flame easier by using a fuel-rich pre-burner [22].

Utilizing a fuel-rich pre-burner ensures stable combustion during the variation of the particles size and reduces the NO_x formation by starving the flame for oxygen [29]. As previously mentioned, the curtailed sensitivity of metal fuels to the oxygen concentration (Figure 2.5) permits diminishing NO_x production through oxygen starvation. Further, Goroshin et al. proved that, as for carbon-air flames, preheating the metal/air mixture allows increasing the initial temperature of the metal-fuel flame, and thus the flame speed and the combustion rate [22]. The high radiative emissivity obtained from metal flames can be helpful to pre-heat the metal-air mixture when absorbed by reactants, thus increasing the flame speed [25]. One last critical parameter to consider during flame burning is the turbulence level as almost all combustion systems employ turbulent flows to increase the burning rates. In this context, Julien et al. observed a significant increase in the flame burning rate caused by some residual turbulence created by the metal-fuel dispersal process [27].

2.2.3 Designing Metal-Fueled Engines

The design of a metal-fueled heat engine must take into consideration both the combustion properties of the metal particles and the ability to collect the produced oxides to be regenerated. The combustion of metal powders in an internal combustion engine (ICE) is unfavorable, as the produced

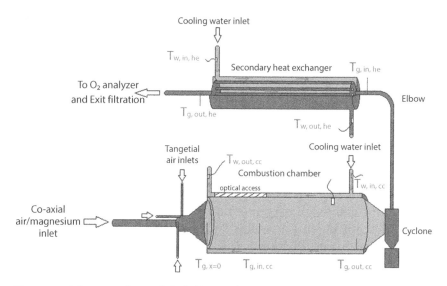

Figure 2.5 Schematic of a metal-fueled burner for clean energy production [32].

metal oxides provoke wear of the valve disks which is a critical concern in long term [30]. Instead, metal powders can be burned in metal-fueled flame combustors that can be simply coupled with Rankine-cycle steam external combustion engines (ECEs) [31]. Figure 2.5 presents the proposed design of a metal-fueled combustor (swirl-stabilized metal-air flame burner) coupled with a cyclone to separate the produced oxides and a heat exchanger to recover the produced energy [32]. This design allows burning metal particles as suspensions in a fuel-rich turbulent chamber giving high power densities. The high-power performance is reached by burning the metal particles at high flame speed and temperature. Modern phase-separation cyclone technology allows an efficient collection of the produced oxides having a particle size bigger than several microns, while nano-oxides collection requires the use of HEPA-type and electrostatic filters [33].

In 2015, Lomba *et al.* studied the combustion regimes of micro-sized aluminum particles demonstrating that the burning velocity decreases as the particle size increases for both laminar and turbulent flames. For laminar flame, the burning velocity decreases from around 40 cm s^{-1} for 7.1 μm particles to 30 cm s^{-1} for 11.7 μm particles and 10 cm s^{-1} for 17.9 μm particles [34]. Further, the flame temperature decreases dramatically from around 4000 K for 17.9 μm particles to around 3000 K for particles smaller than 12 μm. They found that the combustion efficiency is dependent on both the particle size and the pressure. For 17.9 μm particles, it increased from around 40% at 0.8 bar to more than 60% at 1.5 bar, whereas for particles smaller than 12 µm, it increased from 60% to around 80% at 1.5 bar [34]. Later on, under a collaboration between PSA group (now Stellantis), AVL group and ICARE-CNRS laboratory, researchers were able to stabilize an aluminum-air flame (7 µm Al particles) in a Bunsen burner proving that the flame speed and temperature vary very slightly from the aluminum/oxygen equivalence ratio, with average values of 34 cm s^{-1} and 3146 K, respectively [31]. They concluded that using such metal-fueled flame burner, aluminum should be reactive as conventional hydrocarbons with similar specific energies and burning velocities when using 7 µm Al particles. Such a burner has an estimated power output of about 8 kW [35]. Recently, the dilution of oxygen in various diluent gases (N_2, Ar, He) was investigated proving that it reduces the flame temperature and the aluminum combustion rate resulting in a flame closer to the droplet and a longer combustion time [36].

Similar studies were performed for magnesium-air flames showing conversion efficiencies of 80% at atmospheric pressure for Mg particles of 23 to 38 µm, higher by 20% than those obtained for aluminum particles [37]. Further, similar to Al particles, the burning velocity decreases as the Mg particle size increases (35 cm s^{-1} for 23 µm to 20 cm s^{-1} for 38 µm particles). Contrarily to aluminum, the Mg-air flame temperatures are less affected by the particle size variation, with values around 2500 K [37]. The produced oxides consist of crystalline cubic MgO particles and polymorphous/amorphous spherical Al_2O_3 droplets with a particle size of around 200 nm. The formation of those fine particles is coherent with the presence of a vapor-phase flame as suggested from AlO and MgO emissions [37]. Recently, under a collaboration between PSA group (now Stellantis), AVL group and LGRE-UHA laboratory, Laraqui et al. designed an innovative power generation system where 50 to 70 µm Mg particles were burned in a swirled-stabilized metal-air flame combustor and 98% of the produced submicron magnesia (10–1000 nm particles) were trapped and collected. The system proved to have an optimal heat-to-mechanical conversion efficiency where 80% of the power generated from combustion is recovered in the power generation system [32]. The levels of NO_x emissions produced from such metal-fueled combustors were similar to those of current ICEs [38]. Based on those advancements in the development of metal-fueled combustors, some researchers of the Stellantis group suggested coupling a metal-fueled flame burner with an external combustion gas-turbine (ECGT) system as an auxiliary power unit (APU) for future extended-range hybrid electric vehicles (EREV) [39].

2.3 Regeneration of Metal Fuels Through the Solar Reduction of Oxides

The major progress achieved in the use of metal powders as transport fuels in metal-fueled burners makes it necessary to consider new innovative and sustainable processes for the production and regeneration of those fuels. In this context, a collaboration was formed between our group (PROMES-CNRS laboratory) with ICARE-CNRS laboratory, LGRE-UHA laboratory and Stellantis. This collaboration came under the STELLAR (SusTainable mEtaL fueLs for future trAnspoRtation) project funded by French National Research Agency (ANR) in 2018. As shown in Figure 2.3, this project aims to prove the efficiency of the combustion/reduction cycles for the production of recyclable, sustainable and clean solar metal fuels for transportation. In this section, we review the parameters affecting the solar carbothermal reduction of magnesia and alumina as a recycling process for the regeneration of magnesium and aluminum metal fuels.

2.3.1 Thermodynamics and Kinetics of Oxides Reduction

It is well known that the reduction of metal oxides can be promoted by using carbon or methane as reducing agents, as the equilibrium temperature of the reaction is decreased [40]. Despite that using methane is thermodynamically more advantageous, the use of carbon is more economically attractive especially when obtained from biomass pyrolysis as for charcoals. Furthermore, the reaction would occur at a lower temperature when decreasing the operating pressure. Puig and Balat-Pichelin proved that the starting temperature for the magnesia carbothermal reduction decreases from 1800 K at atmospheric pressure to 1500 K and 1300 K at 1000 Pa and 10 Pa respectively. Similarly for alumina, the onset temperature decreases from 2200 K at atmospheric pressure to 1800 K or 1600 K at 1000 Pa or 10 Pa, respectively [41].

At atmospheric pressure, the reaction kinetics of the magnesia carbothermal reduction at temperatures lower than 1750 K (low conversion rates) were well correlated with the MgO/C solid-solid phase-boundary reaction (2.1) with an activation energy of 208 kJ mol^{-1} [42]. At higher temperatures (high conversion rates), a good correlation was found with the solid/gas reaction with an activation energy of 374 kJ mol^{-1}. In this case, the MgO/CO solid-gas reaction (2.2) is favorable, with an activation energy of 470 kJ mol^{-1}, due to the loss of the MgO/C contact caused by the decrease of the reactants surface area, the MgO sintering and densification [42, 43].

The CO production through the Boudouard gasification reaction (2.3), with an activation energy of 180 kJ mol^{-1}, grants the sustainability of the MgO/CO solid/gas reaction [42]. However, under low pressures, the CO partial pressure is reduced and the solid-gas reaction is limited, thus promoting the solid-solid reaction [44]. Indeed, the reaction kinetics are dependent on the reaction stoichiometry λ (see (2.4) and eq. 2.5) that increases when the pressure decreases, thus lowering the CO_2 production and promoting the solid-solid reaction. This was observed by the change of the rate-determining step of the reaction as a function of the pressure as the activation energy reduces from 325 kJ mol^{-1} (solid-gas reaction) at atmospheric pressure to around 200 kJ mol^{-1} (solid-solid phase-boundary reaction) at 100 to 1000 Pa [43].

$$MgO_{(s)} + C_{(s)} \leftrightarrow Mg_{(g)} + CO_{(g)} \qquad (2.1)$$

$$MgO_{(s)} + CO_{(g)} \leftrightarrow Mg_{(g)} + CO_{2(g)} \qquad (2.2)$$

$$C_{(s)} + CO_{2(g)} \leftrightarrow 2\,CO_{(g)} \qquad (2.3)$$

$$MgO_x + \lambda\,C \leftrightarrow MgO_{(x-1)} + (2\lambda - 1)\,CO + (1 - \lambda)\,CO_2 \qquad (2.4)$$

$$\text{Where } \lambda = \frac{(N_{Ci} - N_{Cf})}{(N_{MgOi} - N_{MgOf})} \qquad (2.5)$$

Considering alumina carbothermal reduction under atmospheric pressure, the reaction (2.6) occurs through intermediate reactions (2.7), (2.8) and (2.9) at progressively higher temperatures ranging from 1973 to 2470 K. Hence, operating under lower pressures mitigates the formation of undesired by-products (Al_4C_3, Al_4O_4C, Al_2OC) and improves the aluminum production by decreasing the onset temperature of the reaction [45]. Indeed, under low pressures, the reduction yield is enhanced as the CO partial pressure decreases and the P_{Al2O}/P_{CO} ratio increases, thus promoting reactions (2.15) and (2.16) rather than reactions (2.10) and (2.11) [46].

$$Al_2O_3 + 3\,C \leftrightarrow 2\,Al + 3\,CO \qquad (2.6)$$

$$2\,Al_2O_3 + 3\,C \leftrightarrow Al_4O_4C + 2\,CO \qquad (2.7)$$

$$Al_4O_4C + 6\,C \leftrightarrow Al_4C_3 + 4\,CO \qquad (2.8)$$

$$Al_4O_4C + Al_4C_3 \leftrightarrow 8\,Al + 4\,CO \qquad (2.9)$$

$$Al_2O_{3(s)} + 2\,C_{(s)} \leftrightarrow Al_2O_{(g)} + 2\,CO_{(g)} \quad P_{Al2O}/P_{CO} = 0.5 \qquad (2.10)$$

$$2\,Al_2O_{(g)} + 5\,C_{(s)} \leftrightarrow Al_4C_{3(s)} + 2\,CO_{(g)} \qquad (2.11)$$

$$5\,Al_2O_{3(s)} + 2\,Al_4C_{3(s)} \leftrightarrow 9\,Al_2O_{(g)} + 6\,CO_{(g)} \quad P_{Al2O}/P_{CO} = 1.5 \qquad (2.12)$$

$$3\,Al_2O_{(g)} + Al_4C_{3(s)} \leftrightarrow 10\,Al_{(s)} + 3\,CO_{(g)} \qquad (2.13)$$

The reaction kinetics of the alumina carbothermal reduction under low vacuum conditions (P_{CO} < 5 Pa) correlates well with the phase boundary and nucleation kinetics at the beginning of the reaction (first 50 s) with an activation energy of 101 ± 10 kJ mol^{-1}, whereas a poor correlation is obtained with the diffusion kinetics with an activation energy of 175 ± 20 kJ mol^{-1}. Nevertheless, as the reaction proceeds with time, none of the simple kinetics correlates well with the experimental results, meaning that several mechanisms are involved [47].

2.3.2 Effect of Some Parameters on the Reduction Yield

2.3.2.1 Carbon-Reducing Agent

One of the critical parameters to consider when performing the carbothermal reduction is the type and properties of the carbon reducing agent. Besides being obtained from renewable sources (biomass pyrolysis), charcoals proved to be more reactive than graphite as they provide a higher contact area with MgO due to their higher surface area [48]. Further, the amorphous structure of charcoals improves their reactivity, which suggests milling graphite to generate an amorphous phase [49]. Those assets make charcoals the most attractive carbon reducing agents, particularly when operating under low-pressure conditions where the solid-solid reaction is the rate-determining reaction [50]. For gas-solid reaction kinetics, petcock outperforms wood charcoal as it contains metal impurities that can catalyze the reaction [51].

An excess of carbon content is inevitable and surely beneficial to improve the magnesia reduction yield [42]. However, one must consider reducing carbon consumption as much as possible. For alumina reduction under low pressures, this parameter is not as much as critical where a C/Al_2O_3 stoichiometric ratio is necessary and sufficient. Under 10 Pa, the reduction yield increases from 59% for a C/Al_2O_3 ratio of 1 to 99.3% for a ratio of 3 or 6 [52]. Indeed, during alumina reduction, an excess of carbon can lead to an additional formation of unwanted Al-oxycarbide by-products.

2.3.2.2 Catalysts and Additives

Under atmospheric pressure, metal catalysts (Cu, Co, Ni and Fe) accelerate the reaction rate of the MgO/CO gas-solid reaction [53]. Similarly, additives, such as $Co(NO_3)_2$, $Ni(NO_3)_2$, $Fe(NO_3)_3$, $CaCl_2$, KNO_3, K_2CO_3 or CaF_2, catalyze the gasification reaction and improves the reaction rate by around 2.4 times. Nevertheless, under lower pressures (100 Pa), the solid-solid reaction becomes dominant and the catalytic effect is unnoticeable. Further, those additives may even present an adverse effect on the reaction rate by blocking the MgO/C interparticle contact [54, 55].

Even so, under 30 to 100 Pa, calcium binder has shown a catalytic effect when used as a binder for C/MgO particles, with an optimal quantity of 5% CaF_2 [56]. Contrarily, a lower aluminum yield is observed when adding polyvinyl alcohol (PVA) binder. This can be attributed to the higher emission of CO at beginning of the reaction, promoting the formation of undesired Al-oxycarbide by-products [47]. Moreover, naphthalene acts as a blowing agent for C/Al_2O_3 pellets, thus rising their porosities and improving the reduction yield [57].

2.3.2.3 Mechanical Milling

The mechanical milling of carbon and magnesia particles allows reducing the onset temperature of the carbothermal reduction of magnesia, thus improving the reduction yield. This enhancement is correlated directly to the milling time and indirectly to the MgO particles size and the specific surface area of the C/MgO sample. The weight loss of the sample increases from 24% without milling to 67% with 8 h of milling, correlated with a diminution of the MgO particles size from 50 to 15 nm and an increase of the C/MgO surface area from 5 to 110 m^2 g^{-1} [58]. Indeed, under atmospheric pressure, the mechanical milling changes the kinetics of the reaction. For non-milled samples, a shift from the C/MgO solid-solid reaction (first stage) to the MgO/CO gas-solid reaction (second stage) occurs at 1750 K, whereas this shift occurs at 1670 K when the C/MgO particles are milled for 1 to 2 hours. When milling for 4 to 8 hours, the two-stages mechanism disappears and the solid-solid reaction is predominant along the reaction duration [58]. A similar favorable effect of the mechanical milling was highlighted during magnesia reduction under 10 kPa and can be attributed to the mixing and aggregate attrition of C/MgO particles [54].

2.3.2.4 CO Partial Pressure

As previously discussed in section 2.3.1, lowering the total pressure of the reactor, and thus the CO partial pressure reduces the required temperature of the reaction. For example, the reduction of alumina is complete at 1770 K for a P_{CO} of 20 to 50 Pa while it requires 1970 K or 2073 K for a P_{CO} of 80 to 170 Pa or 350 Pa, respectively [59–61]. Furthermore, the formation of undesired by-products is mitigated as the P_{Al2O}/P_{CO} ratio increases promoting the production of aluminum. During magnesia reduction, the solid-solid reaction becomes the rate-determining step under low CO partial pressure as the CO/MgO gas-solid reaction is limited. This means that under low P_{CO}, the C/MgO properties acquire a huge interest being critical to improve the kinetics of the magnesia reduction.

2.3.2.5 Carrier Gas

The diffusion mechanism of gaseous products (Al, Al_2O, CO, CO_2) may affect the reaction yield of the alumina reduction under low pressures (50 Pa). Thus, the choice of the carrier gas is important as the diffusivities of those elements are higher in He or H_2 (around 10 cm² s⁻¹) than in Ar (< 1 cm² s⁻¹). This is true for temperatures lower than 1973 K as the reaction extent in He or H_2 is much higher than that in Ar (80% compared to 30%). However, when operating at higher temperatures, the extent becomes similar in all environments [57].

2.3.2.6 Fast Preheating

Besides being a renewable energy source, high temperatures are reached very rapidly when using concentrated solar power for the reduction of oxides. This is advantageous for the magnesia reduction as slow preheating decreases the starting and average temperatures of the reaction, thus necessitating longer reaction times meaning more magnesia sintering. Likewise, the fast preheating (150–200 K min⁻¹) to reach the onset temperature of the alumina reduction is necessary to limit the formation of Al-oxycarbide by-products [62, 63].

2.3.2.7 Progressive Heating

A gradual increase of the temperature, after reaching the onset temperature of the reduction, is preferable to increase the production yields of metal powders. This can be attributed to the distribution of the CO gaseous emission along the reaction duration that decreases the CO partial pressure instantaneously. As a confirmation of this suggestion, a higher Al yield was obtained and the Al_2OC formation was limited during the vacuum-assisted (900 Pa) magnesia reduction at temperatures up to 2150 K. Similar conclusions were obtained during the vacuum-assisted carbothermal reduction of magnesia [64].

2.3.3 Reverse Reoxidation of the Produced Metal Powders

During the condensation of the produced Mg powders, reverse re-oxidation may occur reducing the purity of the metal fuels and consequently the reduction yield. One additional advantage of operating under low-pressure conditions, meaning low P_{CO}, is the prevention of this phenomenon [65]. This is attributed to the minimization of the magnesium vapor pressure allowing its condensation, at temperatures lower than 973 K, with a rate faster than the oxidation rate of the condensed surface, thus yielding high purity powders for a $P_{CO} < 300$ Pa [66]. In this context, Vishnevetsky et al. proved that the purity of the Mg powders, at deposit sites lower than 730 to 870 K, decreases by around 10% to 15% when the CO partial pressure increases from 7–10 Pa to 20–70 Pa [62].

Nevertheless, even at low pressures, one important parameter to consider is the temperature of the deposit sites as the reverse reaction can be prevented only at temperatures lower than 900 K, as both the pressure and temperature of the vapors determine the crystal growth rate, purity and morphology of the produced magnesium [65]. However, very low condensation temperatures should be avoided, as in this case, the number of atomic collisions and the nucleation rate increase rapidly leading to the formation of small powders (< 20 μm) that can be oxidized more easily [67]. Further, as previously mentioned in section 2.3.2.1, an excess of carbon is needed but should be limited to prevent the reverse re-oxidation through reaction (2.14) [68].

$$6\,Mg_{(g)} + 4\,CO_{(g)} \leftrightarrow 4\,MgO_{(s)} + Mg_2C_{3(s)} + C_{(s)} \qquad (2.14)$$

Similarly, low P_{CO} (< 10 Pa) and deposit temperatures (730–870 K) are compulsory during the carbothermal reduction of alumina to avoid the re-oxidation and the carbonization of the condensate aluminum [63]. Indeed, the transition region between the hot reaction zone and the cold deposit site should be minimized to suppress the backward reaction [69]. Further, as for magnesium, the morphology of the products depends on the temperature of the deposit sites. Experiments have shown that nanocrystalline powders were produced in the cold zone, whereas solidified liquid Al condensates, covered with an insulating layer of Al-oxycarbides, were obtained in the hot zone where the temperature is 1373 K much higher than the melting point of aluminum (933 K) [60].

2.3.4 Reduction of Oxides Using Concentrated Solar Power

As discussed in the introduction, the use of renewable energy sources as the concentrated solar power (CSP) to regenerate the metal powders makes their use profitable as transport metal fuels. Further, as highlighted in section 2.3.2.6, fast preheating is beneficial to improve the reduction yield and can be provided using CSP. Therefore, in this section, we review the major studies probing the reduction of oxides through solar thermochemical processes [16].

Historically, the main concern of the researchers was to develop solar processes allowing to provide the necessary heat for the reduction experiments and to control the reaction environment as the reaction pressure, which is a critical parameter. In this context, the first trials performed by Steinfeld and Fletcher in 1991 investigate the invention of a solar receiver-reactor for the carbothermal reduction of iron oxide [70]. Some years later, under a collaboration between the University of Minnesota (USA) and the Paul Scherrer Institute (PSI, Switzerland), two solar furnaces were designed and used to evaluate the carbothermal reduction of various metal oxides as SiO_2, Al_2O_3, TiO_2, ZrO_2, MgO, ZnO, and CaO. At the University of Minnesota, experiments were performed in a spherical solar reactor (Figure 2.6a) under argon atmosphere using a 7-kW solar furnace consisting of a solar-tracking heliostat and a stationary parabolic concentrator with a concentration ratio of 7000 (Figure 2.6b). At PSI, experiments were performed in a cylindrical solar reactor (Figure 2.6c) under nitrogen atmosphere using a 15-kW solar furnace similar to that used in Minnesota

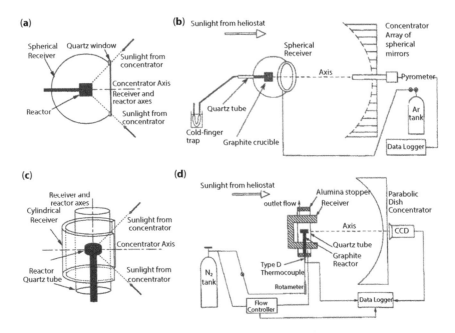

Figure 2.6 Concept of the solar furnaces and reactors used for the reduction experiments at the University of Minnesota (a and b) and at the Paul Scherrer Institute (c and d) [71].

(Figure 2.6d) [71]. The design of those solar reactors inspired the development of new concepts for the regeneration of metal powders through the solar reduction processing of oxides.

In 2004, under a collaboration between PSI, ETH Zurich and PROMES-CNRS laboratory, Osinga *et al.* proposed an innovative shrinking packed-bed solar reactor for the carbothermal reduction of ZnO. The reactor consists of two cavities where the first acts as a solar absorber while the second as a reaction chamber [72]. Later on, when designing a new solar reactor, researchers considered the continuous processing of ZnO using a horizontal axis rotating driving gear to insert the reactant particles in the hot reaction zone submitted to the concentrated radiative flux [73]. In 2010, researchers at PSI and ETH Zurich took into consideration the tightness of the solar reactor during the carbothermal reduction of silica when operating under low vacuum conditions (300 Pa). The solar reactor proved its endurance when exposed to a radiative flux equivalent to 6500 suns with high temperatures up to 2260 K [74]. In the following sections, we review the experimental results of the carbothermic reduction of magnesia and alumina under low vacuum conditions using novel solar reactors and furnaces.

2.3.5 Solar Carbothermal Reduction of Magnesia

In 2015, researchers at PROMES-CNRS investigated the carbothermal reduction of magnesia as one step of the redox cycle for the production of CO or H_2 solar fuels [75]. The experiments were performed in a solar thermogravimeter reactor with a hemispherical glass window that allows the passage of the concentrated solar radiation, as shown in Figure 2.7. They observed that the reaction rate was faster using activated charcoal compared to carbon black at the beginning of the reduction, then it follows the same trend for both carbons. As discussed previously, this can be attributed to the change of the reaction kinetics, under the employed conditions (atmospheric pressure or 15 kPa), from the solid-solid phase boundary reaction (affected by the C/MgO contact and the carbon properties) to the gas-solid diffusion reaction. Nevertheless, under the employed conditions, the reverse reaction was significant resulting that the weight loss did not exceed 30% at atmospheric pressure and 40% at 15 kPa for temperatures up to 1800 K [75]. More recently, those researchers designed a new 1.5 kW solar reactor based on the concept of a directly-irradiated cavity-type receiver with a hemispherical glass window (Figure 2.8), allowing it to operate under lower pressures and higher temperatures [76]. Using this design, they were able to achieve a high conversion yield of 97.8% under 11 kPa using 50% excess activated charcoal as reducing agent. However, the produced Mg powders were highly pyrophoric spherical nano-particles that oxidize directly in

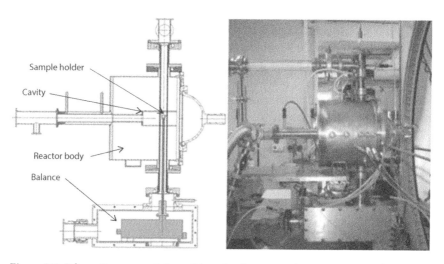

Figure 2.7 Schematic representation of the solar thermogravimeter reactor used at PROMES-CNRS for the carbothermal reduction of oxides [75].

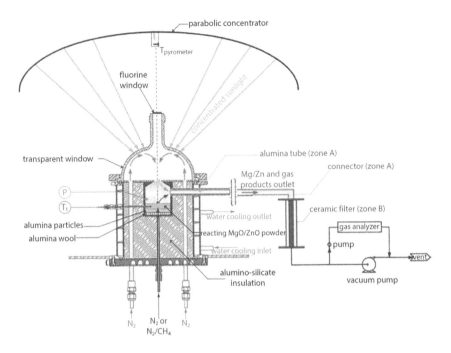

Figure 2.8 Design of the proposed directly-irradiated cavity-type solar reactor for the carbothermal reduction of magnesia under low pressures [76].

native air when the vacuum is broken. Thus, their use as transport metal fuels is restricted due to their instability and low purity (< 33% Mg) [76].

Another group of researchers, at the Weizmann Institute of Science (WIS) started investigating the vacuum-assisted solar carbothermal reduction of magnesia in 2015 using a stationary parabolic dish (coupled with a flat deflecting mirror and a solar tracking heliostat) that concentrates the solar radiation up to 5000 suns [62]. Based on numerical thermal and stress simulations, they optimized the design of a spherically shaped quartz solar reactor with improved optical and powders collection components [77]. Further, those simulations allowed understanding the mass transfer, energy transport and reaction kinetics during a continuous reduction process [78]. Solar experiments proved, as mentioned previously in section 2.3.3, that the reverse reoxidation reaction is limited for a CO partial pressure lower than 15 Pa. Using this solar reactor, pure Mg powders were produced with a maximal yield of 62% of the loaded oxide (equivalent to 95% yield of the reacted oxide) at temperatures of around 1850 K and under a P_{CO} of 3 to 7 Pa [62].

Our group, at PROMES-CNRS laboratory, started in 2016 to study the carbothermal reduction of magnesia under low-pressure conditions using a 1.5-kW solar furnace concentrating the solar power up to 15,000 suns [41]. Preliminary experiments were performed in the Heliotron reactor yielding about 45% Mg, with 60% to 80% purity, when using an excess of graphite or carbon black (C/MgO = 2). This low yield was impacted by the magnesia sintering preventing the completion of the reaction and by the reverse reoxidation of the produced metal powders [41]. The next year, the Sol@rmet reactor was developed with improved optical performance, a better air-tightness and smaller volume allowing it to operate under lower pressures (< 900 Pa) and limiting the reverse re-oxidation [79]. Using this solar reactor, as discussed in section 2.3.2.7, tests proved that gradual heating from 1770 K to 2190 K along the reaction improves the Mg yield [64]. Further, solar experiments proved that a C/MgO ratio of 1.25 is sufficient but necessary for the reaction with a better yield obtained using biochar compared to carbon black [80]. The produced powders consisted of highly pure (94% Mg) stable micrometric agglomerates (20–250 μm) of Mg nanoparticles and crystals with maximal Mg yield of 55% to 60% of loaded oxide (equivalent to 75% yield of reacted oxide) [79].

Recently, the gas circulation in the Sol@rmet reactor was examined through numerical simulations using ANSYS-CFX software aiming to optimize the reactor design and consequently to improve the reduction extent. Results proved that a good swirl circulation was observed when using several argon entries allowing to purge the produced products and thus promote the reaction [81]. Those results were validated experimentally, using a 1.5-kW solar furnace and birch charcoal (C/MgO = 1.25), showing that the Mg yield increases by around 15% when using a double-entry of argon entry from the upper and bottom parts of the reactor. Furthermore, we confirmed the effect of the mechanical milling, mentioned in Section 2.3.2.3, as the MgO particles size decreased and the C/MgO contact area increased leading to a higher reduction yield (by around 15%) [82]. Bentonite was used as a binder for the C/MgO pellets and revealed to have a catalytic effect rising the Mg yield up to 96%, when heating progressively under 830 Pa, of highly pure (96% purity) agglomerates of submicron Mg particles and crystals [83], whereas metal catalysts (Fe, Ni, and Fe-Ni) have an adverse effect due to the rapid consumption of the carbon particles at the beginning of the reaction, thus boosting the MgO sintering [81]. Besides the effects of the charcoal source, synthesis (pyrolysis conditions) and properties were evaluated on the magnesia reduction demonstrating that a fixed carbon content higher than 80% is necessary and sufficient for the reaction when using wood-based, starch-based, and cellulose-based charcoals [84]. The reduction is not efficient when using sugar-based charcoals, even though the carbon content is higher than 80%.

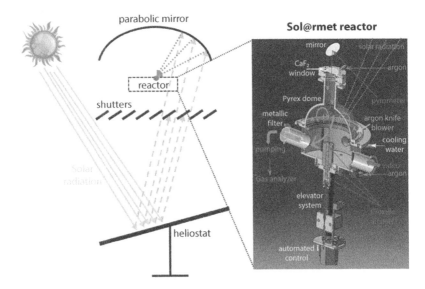

Figure 2.9 Schematic layout of the 1.5-kW solar furnace composed of a solar-tracking heliostat, a parabolic concentrator and the Sol@rmet reactor.

To reach such high carbon content, the pyrolysis should be performed at low heating rates (2 K min^{-1} preferred over 10 K min^{-1}) and high temperatures (1083 K preferred over 783 K) [84]. Currently, the Sol@rmet reactor is modified, as shown in Figure 2.9, to operate semicontinuously and process a higher quantity of magnesia powders, assuring, thus, the continuity of the combustion/reduction cycles and the reliability of the use of Mg metal powders as transport fuels [11].

2.3.6 Solar Carbothermal Reduction of Alumina

Under the ENEXAL project, an assessment of the vacuum-assisted carbothermal reduction process, compared to the conventional Hall-Héroult process, proved that this process provides an energy saving of 21%, a GHGs emission reduction of 52% and an exergy efficiency improvement of 10% [85]. Further, under low pressures, the use of solar power assures a lower overall exergy cost compared to the traditional heating methods as for the shaft-EAF (electric arc furnace) reduction process. Therefore, the WIS and ETH Zurich coupled the low-pressure and high-temperature concepts in an attractive solar vacuum-assisted carbothermal reduction process [85].

Historically, Murray *et al.* were the first researchers who investigated the solar carbothermal reduction of alumina under various conditions [71].

AlN was the main product during the solar (15-kW solar furnace at PSI) reduction at 1773 K under atmospheric pressure of N_2, with no formation of aluminum or Al-oxycarbide by-products, whereas no conversion occurs at 2300 K under atmospheric pressure of Ar using a 7-kW solar furnace at the University of Minnesota. Yet, lowering the pressure to 2×10^4 Pa allows the formation of a small amount of aluminum with a major production of Al-oxycarbide by-products (Al_4O_4C, Al_2OC, Al_4C_3) [71]. A further decrease of the pressure to 350 to 1200 Pa allows reaching a 19% Al yield at 1900 K with a reaction extent of 47%. Those results were obtained using a high-flux solar simulator (HFSS) at PSI with concentration ratios up to 11000 suns and heating rates up to 1000 K s^{-1}, whereas at temperatures lower than 1660 K, only Al_4C_3 and Al_4O_4C by-products were produced with a reaction extent close to 30%, with no aluminum formation [86]. Thus, high temperatures are required and can be achieved by increasing the number of the xenon lamps, opening the venetian shutter and positioning the sample closer to the focal plane of the ellipsoidal reflectors (see Figure 2.10) [87].

Motivated by the experiments performed in a 25-kW fast induction heater proving the production of pure Al powders (71% purity) at 1800 K under a P_{CO} of 10 Pa [61], Vishnevetsky et al. designed the solar reactor, shown in Figure 2.11, and tested experimentally the carbothermal reduction of alumina when employing a radiative flux up to 5000 suns. Using this prototype, they produced highly pure Al powders (90% purity) with conversion up to 90% at 1873 K and under CO partial pressures lower than 7 Pa [60].

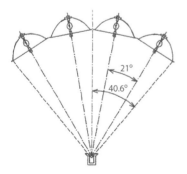

Figure 2.10 High-flux solar simulator consisting of an array of Xe-arc lamps, each close-coupled to a truncated ellipsoidal reflector [87].

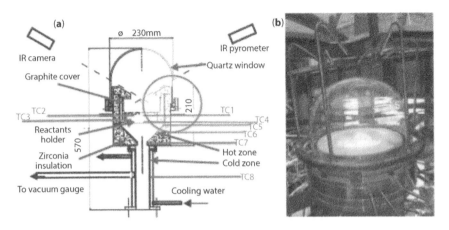

Figure 2.11 Schematic configuration of the solar reactor used at the Weizmann Institute of Science for the vacuum-assisted carbothermal reduction of alumina [60].

In 2016, Puig and Balat-Pichelin performed the carbothermal reduction of alumina in the Heliotron reactor under 1000 to 1600 Pa using a 1.5-kW solar furnace and showed that the flux distribution decreases and becomes less uniform when moving away from the focus of the parabola in all directions (x, y, and z) [41]. Experiments were conducted at stoichiometric conditions using graphite and carbon black giving similar results, with a maximal conversion yield of 52%. The produced powders are microsized agglomerates of nanoparticles (40–200 µm) with low Al purity of 60% [41]. Later on, experiments were performed in the Sol@rmet reactor, using biochar (C/Al_2O_3 = 3) under 900 Pa, proving the beneficial effect of the gradual heating leading to 60% Al yield with 82% purity (compared to 28% yield and 69% purity at fixed temperature). This can be explained by the spread of the CO emissions along the reaction instead of an enormous instant peak [79]. Further, the Al_2OC formation was limited during the gradual heating, even at very high temperatures, whereas for experiments at a fixed temperature, the Al_2OC production increases when the temperature rises due to the formation of $Al_2O_{(g)}$ as suggested by thermodynamic studies [79]. Experiments performed at lower pressures (380 Pa) enable attaining similar yields (61%) but better powders purity of 91%. A further decrease of the pressure to 160-Pa grants to nearly avoid the presence of Al_4C_3 and Al_2O_3 by-products [64]. Very recently, a 68% yield of highly pure (91% purity) submicronic agglomerates of nano-particles/crystals was achieved at 280 Pa [47].

2.4 Conclusions

The use of metal powders as transport fuels becomes attractive due to the possibility of their regeneration using renewable energy sources allowing their sustainable employment through combustion/reduction cycles. Mg/MgO and Al/Al$_2$O$_3$ were the best candidates for those cycles. Thus, we reviewed the recent developments of the metal-fueled combustors and solar recycling processes allowing us to improve the efficiency of the cycle and consequently to consider metal fuels as clean, recyclable, and sustainable substitutes to conventional fossil fuels in the transportation sector.

In this study, we highlighted the historical use of metal powders as combustible for rocket engines, and we probed how recent combustion engineering methods allow their direct combustion in external combustion engines. The development of a metal-fueled engine to be used in future transportation is seriously considered by some important automotive groups. Recently, a swirled-stabilized magnesia-air flame burner was developed with a heat-to-mechanical conversion efficiency of 80% and an ability to collect 98% of the produced submicron magnesia products. Further, when burned in an 8-kW Bunsen flame burner, aluminum proved to be as reactive as conventional hydrocarbons with similar specific energies and burning velocities.

Reviewing the solar carbothermal reduction of magnesia demonstrates that the thermodynamics and kinetics of the reaction depend on the operating parameters and conditions. Thus, we discussed the effect of the total pressure, CO partial pressure, reaction temperature, gradual heating, gas circulation in the Sol@rmet reactor, carbon reducing agent, C/MgO properties, mechanical milling of particles, catalysts, and binders. Further, we inspected the methods that allow limiting the reverse reoxidation reaction of the produced Mg powders. In conclusion, we were able recently to produce highly pure agglomerates (96% purity) of nanocrystalline and submicron Mg particles with conversion yields up to 96%.

During the alumina carbothermal reduction, the main problem was the formation of unwanted Al-oxycarbide by-products. The use of concentrated solar power (CSP), as a heating source, grants achieving fast preheating rates (150–200 K min^{-1}), thus limiting the production of by-products. Pure Al powders (91% purity) were achieved with a maximal yield of 68% when operating under low pressures (280 Pa), meaning low CO partial pressure (< 20 Pa), as the production of undesired by-products is restricted.

Currently, we focus on developing a semicontinuous set up to magnify the solar treatment of oxides and thus rise the production of metal fuels.

Alongside, our collaborators, under the STELLAR project, are optimizing the design of the metal-fueled burners to improve the efficiencies of fuel burning and oxide particles collection.

Acknowledgments

This study is funded under the STELLAR project by the French National Research Agency (ANR), under contract ANR-18-CE05-0040-02. This work was supported by the French "Investments for the future" program managed by the French National Research Agency (ANR), under contract ANR-10-EQPX-49-SOCRATE (Equipex SOCRATE).

References

1. Höök, M. and Tang, X., Depletion of fossil fuels and anthropogenic climate change—a review. *Energy Policy*, 52, 797–809, 2013.
2. Capellán-Pérez, I., Mediavilla, M., de Castro, C., Carpintero, Ó., Miguel, L.J., Fossil fuel depletion and socio-economic scenarios: An integrated approach. *Energy*, 77, 641–66, 2014.
3. Ritchie, H. and Roser, M., Emissions by sector, 2016. https://ourworldindata.org/emissions-by-sector.
4. Dussud, F.X., Joassard, I., Wong, F., Duvernoy, J., Morel, R., *Key Climate Figures of France and the World-2016 Edition (Presented at the UN Climate Change Conference: Chiffres clés du climat France et Monde-Édition 2016)*, Paris, 2016, https://www.connaissancedesenergies.org/sites/default/files/pdf-actualites/rep_-_chiffres_cles_du_climat_2016.pdf.
5. Monthly Energy Review. U.S. Energy Information Administration (EIA), April 2020. https://www.eia.gov/totalenergy/data/monthly/archive/00352004.pdf.
6. Su, B., Heshmati, A., Geng, Y., Yu, X., A review of the circular economy in China: moving from rhetoric to implementation. *J. Clean. Prod.*, 42, 215–27, 2013.
7. Mutezo, G. and Mulopo, J., A review of Africa's transition from fossil fuels to renewable energy using circular economy principles. *Renew. Sust. Energ. Rev.*, 137, 110609, 2021.
8. Shafirovich, E. and Goldshleger, U., Combustion of magnesium particles in carbon dioxide and monoxide. *31st Jt. Propuls. Conf. Exhib*, American Institute of Aeronautics and Astronautics, San Diego, CA, U.S.A, 1995, https://doi.org/10.2514/6.1995-2992.
9. Shkolnikov, E.I., Zhuk, A.Z., Vlaskin, M.S., Aluminum as energy carrier: Feasibility analysis and current technologies overview. *Renew. Sust. Energ. Rev.*, 15, 4611–23, 2011.

10. Utgikar, V.P., Lattin, W., Jacobsen, R.T., Nanometallic fuels for transportation: A well-to-wheels analysis. *Int. J. Energy Res.*, 31, 99–108, 2007.
11. Berro, Y. and Balat-Pichelin, M., Metal fuels production for future long-distance transportation through the carbothermal reduction of MgO and Al_2O_3: A review of the solar processes. *Energ. Convers. Manag.*, 251, 114951, 2022.
12. Bergthorson, J.M., Recyclable metal fuels for clean and compact zero-carbon power. *Prog. Energy Combust. Sci.*, 68, 169–96, 2018.
13. Kuschner, M., The carcinogenicity of beryllium. *Environ. Health Perspect.*, 40, 101–5, 1981.
14. King, M.K., Ignition and combustion of boron particles and clouds. *J. Spacecr. Rockets*, 19, 294–306, 1982.
15. Yuasa, S. and Isoda, H., Carbon dioxide breathing propulsion for a Mars airplane. *25th Jt. Propuls. Conf.*, American Institute of Aeronautics and Astronautics, Monterey, CA, U.S.A, 1989, https://doi.org/10.2514/6.1989-2863.
16. Steinfeld, A. and Palumbo, R., Solar thermochemical process technology, in: *Encyclopedia of Physical Science & Technology*, vol. 15, R.A. Meyers (Ed.), pp. 237–56, Academic Press, San Diego, 2003.
17. Rezan, S.A., Zhang, G., Ostrovski, O., Effect of gas atmosphere on carbothermal reduction and nitridation of titanium dioxide. *Metall. Mater. Trans. B*, 43, 73–81, 2012.
18. Bergthorson, J.M., Goroshin, S., Soo, M.J., Julien, P., Palecka, J., Frost, D.L., Jarvis, D.J., Direct combustion of recyclable metal fuels for zero-carbon heat and power. *Appl. Energy*, 160, 368–82, 2015.
19. Yetter, R.A. and Dryer, F.L., Metal particle combustion and classification, in: *Microgravity Combustion: Fire in Free Fall*, vol. 6, H.D. Ross (Ed.), pp. 419–78, Academic Press, New York, 2001.
20. Goroshin, S., Fomenko, I., Lee, J.H.S., Burning velocities in fuel-rich aluminum dust clouds. *Symp. Int. Combust.*, 26, 1961–7, 1996.
21. Soo, M., Julien, P., Goroshin, S., Bergthorson, J.M., Frost, D.L., Stabilized flames in hybrid aluminum-methane-air mixtures. *Proc. Combust. Inst.*, 34, 2213–20, 2013.
22. Goroshin, S., Higgins, A., Kamel, M., Powdered metals as fuel for hypersonic ramjets. *37th Jt. Propuls. Conf. Exhib*, American Institute of Aeronautics and Astronautics, Salt Lake City, UT, U.S.A, 2001, https://doi.org/10.2514/6.2001-3919.
23. Goroshin, S., Tang, F.-D., Higgins, A.J., Lee, J.H.S., Laminar dust flames in a reduced-gravity environment. *Acta Astronaut.*, 68, 656–66, 2011.
24. Wang, S., Corcoran, A.L., Dreizin, E.L., Combustion of magnesium powders in products of an air/acetylene flame. *Combust. Flame*, 162, 1316–25, 2015.
25. Julien, P., Vickery, J., Goroshin, S., Frost, D.L., Bergthorson, J.M., Freely-propagating flames in aluminum dust clouds. *Combust. Flame*, 162, 4241–53, 2015.
26. Huang, Y., Risha, G.A., Yang, V., Yetter, R.A., Effect of particle size on combustion of aluminum particle dust in air. *Combust. Flame*, 156, 5–13, 2009.

27. Julien, P., Vickery, J., Whiteley, S., Wright, A., Goroshin, S., Bergthorson, J.M., Frost, D.L., Effect of scale on freely propagating flames in aluminum dust clouds. *J. Loss Prev. Process Ind.*, 36, 230–6, 2015.
28. Goroshin, S., Bidabadi, M., Lee, J.H.S., Quenching distance of laminar flame in aluminum dust clouds. *Combust. Flame*, 105, 147–60, 1996.
29. Watson, G.M.G., Munzar, J.D., Bergthorson, J.M., Diagnostics and modeling of stagnation flames for the validation of thermochemical combustion models for NO_x predictions. *Energy Fuels*, 27, 7031–43, 2013.
30. Beach, D.B., Rondinone, A.J., Sumpter, B.G., Labinov, S.D., Richards, R.K., Solid-state combustion of metallic nanoparticles: New possibilities for an alternative energy carrier. *J. Energy Resour. Technol.*, 129, 29–32, 2007.
31. Laboureur, P., Lomba, R., Halter, F., Chauveau, C., Dumand, C., *The use of solid combustible in thermal engines (presented at "9ème édition du COlloque FRancophone en Energie, Environnement, Economie et Thermodynamique": "Utilisation de la combustion solide dans les machines thermiques")*, Strasbourg, 2018, https://hal.archives-ouvertes.fr/hal-01860128.
32. Laraqui, D., Leyssens, G., Schonnenbeck, C., Allgaier, O., Lomba, R., Dumand, C., Brilhac, J.-F., Heat recovery and metal oxide particles trapping in a power generation system using a swirl-stabilized metal-air burner. *Appl. Energy*, 264, 114691, 2020.
33. Wang, H., Zhang, Y., Wang, J., Liu, H., Cyclonic separation technology: Researches and developments. *Chin. J. Chem. Eng.*, 20, 212–9, 2012.
34. Lomba, R., Halter, F., Chauveau, C., Bernard, S., Gillard, P., Mounaim-Rousselle, C., Gillard, P., Experimental characterization of combustion regimes for micron-sized aluminum powders. *53rd AIAA Aerosp. Sci. Meet*, American Institute of Aeronautics and Astronautics, Kissimmee, Florida, 2015, https://doi.org/10.2514/6.2015-0925.
35. Lomba, R., Laboureur, P., Dumand, C., Chauveau, C., Halter, F., Determination of aluminum-air burning velocities using PIV and laser sheet tomography. *Proc. Combust. Inst.*, 37, 3143–50, 2019.
36. Braconnier, A., Chauveau, C., Halter, F., Gallier, S., Experimental investigation of the aluminum combustion in different O_2 oxidizing mixtures: Effect of the diluent gases. *Exp. Therm. Fluid Sci.*, 117, 110110, 2020.
37. Lomba, R., Bernard, S., Gillard, P., Mounaïm-Rousselle, C., Halter, F., Chauveau, C., Tahtouh, T., Guézet, O., Comparison of combustion characteristics of magnesium and aluminum powders. *Combust. Sci. Technol.*, 188, 1857–77, 2016.
38. Laraqui, D., Allgaier, O., Schönnenbeck, C., Leyssens, G., Brilhac, J.-F., Lomba, R., Dumand, C., Guézet, O., Experimental study of a confined premixed metal combustor: Metal flame stabilization dynamics and nitrogen oxides production. *Proc. Combust. Inst.*, 37, 3175–84, 2019.
39. Reine, A. and Nader, W.B., Fuel consumption potential of different external combustion gas-turbine thermodynamic configurations for extended range electric vehicles. *Energy*, 175, 900–13, 2019.

40. Steinfeld, A., Kuhn, P., Tamaura, Y., CH_4-utilization and CO_2-mitigation in the metallurgical industry via solar thermochemistry. *Energy Convers. Manage.*, 37, 1327–32, 1996.
41. Puig, J. and Balat-Pichelin, M., Production of metallic nanopowders (Mg, Al) by solar carbothermal reduction of their oxides at low pressure. *J. Magnes. Alloys*, 4, 140–50, 2016.
42. Rongti, L., Wei, P., Sano, M., Kinetics and mechanism of carbothermic reduction of magnesia. *Metall. Mater. Trans. B*, 34, 433–7, 2003.
43. Chubukov, B.A., Palumbo, A.W., Rowe, S.C., Hischier, I., Groehn, A.J., Weimer, A.W., Pressure dependent kinetics of magnesium oxide carbothermal reduction. *Thermochim. Acta*, 636, 23–32, 2016.
44. Xiong, N., Tian, Y., Yang, B., Xu, B., Dai, T., Dai, Y., Results of recent investigations of magnesia carbothermal reduction in vacuum. *Vacuum*, 160, 213–25, 2019.
45. Halmann, M., Frei, A., Steinfeld, A., Carbothermal reduction of alumina: Thermochemical equilibrium calculations and experimental investigation. *Energy*, 32, 2420–7, 2007.
46. Balomenos, E., Panias, D., Paspaliaris, I., Theoretical investigation of the volatilization phenomena occurring in the carbothermic reduction of alumina. *World Metall.-ERZMETALL*, 64, 6, 312–20, 2011.
47. Puig, J. and Balat-Pichelin, M., Experimental carbothermal reduction of Al_2O_3 at low pressure using concentrated solar energy. *J. Sustain. Metall.*, 6, 161–73, 2020.
48. Fruehan, R.J. and Martonik, L.J., The rate of reduction of MgO by carbon. *Metall. Trans. B*, 7, 537–42, 1976.
49. Jiang, Y., Ma, H.W., Liu, Y.Q., Experimental study on carbothermic reduction of magnesia with different carbon materials. *Adv. Mater. Res.*, 652–654, 2552–5, 2013.
50. Rongti, L., Wei, P., Sano, M., Li, J., Kinetics of reduction of magnesia with carbon. *Thermochim. Acta*, 390, 145–51, 2002.
51. Gálvez, M.E., Frei, A., Albisetti, G., Lunardi, G., Steinfeld, A., Solar hydrogen production via a two-step thermochemical process based on MgO/Mg redox reactions—Thermodynamic and kinetic analyses. *Int. J. Hydrog. Energy*, 33, 2880–90, 2008.
52. Halmann, M., Frei, A., Steinfeld, A., Vacuum carbothermic reduction of Al_2O_3, BeO, MgO-CaO, TiO_2, ZrO_2, HfO_2+ZrO_2, SiO_2, $SiO_2+Fe_2O_3$, and GeO_2 to the metals. A thermodynamic study. *Miner. Process. Extr. Metall. Rev.*, 32, 247–66, 2011.
53. Rongti, L., Wei, P., Sano, M., Li, J., Catalytic reduction of magnesia by carbon. *Thermochim. Acta*, 398, 265–7, 2003.
54. Chubukov, B.A., Palumbo, A.W., Rowe, S.C., Wallace, M.A., Weimer, A.W., Enhancing the rate of magnesium oxide carbothermal reduction by catalysis, milling, and vacuum operation. *Ind. Eng. Chem. Res.*, 56, 13602–9, 2017.

55. Chubukov, B.A., Palumbo, A.W., Rowe, S.C., Wallace, M.A., Sun, K.Y., Weimer, A.W., Design and fabrication of pellets for magnesium production by carbothermal reduction. *Metall. Mater. Trans. B*, 49, 2209–18, 2018.
56. Tian, Y., Qu, T., Yang, B., Dai, Y.-N., Xu, B.-Q., Geng, S., Behavior analysis of CaF_2 in magnesia carbothermic reduction process in vacuum. *Metall. Mater. Trans. B*, 43, 657–61, 2012.
57. Ostrovski, O., Zhang, G., Kononov, R., Dewan, M.A.R., Li, J., Carbothermal solid state reduction of stable metal oxides. *Steel Res. Int.*, 81, 841–6, 2010.
58. Nusheh, M., Yoozbashizadeh, H., Askari, M., Kuwata, N., Kawamura, J., Kano, J., Saito, F., Kobatake, H., Fukuyama, H., Effect of mechanical milling on carbothermic reduction of magnesia. *ISIJ Int.*, 50, 668–72, 2010.
59. Vishnevetsky, I., Ben-Zvi, R., Epstein, M., Solar metal oxides reduction under vacuum, experimental investigation of the alumina case. *SolarPACES2012*, Marrakech, Morocco, 2012.
60. Vishnevetsky, I., Epstein, M., Rubin, R., Solar carboreduction of alumina under vacuum. *Energy Proc.*, 49, 2059–69, 2014.
61. Halmann, M., Steinfeld, A., Epstein, M., Guglielmini, E., Vishnevetsky, I., Vacuum carbothermic reduction of alumina. *Proc. ECOS 2012*, Perugia, Italy, vol. 14, pp. 1–8, 2012.
62. Vishnevetsky, I. and Epstein, M., Solar carbothermic reduction of alumina, magnesia and boria under vacuum. *Sol. Energy*, 111, 236–51, 2015.
63. Vishnevetsky, I., Solar thermal reduction of metal oxides as a promising way of converting CSP into clean electricity on demand. *Proc. ISES Sol. World Congr. 2015*, International Solar Energy Society, Daegu, Korea, pp. 1–12, 2016.
64. Puig, J., Balat-Pichelin, M., Beche, E., Solar metallurgy for the production of Al and Mg particles. *Solar PACES 2017, Santiago, Chile, AIP Conference Proceedings*, vol. 2033, p. 140002, 2018.
65. Yang, C., Tian, Y., Qu, T., Yang, B., Xu, B., Dai, Y., Analysis of the behavior of magnesium and CO vapor in the carbothermic reduction of magnesia in a vacuum. *J. Magnes. Alloys*, 2, 50–8, 2014.
66. Hischier, I., Chubukov, B.A., Wallace, M.A., Fisher, R.P., Palumbo, A.W., Rowe, S.C., Groehn, A.J., Weimer, A.W., A novel experimental method to study metal vapor condensation/oxidation: Mg in CO and CO_2 at reduced pressures. *Sol. Energy*, 139, 389–97, 2016.
67. Xiong, N., Tian, Y., Yang, B., Xu, B.-Q., Liu, D.-C., Dai, Y.-N., Volatilization and condensation behaviors of Mg under vacuum. *Vacuum*, 156, 463–8, 2018.
68. Tian, Y., Xu, B., Yang, C., Yang, B., Liu, D., Qu, T., Dai, Y., Study on mechanism of magnesia production by reversion reaction process in vacuum, in: *Magnesium Technology 2016*, A. Singh (Eds.), pp. 61–6, John Wiley & Sons, Inc., Hoboken, NJ, USA, 2016.

69. Vishnevetsky, I. and Epstein, M., Metal oxides reduction in vacuum: Setup development and first experimental results. *Solar PACES 2011*, Granada, Spain, 2011.
70. Steinfeld, A. and Fletcher, E.A., Theoretical and experimental investigation of the carbothermic reduction of Fe_2O_3 using solar energy. *Energy*, 16, 1011–9, 1991.
71. Murray, J., Steinfeld, A., Fletcher, E., Metals, nitrides, and carbides via solar carbothermal reduction of metal oxides. *Energy*, 20, 695–704, 1995.
72. Osinga, T., Olalde, G., Steinfeld, A., Solar carbothermal reduction of ZnO: Shrinking packed-bed reactor modeling and experimental validation. *Ind. Eng. Chem. Res.*, 43, 7981–8, 2004.
73. Abanades, S., Charvin, P., Flamant, G., Design and simulation of a solar chemical reactor for the thermal reduction of metal oxides: Case study of zinc oxide dissociation. *Chem. Eng. Sci.*, 62, 6323–33, 2007.
74. Loutzenhiser, P.G., Tuerk, O., Steinfeld, A., Production of Si by vacuum carbothermal reduction of SiO_2 using concentrated solar energy. *JOM*, 62, 49–54, 2010.
75. Levêque, G. and Abanades, S., Investigation of thermal and carbothermal reduction of volatile oxides (ZnO, SnO_2, GeO_2, and MgO) via solar-driven vacuum thermogravimetry for thermochemical production of solar fuels. *Thermochim. Acta*, 605, 86–94, 2015.
76. Chuayboon, S. and Abanades, S., Solar metallurgy for sustainable Zn and Mg production in a vacuum reactor using concentrated sunlight. *Sustainability*, 12, 6709, 2020.
77. Ben-Zvi, R., Numerical simulation and experimental validation of a solar metal oxide reduction system under vacuum. *Sol. Energy*, 98, 181–9, 2013.
78. Ben-Zvi, R., Transient one-dimensional reactive pellet simulation. *Sol. Energy*, 115, 10–5, 2015.
79. Puig, J. and Balat-Pichelin, M., Production of Mg and Al using concentrated solar energy for future fuel applications. *Proc. EMC 2017 (European Metallurgical Conference)*, Leipzig, Germany, vol. 2, pp. 817–32, 2017.
80. Puig, J. and Balat-Pichelin, M., Experimental carbothermal reduction of MgO at low pressure using concentrated solar energy. *J. Min. Metall. B Metall.*, 54, 39–50, 2018.
81. Berro, Y., Masse, R., Puig, J., Balat-Pichelin, M., Improving the solar carbothermal reduction of magnesia for metallic fuels production through reactor designing, milling and binders. *J. Clean. Prod.*, 315, 128142, 2021.
82. Berro, Y., Puig, J., Balat-Pichelin, M., Improving the process of metallic fuels production through the solar carbothermal reduction of magnesia. *Proc. 7th World Congr. Mech. Chem. Mater. Eng. MCM21*, Virtual Conference, 2021, https://doi.org/10.11159/mmme21.102.
83. Berro, Y., Puig, J., Balat-Pichelin, M., Improving the solar carbothermal reduction of magnesia as a production process of metal fuels. *Int. J. Min. Mater. Metall. Eng.*, 7, 22–29, 2021. https://doi.org/10.11159/ijmmme.2021.003.

84. Berro, Y., Kehrli, D., Brilhac, J.-F., Balat-Pichelin, M., Metal fuel production through the solar carbothermal reduction of magnesia: Effect of the reducing agent. *Sustain. Energy Fuels*, 5, 6315–27, 2021.
85. Balomenos, E., Panias, D., Paspaliaris, I., Jaroni, B., Friedrich, B., Carbothermic reduction of alumina: A review of developed processes and novel concepts. *Proc. EMC 2011 (European Metallurgical Conference)*, Düsseldorf, Germany, vol. 3, pp. 729–43, 2011.
86. Kruesi, M., Galvez, M.E., Halmann, M., Steinfeld, A., Solar aluminum production by vacuum carbothermal reduction of alumina—thermodynamic and experimental analyses. *Metall. Mater. Trans. B*, 42, 254–60, 2011.
87. Petrasch, J., Coray, P., Meier, A., Brack, M., Häberling, P., Wuillemin, D., Steinfeld, A., A novel 50kW 11,000 suns high-flux solar simulator based on an array of xenon arc lamps. *J. Sol. Energy Eng.*, 129, 405–11, 2007.

3

Design Optimization of a Solar Fuel Production Plant by Water Splitting With a Copper-Chlorine Cycle

Samane Ghandehariun[1]*, Shayan Sadeghi[1] and Greg F. Naterer[2]

[1]*Sustainable Energy Research Group (SERG), School of Mechanical Engineering, Iran University of Science and Technology, Tehran, Iran*
[2]*Faculty of Sustainable Design Engineering, University of Prince Edward Island, Charlottetown, PE, Canada*

Abstract

Due to growing concerns about climate change, the production of renewable fuels has attracted the attention of many researchers recently. Solar energy is an abundant energy resource and a good alternative for the transition toward a more sustainable energy future. Solar energy can be used in many ways to produce solar fuels, for instance, by cracking of hydrocarbons or splitting of water into hydrogen and oxygen. Among these methods, the utilization of thermochemical cycles to produce hydrogen from water is one of the most promising ways of producing solar fuel. The copper-chlorine (Cu-Cl) high-performance thermochemical cycle requires relatively lower temperatures compared to other thermochemical cycles and is a good choice for integration with solar power tower systems. In this study, the design and optimization of a standalone plant for hydrogen generation, powered by solar energy, is investigated. To supply the required thermal energy of the Cu-Cl cycle, and also, to store the thermal energy throughout the night, a high-temperature carbonate molten salt (LiNaK) is used. Also, a three-stage super-critical steam Rankine cycle with four feedwater heater subsystems is also integrated into the solar production plant to provide the required electrical needs of the system. Therefore, there is no need for any auxiliary fuel or electricity to have a stable and continuous operation of the solar fuel production plant. Also, a modified heat exchanger network is utilized for the Cu-Cl cycle to use the available waste heat of the cycle and improve the system overall thermal efficiency.

---

*Corresponding author: samane_ghandehariun@iust.ac.ir

The technical and economic performance of the integrated system is investigated, and energy and exergy efficiencies, along with the investment and hydrogen production costs are evaluated. Based on the thermodynamic results, an efficiency of 40.4% for the Cu-Cl cycle and 45.0% for the supercritical Rankine cycle is obtained. The overall energy efficiency of the system is 28.6%, while the exergy efficiency of the overall system is 29.5%. An economic analysis showed that the investment cost of the solar thermochemical plant for a plant capacity of 1524 kg of hydrogen per hour is 516 million dollars. Using the Non-dominated Sorting Genetic Algorithm-II (NSGA-II) for multicriteria optimization, the optimum cost of producing hydrogen will be $2.98/kg H_2 while maintaining solar to hydrogen and exergetic efficiencies of 29.2% and 30.1%, respectively. Also, a comparison of the optimized proposed system with other renewable and nonrenewable hydrogen production systems indicates that this method is quite competitive among the other available methods.

***Keywords*:** Multiobjective optimization, hydrogen production, solar energy, solar fuel, thermochemical water dissociation

Nomenclature

A	Area (m^2)
C	Cold stream
h	Specific enthalpy (kJ/kg)
H	Hot stream/height (m)
i	Discount rate (%)
$\dot{m}$	Mass flow rate (kg/s)
n	Plant lifetime (years)/number of heliostat mirrors
P	Pressure (kPa)
$\dot{Q}$	Heat transfer rate (W)
T	Temperature (K)
V	Volume (m^3)
$\dot{W}$	Power (W)
Z	Purchased equipment cost ($)

Subscripts

ce	Common equity
CI	Capital investment
Cond	Condenser
d	Debt
Elec	Electrolyzer
K	kth component
L	Levelized/loss
ps	Preferred stock

Abbreviations

BPV	Byproduct value
CFWH	Close feedwater heater
CRF	Capital recovery factor
CuCl	Copper chloride
DNI	Direct normal irradiance
FC	Fuel costs
HX	Heat exchanger
ITX	Income taxes
LCOH	Levelized cost of produced hydrogen
LHV	Lower heating value
LINMAP	Linear programming technique for multidimensional analysis of preference
MPQ	Main product quantity
NSGA-II	Nondominated sorting genetic algorithm-II
OFWH	Open feedwater heater
OMC	Operation and maintenance costs
OTXI	Other taxes and insurance
ROI	Return of investment
SRC	Steam Rankine cycle
TCR	Total capital recovery
TES	Thermal energy storage
TRR	Total revenue requirement

3.1 Introduction

The world primary energy consumption is increasing every year. With the decline of natural resources, and the increasing concern of climate change, it is necessary to limit the usage of fossil fuels and use more clean and available energy sources, manage energy consumption/production more effectively, and use more efficient energy systems. Considering a cleaner replacement for conventional fuels is essential for moving toward a sustainable energy future.

Hydrogen can be considered as a promising energy carrier for industrial and transportation sectors due to its high-energy content and no carbon footprint. According to the International Renewable Energy Agency, more than two-thirds of the annual global hydrogen produced in the world is used to produce ammonia and refine petroleum [1]. In the petroleum industry, the hydrogen is used to refine and upgrade crude petroleum through various processes, such as hydrotreating and hydrocracking. In the chemical and petrochemical industries, hydrogen is used to produce various compounds, such as ammonia, methanol, resins, polymers, amines, and hydrogen peroxide. But most of the hydrogen is used to produce ammonia and methanol. Total hydrogen demand in the petroleum industry is over 70% of the total global hydrogen production [1]. In the transportation and energy sectors, hydrogen is used in hydrogen fuel cell vehicles, as well as internal combustion engines [2].

In its stable form, hydrogen cannot be found in nature; therefore, it is extracted/produced from chemical feedstock that contains hydrogen atoms. These materials include hydrocarbons, water, sugars, etc. [1]. Currently, hydrogen is mostly produced from fossil fuels [1]. About 95% of the worldwide hydrogen demand is produced from natural gas, oil, and coal [3]. Hydrogen generation from fossil fuels is cost-effective; however, it emits significant quantities of carbon dioxide and other greenhouse gases [4]. Researchers have suggested alternative methods to produce hydrogen via water dissociation using electricity, heat, or both [5].

Supplying the required energy for the large-scale hydrogen generation from water is a major challenge [6]. Solar energy may be utilized to provide the energy required for the production of hydrogen [7]. There are many methods to harvest solar energy and produce heat or power. Some of the most common methods include photovoltaic (PV) or concentrated solar power (CSP) systems. Both these systems can be used to produce heat and electricity. However, photovoltaics combined with solar collectors that are called PV/Thermal systems can only provide heat up to 80°C but the CSP systems, depending on their heat transfer fluid (HTF), can produce heat for temperatures up to 2,000°C [8].

The various types of CSP systems include parabolic trough collectors (PTC), linear Fresnel reflectors (LFR), solar power towers (SPT), and parabolic dish collectors (PDC) [9]. PTCs and LFRs produce lower temperatures compared to SPTs and PDCs, require a higher amount of land, need separate HTF lines for each assembly, and have lower progression perspectives. For the PDCs, although they produce higher temperatures compared to the mentioned two methods, they need separate HTF lines per assembly, and also, come in small capacities. On the other hand, SPTs can work with various HTFs, does not need separate lines per assembly, produce high temperatures, will not require a lot of land, and have high progression perspectives [9]. Also, they can easily be integrated into any cycles to produce power/heat or hydrogen. One of the main challenges of this technology, or solar energy, is its intermittent nature. To overcome this problem, it is necessary to use energy storage devices to store the energy of the sun, and use it when it is nighttime or the skies are cloudy. In this way, it is possible to achieve continuous operation of the whole integrated system. Sensible or latent thermal energy storage (TES) systems can be integrated with CSPs. Today, most of the CSP plants worldwide utilize molten salt as both their HTF and TES at the same time. The cost of the plant is lower with this approach [9].

Thermochemical or hybrid cycles, low-/high-temperature water electrolysis, and photo-electrochemical cells can be considered as promising methods for water splitting hydrogen production [10, 11]. The use of algae and photoelectrochemical cells for hydrogen production are still in the initial stages of research and development, however, the electrolysis of water and thermochemical cycles are more common than these two mentioned methods. In electrolysis, water dissociates to hydrogen and oxygen under the effect of direct electrical current. The following reaction describes this water splitting process [12]:

$$2H_2O \rightarrow 2H_2 + O_2 \qquad (3.1)$$

Three main types of electrolysis can be used for this process. These include alkaline water electrolysis, proton exchange membrane water electrolysis, and solid oxide water electrolysis [12]. Some studies examined the solar-based hydrogen generation using water electrolysis from technical, environmental, and economic viewpoints. Wang et al. [13] studied a hydrogen generation system using photovoltaic/thermal panels and a proton exchange membrane (PEM) water electrolyzer. An exergy efficiency of 20.3% and a solar-to-hydrogen efficiency of 17.3% was reported [13]. Hydrogen generation through high-temperature solid oxide water electrolysis using parabolic solar dishes was studied by Mastropasqua et al. [14]. Considering an efficiency of 80% for the solid oxide water electrolyzer, a solar-to-hydrogen efficiency of around 30% was reported [14]. Also, the cost of hydrogen is estimated to be in the range of €5.9 to 9.1/kg [10].

In another study, various solar-powered hydrogen generation systems were investigated for South Africa [15]. The methods include PV low-temperature electrolysis, concentrating solar power (CSP) high-temperature electrolysis, CSP low-temperature electrolysis, and solar-assisted steam methane reforming (SMR). Among the investigated methods, the lowest hydrogen production cost was $1.5/kg for the solar SMR and the highest one was $4.69/kg for the CSP low-temperature electrolysis [15].

Splitting the water into hydrogen and oxygen, by only suppling thermal energy, is called thermolysis or single step thermal dissociation of water. For this procedure, extremely high temperatures are required [16, 17]. For this reaction to proceed close to the completion or maximum degree of dissociation, the heat source must exceed 2,500 K [18]. Another major issue of this process is the separation of the product gasses (oxygen and hydrogen), which can easily react with each other in the reaction mixture and produce water [18]. To overcome these issues, researchers proposed the use of multistep thermochemical/hybrid cycles to produce hydrogen from water.

The history of thermochemical cycles dates back to 1970s, where they were first proposed as an alternative way to produce hydrogen by splitting the water molecules [5]. Most of the efforts to develop thermochemical cycles were reported by the nuclear sector to utilize the very high temperature produced by the nuclear reactors. Most of the results associated with development of thermochemical cycles were reported by the Joint Research Center of the European Union in Ispra, Italy, General Atomics, Westinghouse in the U.S., and also, the Japanese Atomic Energy Research Institute [19].

In the U.S., a working group consisting of different institutions, such as the Department of Energy, Sandia National Laboratory, University of Kentucky, and General Atomics, was engaged in determining the most suitable thermochemical method to produce hydrogen from 115 different cycles as a part of a Nuclear Energy Research Institute (NERI) study [20]. In that program, the sulfur-iodine cycle was proposed as the most promising cycle that can be coupled with high-temperature reactors [20]. The Japanese Atomic Energy Agency (JAEA) chose the sulfur-iodine cycle for further research and development. Also, the Solar Thermochemical Cycles for Hydrogen Production (STCH) R&D project in the U.S. determined the most suitable thermochemical cycles that can be coupled with concentrated solar power for hydrogen production [21]. The initial selection of 350 possible cycles was studied, but only nine cycles were chosen for further research and development [21]. In Canada, the Copper-Chlorine (Cu-Cl) cycle was identified as the most promising thermochemical cycle for coupling with Super Critical Water Reactors by Atomic Energy of Canada Limited (AECL) [22].

Thermochemical cycles utilize a series of chemical reactions to split water molecules at reasonable temperatures (lower than thermolysis). All of the chemicals used in the process can be recycled. If the decomposition of water is supplied by thermal energy only, it is referred to as thermochemical water splitting and if both thermal and electrical power are used for the process, it is called a hybrid thermochemical process [5].

Thermochemical cycles have two major advantages over typical thermolysis processes [5]:

I. The operating temperature is lower than that of thermolysis;
II. No membranes are required to separate the explosive hydrogen and oxygen mixture because they are produced in different reactions.

There are more than 3000 different thermochemical cycles available in the literature [18]. But presently, the sulfur-iodine (S-I) thermochemical cycle and copper-chlorine (Cu-Cl) hybrid cycle are considered as the most technically viable, common and well-developed methods to produce hydrogen [5]. Wang *et al.* presented a comparative investigation of these two thermochemical cycles [23]. The two cycles have similar thermal energy requirements, thermal efficiencies, and hydrogen generation costs. However, since the Cu-Cl cycle requires heat at a lower temperature, it may be integrated with available energy sources more easily [23].

The first Cu-Cl cycle that comprises an electrochemical reaction was proposed by the National Chemical Laboratory for Industry in Japan in 1976 as a two-step cycle [24]. Later, a four step Cu-Cl cycle was investigated by the Argonne National Laboratory in the United States, and also, in Canada by AECL [23, 24]. The inputs of this cycle are water, heat, and electricity while producing hydrogen and oxygen. Because electricity and heat are both used in this process, this cycle is a hybrid method. Currently, seven types of different Cu-Cl cycles are proposed in the literature. These include one 2-step cycle, three 3-step cycles, two 4-step cycles, and one

5-step cycle [24]. Some of these cycles are more efficient compared to the other options. Ishaq and Dincer [25] investigated the thermal and exergetic performance of three of these cycles and observed that the four-step Cu-Cl cycle is the most efficient one [25].

Different aspects of the copper-chlorine cycle have been studied previously. Ozbilen et al. [26] proposed a heat exchanger network (HEN) for heat recovery within the Cu-Cl cycle and reported an overall efficiency of 35.7% [26]. Ghandehariun et al. [27] conducted a pinch analysis to improve the overall thermal efficiency of the cycle with heat recovery. Also experimental investigation of heat recovery from molten salt was performed [28]. A thermal efficiency of 40.5% was reported for a four-step Cu-Cl cycle with heat recovery [27].

Integration of a solar parabolic trough collector (PTC) with the copper-chlorine cycle was studied by Ouagued et al. [29]. An exergy efficiency of 92.3% and an energy efficiency of 48.1% were reported [29]. An integration of a solar power tower system with a four-step Cu-Cl cycle was investigated by Sadeghi et al. [18]. The solar system has a pressurized cavity receiver and uses air as the heat transfer fluid and a phase change material (PCM) for thermal energy storage. The optimal design had an exergy efficiency of 50.1% and a unit product cost of $11.94/GJ. Table 3.1 presents major studies on thermochemical hydrogen generation using solar energy with a focus on the copper-chlorine cycle.

The copper-chlorine cycle may be integrated with a molten-salt-based solar system. The heat required for the thermolysis and hydrolysis reactors of the Cu-Cl cycle is supplied through steam generation. However, the operational temperature of conventional molten salts is in a range of 290°C to 560°C, while the four-step Cu-Cl cycle requires thermal energy at a maximum temperature of about 530°C. Therefore, a high flow rate of molten salt is required to satisfy the thermolysis reactor requirement. By replacing a conventional molten salt with a high-temperature molten salt, the molten salt can be used directly in the copper-chlorine reactors and heat exchangers, and there is no need to produce steam. Therefore, the hydrogen production cost and the system performance will be improved significantly.

It is noted that in the previous studies, the hydrogen production system needed an external electricity source for the electrolysis step. In this chapter, a standalone solar hydrogen plant is proposed, which uses the recovered heat within the cycle and does not require an external energy source.

A solar power tower using a high-temperature carbonate molten salt is linked with a four-step copper-chlorine cycle for hydrogen generation. A ternary mixture of Li_2CO_3(32.1%)-Na_2CO_3(33.4%)-K_2CO_3(34.5%) carbonate salt with a decomposition temperature of 800° and a melting

SOLAR FUEL PRODUCTION BY WATER SPLITTING   105

Table 3.1 Selected studies on solar thermochemical water splitting hydrogen generation.

Study	Year	Type of thermochemical cycle	Type of solar subsystem	System outputs	Solar HTF and TES	Detailed integration	Overall system exergy efficiency (%)	Overall system energy efficiency (%)	Use of external electricity source	Reference
Mehrpooya et al.	2022	Four-step Cu-Cl	PTC	Hydrogen, oxygen, hot water, power	Biphenyl	Yes	34.23	25.8	No	[46]
Sadeghi and Ghandehariun	2022	Four-step Cu-Cl	SPT	Hydrogen, oxygen	Molten salt	Yes	-	29.18	No	[45]
Corumlu and Ozturk	2021	Three-step Cu-Cl	SPT	Liquid hydrogen, oxygen, heating, cooling	Molten salt	Yes	14.02	22.21	No	[44]
Temiz and Dincer	2021	Four-step Cu-Cl	PTC	Hydrogen, oxygen, power, water, heating	Molten salt	Yes	47.1	52.6	No	[43]
Sadeghi et al.	2021	Four-step Cu-Cl	SPT	Hydrogen, oxygen, power, heating	Air/PCM	Yes	49.9	44.9	No	[42]

(Continued)

Table 3.1 Selected studies on solar thermochemical water splitting hydrogen generation. (*Continued*)

Study	Year	Type of thermochemical cycle	Type of solar subsystem	System outputs	Solar HTF and TES	Detailed integration	Overall system exergy efficiency (%)	Overall system energy efficiency (%)	Use of external electricity source	Reference
Sadeghi and Ghandehariun	2020	Four-step Cu-Cl	SPT	Hydrogen, oxygen, power, heating	Air/PCM	Yes	49	45	No	[41]
Ishaq and Dincer	2019	Four-step Cu-Cl	SPT	Hydrogen, oxygen, power, cooling, heating	Molten salt	No	31.5	29.9	No	[39]
Siddiqui et al.	2019	Four-step Cu-Cl	SPT	Hydrogen, oxygen, power, cooling	Molten Salt	No	19.1	19.6	No	[38]
Ishaq and Dincer	2019	Four-step Cu-Cl	SPT	Hydrogen, oxygen	Molten salt	No	33.2	32.7	Yes	[40]
Ouagued et al.	2018	Four-step Cu-Cl	PTC	Hydrogen, oxygen	Thermal oil	No	-	-	Yes	[29]

(*Continued*)

Table 3.1 Selected studies on solar thermochemical water splitting hydrogen generation. (*Continued*)

Study	Year	Type of thermochemical cycle	Type of solar subsystem	System outputs	Solar HTF and TES	Detailed integration	Overall system exergy efficiency (%)	Overall system energy efficiency (%)	Use of external electricity source	Reference
Ishaq et al.	2018	Four-step Cu-Cl	SPT	Compressed hydrogen, oxygen, power, heating	Molten salt	No	48.2	49	No	[37]
Yilmaz and Selbas	2017	Three-step S-I	SPT	Hydrogen, oxygen	Molten salt	No	34.6	32.7	Yes	[35]
Al-Zareer et al.	2017	Five-step Cu-Cl	SPT	Compressed hydrogen, oxygen,	Molten salt	Yes	20.7	12.6	No	[34]
Sayyaadi and Boroujeni	2017	Five-step Cu-Cl	PTC	Hydrogen, oxygen	Molten salt	Yes	58.2	49.8	Yes	[36]
Balta et al.	2014	Three-step Mg-Cl	SPT	Hydrogen, oxygen	Molten salt	No	19.1	18.8	No	[32]
Ozcan and Dincer	2014	Three-step Mg-Cl	SPT	Hydrogen, oxygen, power	Molten salt	Yes	19.9	18.8	No	[33]
Ratlamwala and Dincer	2013	Four-step Cu-Cl	SPT	Hydrogen, oxygen, power	Molten salt	No	49.2	40	Yes	[31]
Ghandehariun et al.	2010	Five-step Cu-Cl	PTC	Hydrogen, oxygen	Molten salt	No	-	49	Yes	[30]

temperature of 400°C is used [47]. The proposed system includes a solar heliostat field and receiver with thermal energy storage (TES), a supercritical steam Rankine cycle (SRC) for electricity generation, and a four-step Cu-Cl cycle for hydrogen production. Energy, exergy, and economic analyses of the proposed system are conducted and the system is optimized from thermodynamic and economic aspects.

3.2 System Description

As mentioned earlier, 7 different types of Cu-Cl cycles have been reported. From these options, three are more promising than others. These three cycles include the Cu-Cl-3A cycle developed by ANL in 2005, which is also known as UNLV-191; the second is Cu-Cl-4B, which was developed by the University of Ontario Institute of Technology (UOIT); and the last is the five step cycle, the Cu-Cl-5 proposed by IGT [24]. Generally, five different chemical reactions occur in a Cu-Cl cycle, namely:

- Hydrogen Chloride (HCl) production
- Oxygen production
- Copper production
- Cupric chloride ($CuCl_2$) drying
- Hydrogen production

In the three step Cu-Cl cycle, copper, hydrogen production and cupric chloride drying steps are combined into one step. In the four step Cu-Cl cycle, the copper production and cupric chloride steps are combined [24].

The Cu-Cl-3A cycle starts with the chlorination step in which hydrogen and molten CuCl are produced from the exothermic reaction of copper and hydrogen chloride gas at 450 °C [24]:

$$2Cu\ (s) + 2HCl\ (g) \rightarrow 2CuCl\ (molten) + H_2\ (g) \quad 450\ °C \quad (3.2)$$

The next step is called a disproportionation step (simultaneous oxidation and reduction), where an electrochemical reaction takes place in the HCl acid at about 30°C to 80°C, and the produced cuprous chloride of the first step turns into solid copper and aqueous cupric chloride ($CuCl_2$) [24]:

$$4CuCl\ (s) \rightarrow 2Cu\ (s) + 2CuCl_2\ (aq) \quad 30-80°C \quad (3.3)$$

The last step is called an oxychlorination step. This is an endothermic reaction occurring at about 530 °C. In this step, molten CuCl, oxygen, and HCl are produced from liquid water and aqueous cupric chloride [24]:

$$CuCl_2(aq) + H_2O(l) \rightarrow 2CuCl(molten) + 2HCl + \frac{1}{2}O_2 \quad 530\,°C \tag{3.4}$$

The four-step Cu-Cl-4B cycle starts the hydrolysis step where solid copper oxychloride and gaseous hydrogen chloride are produced in an endothermic reaction at about 400 °C from the reaction of steam with solid cupric chloride [24]:

$$2CuCl_2(s) + H_2O(g) \rightarrow Cu_2OCl_2(s) + 2HCl(g) \quad 400\,°C \tag{3.5}$$

The second step of this cycle occurs at 500 to 530°C in which solid copper oxychloride decomposes in an endothermic reaction to give oxygen and liquid cuprous chloride [24]:

$$Cu_2OCl_2(s) \rightarrow \frac{1}{2}O_2(g) + 2CuCl(l) \quad 500\text{-}530\,°C \tag{3.6}$$

The third step of this cycle is the electrochemical production of hydrogen gas and aqueous cupric chloride at around 25 °C from the reaction of cuprous chloride and hydrogen chloride gas [24]:

$$2CuCl(aq) + 2HCl(aq) \rightarrow H_2(g) + 2CuCl_2(aq) \quad 25\,°C \tag{3.7}$$

The final step is the physical drying of aqueous cupric chloride, which takes place at temperatures of around 80 °C. Crystallization requires 30°C to 80°C while spray drying requires 100°C to 260°C [24]:

$$CuCl_2(aq) \rightarrow CuCl_2(s) \quad 80\text{-}260\,°C \tag{3.8}$$

The Cu-Cl-5 cycle starts with the chlorination process in which solid copper reacts with hydrogen chloride gas in an exothermic reaction at around 430°C to 450°C to yield hydrogen gas and molten cuprous chloride [24]:

$$2Cu(s) + 2HCl(g) \rightarrow H_2(g) + 2CuCl(l) \quad 430\text{–}450°C \quad (3.9)$$

Solid copper and cupric chloride are produced in the disproportionation step from the electrochemical reaction of cuprous chloride in aqueous solution of HCl at around 25°C to 80°C [24]:

$$4CuCl(aq) \rightarrow 2Cu(s) + 2CuCl_2(aq) \quad 25\text{–}80°C \quad (3.10)$$

The copper produced in the second step is then transported to the first reactor and the cupric chloride is sent to the endothermic drying step to form solid cupric chloride. Again, crystallization requires 80 °C while spray drying requires 100°C to 260°C [24].

$$CuCl_2(aq) \rightarrow CuCl_2(s) \quad 80\text{–}260°C \quad (3.11)$$

In the fourth step, solid cupric chloride reacts with steam at around 375°C to 400°C in an endothermic reaction to produce copper oxychloride and hydrogen chloride [24]:

$$2CuCl_2(s) + H_2O(g) \rightarrow Cu_2OCl_2(s) + 2HCl(g) \quad 375\text{–}400°C \quad (3.12)$$

In the last step of this cycle, copper oxychloride decomposes to oxygen and liquid cuprous chloride in an endothermic reaction in about 500°C to 530°C [24]:

$$Cu_2OCl_2(s) \rightarrow \frac{1}{2}O_2(g) + 2CuCl(l) \quad 500\text{–}530\,°C \quad (3.13)$$

In all of the cycles, the only material, which will not be recycled, is water while all of the other compounds are re-used in the cycle. In this chapter, a four-step Cu-Cl cycle is chosen because of its higher exergetic and thermal efficiencies [25].

Due to the high temperature requirement of the thermolysis reaction (around 530°C), limited energy sources are available that can supply this temperature. Solar energy and only some of the solar energy harvesting systems are capable of supplying this temperature. As solar power tower systems use various HTFs, will not require separate lines per assembly, and because of their ability to supply high temperatures, they are selected for integration with a copper-chlorine cycle [8].

One of the limitations of solar energy is its intermittent nature. For overcoming this issue, it is necessary to use energy storage devices. Thermal energy storage systems can be integrated with solar power towers to achieve continuous unit operation, even at night. Sensible or latent TES can be utilized with SPTs. However, it is common to use solar salt, a binary mixture of KNO_3 and $NaNO_3$, with SPTs. These fluids can be used as both TES and HTF. In some of the solar harvesting facilities, including PTCs, two tank TES systems are used where thermal oils are utilized as the HTF while solar salt is used as TES. In these systems, one extra heat exchanger assembly is utilized. A more efficient way of storing thermal energy in SPTs includes the use of direct two-tank TES systems. In these systems, solar salt or any molten salt can be used as both TES and HTF. Therefore, there is no need to transfer the heat content of HTF to TES media [48].

As mentioned earlier, thermal decomposition of Cu_2OCl_2 occurs at temperatures around 530°C and the maximum operating temperature of solar salt is 560°C [24, 48]. Considering a pinch temperature of 10°C for a heat transfer limitation, the temperature difference for the HTF that supplies the required heat of thermolysis reaction will only be 20°C. This means that very high flow rates of molten salt will be required to provide the thermal energy requirement of the thermolysis reaction. It will result in a low hydrogen production yield, and also will affect the thermal performance of the integrated system significantly. For this reason, a more suitable molten salt is selected as both the HTF and TES of this study. As stated earlier, a ternary mixture of Li_2CO_3(32.1%)-Na_2CO_3(33.4%)-K_2CO_3(34.5%) carbonate salt with a decomposition temperature of 800°C and a melting temperature of 400°C is used [47].

For a stand-alone hydrogen production facility using the Cu-Cl cycle, it is necessary to generate the required power for the electrolysis reaction where hydrogen is produced. Various power generation cycles like the Kalina, conventional steam Rankine cycle (SRC), or even organic Rankine cycle (ORC) can be utilized. However, because a higher temperature molten salt is used in this study, it is possible to utilize a supercritical steam Rankine cycle (S-SRC). A higher performance power generation cycle can drastically improve the overall system thermal and exergetic efficiencies.

Figure 3.1 shows a schematic of the proposed solar hydrogen system. At state 30, molten salt enters the solar receiver tower at 400°C [45]. The molten salt exiting the tower at 650°C flows through the hot tank. Hot molten salt splits into two streams. One stream flows through the shell of the thermolysis reactor at state 32 to supply the heat required. The molten salt enters Heater 1 at state 34 to heat up the $CuCl_2$. At state 35, cooler molten salt flows through the dryer unit to dry up the aqueous $CuCl_2$. At state 37,

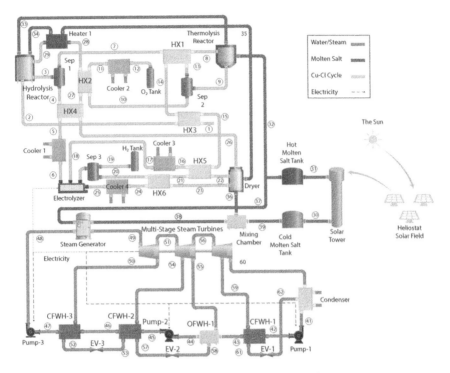

Figure 3.1 Schematic of a standalone solar hydrogen system [45].

the molten salt stream from the hot tank flows through the steam generator of the Rankine cycle to generate superheated steam. Molten salt at state 36 mixes with the one at state 38 in a mixing chamber and then flows to the cold tank at 400°C.

In the copper-chlorine cycle, water at 25°C is heated using the waste heat of the CuCl(I) at state 14. Water vapor at state 2 along with heated $CuCl_2(s)$ at state 29 enter the hydrolysis reactor to produce Cu_2OCl_2 and gaseous HCl. The heat requirement of the hydrolysis reactor is provided by the molten salt (state 33). Cu_2OCl_2 is heated in the first internal heat exchanger (HX1) using some of the waste heat of the molten CuCl leaving the thermolysis reactor. The heated Cu_2OCl_2 at state 8 enters the thermolysis reactor where it is decomposed into molten CuCl and oxygen.

Molten CuCl is used in HX1, HX3, and HW5 to heat Cu_2OCl_2, water, and $CuCl_2$ (aq), respectively. Molten CuCl is cooled further in Cooler 3 before entering the electrolyzer at state 17. Water along with HCl and CuCl enters the electrolyzer to generate hydrogen in an electrochemical cell. $CuCl_2$ (aq) is heated up in HX6 and HX5 before entering the dryer at state

22. Water vapor leaving the dryer unit at state 23 is cooled down in HX6 and Cooler 4 before entering the electrolyzer unit at state 25. $CuCl_2$, leaving the dryer at state 26, enters three heat exchangers and is heated up to about 390 °C before entering the hydrolysis reactor.

As stated earlier, a supercritical Rankine cycle comprising a three-stage turbine is used to produce electricity. The cycle has three closed feedwater heaters (CFWHs) and one open feedwater heater (OFWH). The purpose of the power generation system is to provide the required electricity for the pumps and the electrolysis step of the Cu-Cl cycle, and no excess power is generated. The only inputs of the integrated system are water and solar energy, while the only outputs are hydrogen and oxygen.

3.3 Mathematical Modeling and Optimization

3.3.1 Energy and Exergy Analyses

MATLAB software with the REFPROP 9 library is used for the thermodynamic analysis. The mass, energy, and exergy rate balance equations at steady state are given respectively as follows [49, 50].

$$\sum \dot{m}_i - \sum \dot{m}_e = 0 \qquad (3.14)$$

$$\dot{Q} - \dot{W} = \sum (\dot{m}h)_e - \sum (\dot{m}h)_i \qquad (3.15)$$

$$\dot{Ex}_D = \dot{Ex}_Q - \dot{Ex}_W + \sum (\dot{m}ex)_i - \sum (\dot{m}ex)_e \qquad (3.16)$$

Here, $\dot{m}, \dot{Q}, \dot{W}, \dot{Ex}$ represent the rate of mass, heat, work, and exergy flow, respectively. Subscripts i, e, d, Q, and W denote inlet, outlet, destruction, heat, and work, respectively. Also, h and ex are the specific enthalpy and exergy, respectively. Other terms are calculated as follows [49, 50]:

$$\dot{Ex}_Q = \left(1 - \frac{T_0}{T_i}\right) \dot{Q}_i \qquad (3.17)$$

$$\dot{Ex}_W = \dot{W} \qquad (3.18)$$

$$ex = ex_{ph} + ex_{ch} \tag{3.19}$$

$$ex_{ph} = (h_i - h_o) - T_o(s_i - s_o) \tag{3.20}$$

Here, ex_{ph} and ex_{ch} represent physical and chemical-specific exergy, respectively. The subscript o refers to the reference environment state, which is 25°C and 101 kPa. Chemical exergy is only important in components in which the substance composition changes. Therefore, chemical exergy should be considered in the Cu-Cl cycle analysis.

The Shomate equations are used to evaluate the thermodynamic properties for each stream of the copper-chlorine cycle, The specific heat, enthalpy, and entropy of each stream are calculated as follows, respectively [51]:

$$C_p^o = A + B \times t + C \times t^2 + D \times t^3 + E/t^2 \tag{3.21}$$

$$h^o - h_{ref}^o = A \times t + B \times \frac{t^2}{2} + C \times \frac{t^3}{3} + D \times \frac{t^4}{4} - \frac{E}{t} + F - H \tag{3.22}$$

$$S^o = A \times \ln(t) + B \times t + C \times \frac{t^2}{2} + D \times \frac{t^3}{3} - \frac{E}{2 \times t^2} + G \tag{3.23}$$

$$t = \left(\frac{T}{1000}\right) \tag{3.24}$$

The parameters A to H are the Shomate constants, determined using the NIST database [51].

The heat supplied to the copper-chlorine and Rankine cycle is evaluated considering the thermal energy emitted from the sun, reflected by heliostat mirrors, and absorbed by the heat transfer fluid. The solar energy is given as follows [52]:

$$\dot{Q}_{Sun} = A_h \cdot n_h \cdot DNI \tag{3.25}$$

Here, n_h is the number of heliostat mirrors, A_h is the aperture area of a single heliostat, and DNI is the direct normal irradiance.

The heat reflected from the heliostat mirrors and its corresponding heat loss are evaluated from the following equations [52]:

$$\dot{Q}_{Rec} = \dot{Q}_{Sun} \cdot \eta_{Opt} \qquad (3.26)$$

$$\dot{Q}_0 = \dot{Q}_{Sun} - \dot{Q}_{Rec} \qquad (3.27)$$

Here, η_{Opt} is the optical efficiency of the heliostat mirrors. Heat loss from the receiver and the heat absorbed by the heat transfer fluid are given, respectively as [53]:

$$\dot{Q}_{Loss,rec} = \dot{Q}_{Loss,cond} + \dot{Q}_{Loss,conv} + \dot{Q}_{Loss,em} + \dot{Q}_{Loss,ref} = \dot{Q}_{Rec} - \dot{Q}_{Abs} \qquad (3.28)$$

$$\dot{Q}_{Abs} = \dot{m}_{ms}(h_{ms,e} - h_{ms,i}) \qquad (3.29)$$

The subscript ms represents molten salt. Heat losses of the receiver include heat losses associated with conduction ($\dot{Q}_{Loss,cond}$), natural and forced convection ($\dot{Q}_{Loss,conv}$), emission ($\dot{Q}_{Loss,em}$) and reflection ($\dot{Q}_{Loss,ref}$). Finally, the receiver efficiency is calculated by [52]:

$$\eta_{Rec} = \frac{\dot{Q}_{Abs}}{\dot{Q}_{Rec}} \qquad (3.30)$$

The density, thermal conductivity, viscosity, and specific heat of LiNaK high-temperature molten salt in this study can be calculated from the following equations [54]:

$$\rho = 2.27 - 4.34 \times 10^{-4}\, T[K]\ (g/cm^3) \qquad (3.31)$$

$$k = 0.336 + 2.58 \times 10^{-4}\, T\,[K]\ (W/m.K) \qquad (3.32)$$

$$\mu = 0.0650 \exp\left(\frac{4431.3}{T[K]}\right)(cP) \qquad (3.33)$$

$$c_p = 161 \text{ (J/g.K)} \tag{3.34}$$

The thermal efficiency of the Cu-Cl cycle, overall system energy efficiency, and overall system exergy efficiencies are calculated using the following equations:

$$\eta_{Cu-Cl} = \frac{\dot{m}_{19} \times LHV_{H_2}}{\left(\sum \dot{Q}_{input}\right) + \left(\dfrac{\dot{W}_{ele}}{\eta_{SRC}}\right)} \tag{3.35}$$

$$\eta_{I,total} = \frac{\dot{m}_{19} \times LHV_{H_2}}{\dot{Q}_{Sun}} \tag{3.36}$$

$$\eta_{II,total} = \frac{\dot{m}_{19} \times Ex_{H_2}}{\dot{E}x_{Sun}} \tag{3.37}$$

where $LHV_{H_2}, \dot{Q}_{input}, \dot{W}_{ele}$ and η_{SRC} denote the lower heating value of hydrogen, input heat of the Cu-Cl cycle, required electricity of the electrolysis, and efficiency of steam Rankine cycle, respectively.

3.3.2 Economic Analysis

The total revenue requirement (TRR) method is used for the economic analysis. It includes total capital recovery, return on investment for debt, common equity, and preferred stock, income taxes, other taxes and insurance, fuel costs, and operation and maintenance (O&M) costs [55]:

$$TRR_j = TCR_j + ROI_{j,d} + ROI_{j,ce} + ROI_{j,ps} + ITX_j + OTXI_j + FC_j + OMC_j \tag{3.38}$$

The total revenue requirements of the plant are obtained as follows [55]:

$$TRR_L = CRF \sum_{1}^{n} \frac{TRR_j}{(1+i_{eff})^j} \tag{3.39}$$

Here, i_{eff} is the annual effective discount rate, n is the plant economic lifetime, and TRR_j is the total revenue requirement of the j_{th} year of the plant operation. CRF represents the capital recovery factor, given as follows [55]:

$$CRF = \frac{i_{eff}(1+i_{eff})^n}{(1+i_{eff})^n - 1} \quad (3.40)$$

The levelized costs for the fuel and O&M is evaluated as follows [55]:

$$FC_L = FC_0 CELF = FC_0 \frac{k(1-k^n)}{(1-k)} CRF \quad (3.41)$$

$$OMC_L = OMC_0 CELF = OMC_0 \frac{k(1-k^n)}{(1-k)} CRF \quad (3.42)$$

where k is given as:

$$k = \frac{1+r}{1+i_{eff}} \quad (3.43)$$

Here, r denotes to nominal escalation rate.

Finally, the levelized cost of hydrogen (LCOH) is evaluated as follows [55]:

$$LCOH = \frac{TRR_L - BPV}{MPQ} \quad (3.44)$$

where BPV and MPQ represent the total value of the plant byproduct, and the annual main product quantity, respectively. Table 3.2 gives the input data used in the thermodynamic and economic analyses for the base case study.

Table 3.2 Input data used for the economic and thermodynamic analyses.

Cycle/subsystem	Parameter	Value	Reference
Solar power tower	Aperture area of each heliostat mirror (m²)	100	[18]
	Concentration ratio	1000	[53]
	Number of heliostat mirrors	5000	[18]
	Heliostat field optical efficiency (%)	75	[53]
	Outlet temperature of molten salt (°C)	650	[47]
	Inlet temperature of molten salt (°C)	400	[47]
	View factor	0.8	[53]
	Reflectivity	0.04	[53]
	Emissivity	0.8	[53]
	Tube diameter (m)	0.019	[53]
	Tube thickness (m)	0.00165	[53]
	Number of tubes in the receiver	20	[53]
Cu-Cl cycle	Pinch point (°C)	10	[18]
	Electrolysis reaction temperature (°C)	25	[24]
	Hydrolysis reaction temperature (°C)	390	[24]
	Thermolysis reaction temperature (°C)	530	[24]
	Drying phase temperature (°C)	70	[24]
Rankine cycle	Turbine inlet pressure (kPa)	24000	[50]
	Turbine inlet temperature (°C)	560	[50]

(Continued)

Table 3.2 Input data used for the economic and thermodynamic analyses. (*Continued*)

Cycle/subsystem	Parameter	Value	Reference
	Turbine first bleeding pressure (kPa)	5000	[50]
	Turbine second bleeding pressure (kPa)	1400	[50]
	Turbine third bleeding pressure (kPa)	300	[50]
	Turbine fourth bleeding pressure (kPa)	180	[50]
	Condenser pressure (kPa)	10	[50]
	Isentropic efficiency of pumps (%)	80	[50]
	Isentropic efficiency of steam turbine (%)	86	[50]
Economic analysis	Plant lifetime (years)	35	[56]
	Plant tax life (years)	15	[55]
	Load factor (%)	95	[56]
	Interest rate (%)	4	[55]
	Rate of return for debt, common equity, and preferred stock	10	[55]
	Nominal escalation rate (%)	1	[55]
	Operation and maintenance factor (%)	2	[57]
	Total rate of income tax (%)	30	[55]
	Other taxes and insurance factor (%)	2	[55]

3.3.3 Multiobjective Optimization (MOO) Algorithm

The system will be optimized by a non-dominated sorting genetic algorithm-II (NSGA-II), which is a powerful stochastic algorithm for multiobjective optimization [58]. A series of balanced non-dominated solutions, called the Pareto front, are evaluated based on a genetic algorithm that will satisfy all of the objective functions [58].

In the multiobjective optimization, the ideal point is outside the Pareto Front. Therefore, after generating the non-dominated Pareto solutions, it is necessary to use one of the multiobjective decision-making methods to choose the optimal Pareto solution. A Linear Programming Technique for Multidimensional Analysis of Preference (LINMAP) is selected to assess the optimal Pareto solution [18].

The solutions of the Pareto frontier are normalized by:

$$S_{norm} = \frac{S_i - S_{min}}{S_{max} - S_{min}} \quad (3.45)$$

The optimal solution of the Pareto front by the LINMAP method is the point in which the minimum Distance$_i^+$ is compared to other Pareto solutions [18]:

$$Distance_i^+ = \sqrt[2]{\sum_j^k (S_{ij} - S_j^{ideal})^2} \quad (3.46)$$

where i and k represent the solution number i and number of objectives, respectively. S_j^{ideal} is the ideal value of the j_{th} objective function.

Input design parameters for this study include 14 parameters. Some input parameters related to the solar system, copper-chlorine hydrogen production cycle, and the Rankine cycle can affect the economic and thermal performance of the integrated system significantly. For instance, the outlet temperature of the molten salt from the receiver and the number of heliostats are the major parameters of the solar system. Considering a decomposition temperature of about 800°C for LiNaK, the receiver should be pressurized for a molten salt temperature of above 700 degrees. Therefore, the range of molten salt outlet temperature from the solar receiver is in the range of 620°C to 700°C [59]. The temperature of the thermolysis and the hydrolysis reactors and the pinch point are selected as the input variables for the Cu-Cl cycle. As shown in previous studies, at operating temperature ranges of 510°C to 530°C for the thermolysis reactor and 380°C to 400°C for the hydrolysis reactor, the reactions are close to stoichiometric reactions [25, 29, 36]. The steam generator pressure, the condenser pressure, the turbine inlet temperature, the turbine extraction pressures, and the isentropic efficiencies of turbines/pumps are the variable parameters used for the Rankine cycle.

The chosen range of each parameter for the optimization of this study include: 1) number of heliostats ranging from 1,000 to 10,000; 2) molten salt outlet temperature ranging from 620-700 °C, 3) hydrolysis temperature ranging from 380°C to 400°C, 4) thermolysis temperature ranging from 510-530 °C, 5) pinch point ranging from 8-12 °C, 6) SRC turbine inlet temperature ranging from 500°C to 600°C, 7) The pressure of steam generator ranging from 10 to 25 MPa, 8) SRC turbine first bleeding pressure ranging from 1.9 to 7 MPa, 9) SRC turbine second bleeding pressure ranging from 0.9 to 1.6 MPa, 10) SRC turbine third bleeding pressure ranging from 0.2 to 0.6 MPa, 11) SRC turbine fourth bleeding pressure ranging from 60 to 180 kPa, 12) SRC condenser pressure ranging from 8 to 20 kPa, 13) pumps isentropic efficiencies ranging from 75% to 90%, and 14) turbine isentropic efficiencies ranging from 75% to 90%.

Multiobjective optimization is performed considering the solar-to-hydrogen efficiency and the levelized cost of hydrogen as the two objective functions.

3.4 Results and Discussion

The thermal efficiency of the copper-chlorine cycle is evaluated as 40.4%. The thermal efficiency of the steam Rankine cycle is 45%, and the overall system efficiency is 28.6%. The exergy destruction rate is 118.6 MW, and the exergy efficiency of the system is 29.5%. Also, hydrogen production capacity of the system for the base case is calculated as 1,523.7 kg/h. The economic analysis results show that for the base case, the capital investment cost is 515.98 million dollars. The levelized total revenue requirement of the system is evaluated as 63.91 million dollars. Considering $80 per square meter of heliostats and an oxygen selling price of $0.154/kg, the levelized cost hydrogen is $3.12/kg H_2.

It is important to consider the variations of solar irradiance with time in modeling the solar system [59]. Figure 3.2 shows the results of the dynamic analysis of the solar system for summer data as well as the accumulated thermal energy in the molten salt hot tank for various seasons [45]. Considering the base case, no solar irradiance is available from about 7 P.M. to 5 A.M.,

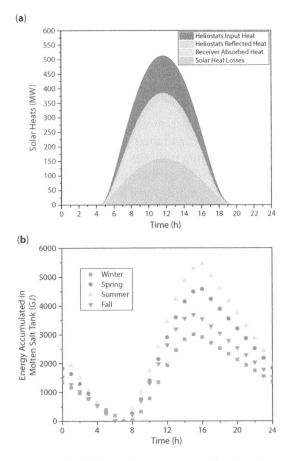

Figure 3.2 (a) Variations of solar heat with respect to time for the solar power tower subsystem, (b) energy accumulated in the molten salt tank for four different seasons [45].

and hence, there is no molten salt flow through the receiver. The solar irradiance peak occurs at about 12 P.M. as the solar heat input to the heliostat mirrors is 514.7 MW. A portion of this thermal energy is lost based on the optical efficiency of the heliostat solar field. The reflected solar energy from the heliostat mirrors to the molten salt receiver is 386 MW at 12 P.M.

The thermal energy absorbed by the molten salt is 355.5 MW for 12 P.M., considering the heat loss associated with the receiver tower. The outlet temperature of the molten salt is a main parameter affecting the amount of heat loss from the receiver tower. For the base case, heat loss from the heliostats are higher than that from the receiver tower. However, if the outlet temperature of molten salt increases, the difference will decrease.

For Figure 3.4 part B, the thermal energy is discharged from the tank, at the start of the day, until it gets empty. In the winter, spring, and fall, the energy content of the tank becomes zero at 7 A.M., while in the summer, the tank gets empty at 6 A.M. Afterward, as the sun comes up, the tank is charged until it gets full. The maximum accumulated energy is 3,001, 4,572, 5,456, and 3,687 GJ for winter, spring, summer, and fall, respectively. TES energy discharge from the hot tank begins as soon as the solar radiations declines at 3:00 to 4:00 pm.

The influence of varying the number of heliostat mirrors and molten salt outlet from the solar receiver on system outputs and performance are shown in Figure 3.3. The total capital investment, hydrogen production capacity, and exergy destruction rate increase from 113.18 million dollars, 301.41 kg/h, and 23.82 MW to 1,007.52 million dollars, 3052.88 kg/h, and 236.97 MW when the number of heliostat mirrors increases from 1,000 to 10,000. For the same increase in the number of heliostat mirrors, the solar to hydrogen and exergetic efficiencies increase from 28.4%, and 29.2% to 28.7% and 29.5%, respectively. As the amount of energy input of the system increases, the ratio of heat losses to heat input of the receiver decreases, and therefore, the performance of the solar system improves. Also, since the hydrogen production rate increases, the unit cost of hydrogen is reduced. When the heliostat number increases from 1,000 to 10,000, the LCOH decreases from \$3.67/kg H_2 to \$3.0/kg H_2, respectively. Increasing the temperature of the solar receiver outlet temperature worsens the system thermal performance including efficiencies and the rate of hydrogen generation.

When the outlet temperature of the molten salt leaving the receiver increases, heat losses are increased drastically, and hence, the efficiency of the receiver is reduced. Also, for a fixed amount of received solar irradiation, by increasing the outlet temperature of the molten salt, the mass flow rate of heat transfer fluid is reduced. Therefore, a reduction in hydrogen generation rate and the efficiencies will happen. As the molten salt outlet temperature increases from 873 to 973 K, the energy efficiency, exergy efficiency, and hydrogen generation rate reduce from 28.9%, 29.7%, and 1,535.08 kg/h to 28.4%, 29.2%, and 1,510.94 kg/h, respectively. Additionally, due to a higher heat loss, the exergy destruction rate increases from 118.2 to 118.98 MW considering a fixed increase in the outlet temperature of the receiver. Decreasing the mass flow rate of molten salt reduces all of the associated costs. Increasing the outlet temperature of the receiver from 873 to 973 K, reduces the levelized cost of hydrogen from \$3.32/kg H_2 to \$3.02/kg H_2.

124 SOLAR FUELS

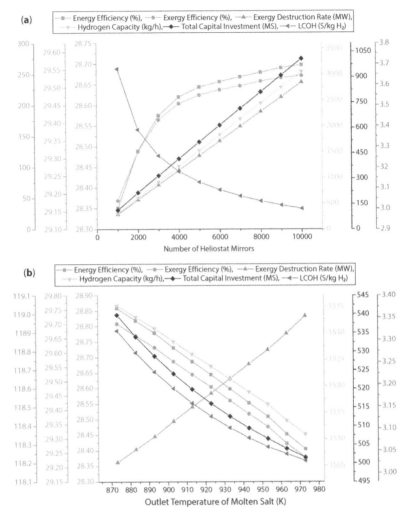

Figure 3.3 Effect of increasing the number of heliostat mirrors (a) and molten salt outlet temperature (b) on system outputs.

Figure 3.4 shows the impact of the turbine extraction pressures on the thermal efficiency of the Rankine cycle. For a condenser pressure of 10 kPa and the boiler pressure of 24,000 kPa, the optimum four extraction pressures are 5,700, 1,500, 300, and 75 kPa.

Figure 3.5 shows the effects of the turbine inlet temperature and the steam generator pressure on the system performance. As can be seen, an increase in pressure of the steam generator improves almost all of the system performance criteria except the total capital investment expenses.

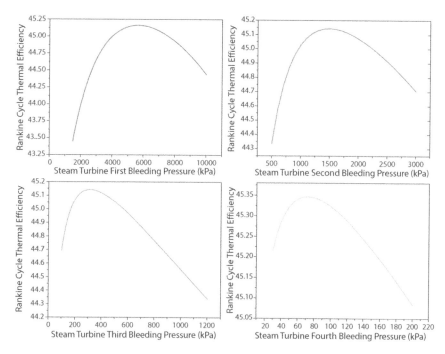

Figure 3.4 Influence of steam turbine extraction pressures on Rankine cycle thermal efficiency.

When the boiler pressure increases from 8 to 25 MPa, the hydrogen production capacity, thermal efficiency, and the exergy efficiency increase from 1,463.04 kg/h, 27.5%, and 28.3% to 1,525.05 kg/h, 28.7%, and 29.5%, respectively. As the thermal efficiency of the Rankine cycle is improved due to an increase in the steam generator pressure, for a fixed heat input, the amount of power output increases. This causes a reduction in the mass flow rate of stream 37 and an increase in the mass flow rate of molten salt through the Cu-Cl cycle. Therefore, the rate of hydrogen generation is improved. However, using a higher-pressure steam generator increases the investment cost but reduces the cost of hydrogen generation. The total investment cost and levelized cost of hydrogen changes from 503.3 million dollars and \$3.22/kg H_2 to 516.35 million dollars and \$3.12/kg H_2, respectively. Moreover, because the energy and exergy efficiencies of the system are improved, the exergy destruction rate is reduced from 120.54 to 118.53 MW when the boiler pressure increases from 8 to 25 MPa.

By increasing the turbine inlet temperature, the average temperature of heat addition to the cycle increases, and therefore, the thermal efficiency of the Rankine cycle increases. Also, a lower mass flow rate of molten salt is

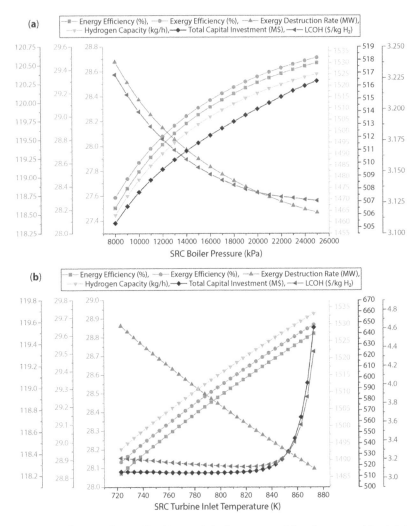

Figure 3.5 Influence of steam Rankine cycle boiler pressure (a) and steam turbine inlet temperature (b) on system outputs.

required for the same amount of hydrogen generation. Therefore, if the total heat input of the cycle remains unchanged, the mass flow rate of stream 32 increases. This improves the hydrogen generation rate. As the turbine inlet temperature increases from 723 to 873 K, the thermal efficiency, exergy efficiency, and hydrogen production capacity increases from 28.1%, 28.9%, and 1,493.2 kg/h to 28.8%, 29.7%, and 1,533.1 kg/h, respectively. Increasing the temperature to about 823 K does not change the total investment cost. However, if the turbine inlet temperature increases from 823 to 873 K, the capital cost increases from 680 to 811 million dollars. This increase in the investment cost is due to the requirement of a higher-performance and hence, a more expensive turbine operating at higher temperatures. Since, the investment cost remains constant and the thermal efficiency improves for inlet turbine temperatures up to 823 K, the levelized cost of hydrogen reaches its optimal value of $3.11/kg H_2. The investment cost is 513.87 million dollars for a turbine inlet temperature of 723 K, and 645.33 million dollars for a turbine inlet temperature of 873 K.

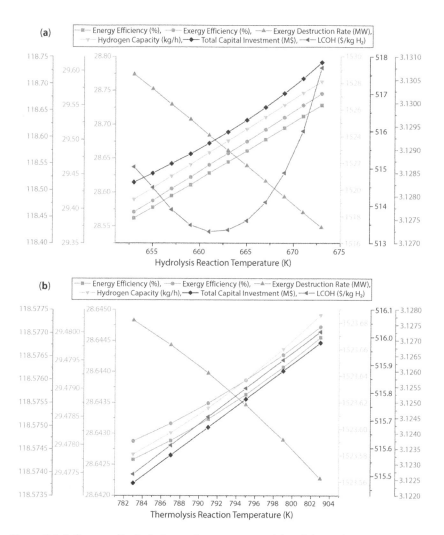

Figure 3.6 Influence of hydrolysis reaction temperature (a) and thermolysis reaction temperature (b) on system outputs.

The influence of hydrolysis and thermolysis reaction temperatures on system outputs are shown in Figure 3.6. As the hydrolysis reaction temperature increase from 653 to 673 K, the energy efficiency, exergy efficiency, hydrogen production capacity, and total capital investment increase from 28.6%, 29.4%, 1,519.3 kg/h, and 514.63 million dollars to 28.7%, 29.6%, 1528.1 kg/h, and 517.84 million dollars, respectively. Moreover, the exergy destruction rate decreases from 118.71 to 118.43 MW for the same amount of increase in hydrolysis reaction temperature. The LCOH will be at its lowest point when the hydrolysis reaction temperature is around 663 K. As can be seen from the other graph, increasing the temperature of the thermolysis reaction has a very slight effect on system performance, and nearly all of the system performance criteria remain constant with changes of thermolysis reaction temperature.

Figure 3.7 shows the Pareto Frontier of optimal solutions. Fourteen variables are considered and three solutions are selected on the Pareto Frontier solutions. Solution A has the lowest levelized cost of hydrogen with the lowest solar-to-hydrogen efficiency. Solution C has the highest efficiency and highest cost. LINMAP method is used to select the solution B, which satisfies both of objective functions. Table 3.3 presents the optimization results for the selected solutions. For solution A, a hydrogen production cost of $2.91/kg H_2 is evaluated while the energy and exergy efficiencies are 29.0% and 29.9%, respectively. Similarly, for solution C, the hydrogen production cost is $3.07/kg H_2 and the energy and exergetic efficiencies are 29.3% and 30.2%, respectively. For the best case scenario of this study, the solution B, LINMAP selected Pareto solution, the hydrogen production

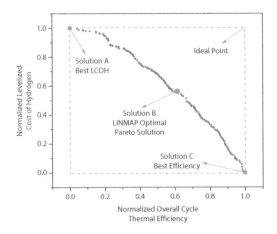

Figure 3.7 Pareto frontier of optimal solutions for the multiobjective optimization.

Table 3.3 Results of the multiobjective optimization for three selected Pareto solutions.

Parameter	Solution A	Solution B	Solution C
Pinch point (K)	11.03	8.50	8.37
Number of heliostat mirrors	8972.08	8982.49	8980.77
Thermolysis step reaction temperature (K)	793.02	793.52	793.99
Hydrolysis step reaction temperature (K)	671.21	672.90	672.33
Turbine inlet temperature (K)	832.23	835.44	839.49
Turbine first extraction pressure (kPa)	4532.30	4545.17	4547.76
Turbine second extraction pressure (kPa)	1270.34	1271.18	1270.23
Turbine third extraction pressure (kPa)	372.57	375.11	376.66
Turbine fourth extraction pressure (kPa)	102.32	102.16	101.60
Condenser pressure (kPa)	8.11	8.01	8.03
Boiler pressure (MPa)	23.90	23.96	23.97
Molten salt temperature exiting solar receiver (K)	954.86	926.99	902.97
Isentropic efficiency of turbine (%)	89.74	89.76	89.76
Isentropic efficiency of pumps (%)	83.21	83.74	83.77
Exergy efficiency (%)	29.87	30.07	30.2
Energy efficiency (%)	29.03	29.21	29.34
LCOH ($/kg H_2)	2.91	2.98	3.07

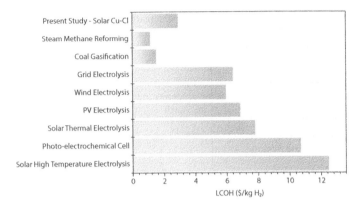

Figure 3.8 Comparison between the LCOH evaluated in this study and other available hydrogen production methods.

cost is evaluated at \$2.98/kg H_2 and the energy and exergy efficiencies are evaluated at 29.2% and 30.1%, respectively.

Figure 3.8 compares the levelized cost of hydrogen evaluated here with those of other hydrogen generation systems [6, 61, 62]. As observed, the cost of hydrogen generation through solar-based copper-chlorine cycle is competitive compared to other available renewable-based hydrogen production methods like wind, CSP, and PV water electrolysis methods. With a hydrogen production cost of \$2.91/kg H_2 for the best economic case of this study, this method stands higher in terms of economic performance compared to \$5.96/kg H_2 of wind electrolysis, \$6.85/kg H_2 of PV electrolysis, \$7.79/kg H_2 of CSP electrolysis, \$10.7/kg H_2 of photoelectrochemical cell, and \$12.48/kg H_2 of solar high-temperature steam electrolysis. Although this method may not be quite competitive with conventional methods such as steam reforming of methane or coal gasification at \$1.13/kg H_2 and \$1.5/kg H_2, it is hoped that with the reduction of capital investment costs of CSP plants, the solar Cu-Cl cycle will also be more competitive with these methods.

3.5 Conclusions

In this study, a standalone solar thermochemical hydrogen generation plant is proposed and investigated. The integrated system includes a four-step copper-chlorine cycle, a supercritical Rankine cycle, and a solar power tower with a high-temperature molten salt, $(LiNaK)_2CO_3$, as a heat transfer fluid and for thermal energy storage. Thermodynamic and economic analyses of the system are performed, and the system is optimized using the NSGA-II, for two objective functions including the levelized cost of produced hydrogen and the solar-to-hydrogen efficiency. The major findings of this study are as follows:

- For the base case, an efficiency of 40.4% for the Cu-Cl cycle and 45.0% for the supercritical Rankine cycle are evaluated. Also, an overall system energy efficiency of 28.6% and overall system exergy efficiency of 29.5% are obtained.
- The rate of hydrogen generation of the integrated system is about 1,524 kg/h.
- The levelized cost of hydrogen is $3.12/kg H_2 while the total investment cost is about 516 million dollars.
- Considering 14 input parameters, the optimization results for the LINMAP-based Pareto solution, gives the solar-to-hydrogen efficiency of 29.2%, exergetic efficiency of 30.1%, and the levelized cost of hydrogen of $2.98/kg H_2.
- A comparison of economic performance of the current model with other available hydrogen production methods indicated that the proposed solar Cu-Cl cycle is more competitive in terms of hydrogen production cost compared to other renewable-based methods like solar PV, solar CSP, or wind water electrolysis methods.

References

1. IRENA, *Hydrogen: A Renewable Energy Perspective*, International Renewable Energy Agency, Abu Dhabi, 2019.
2. Sadeghi, S. and Ghandehariun, S., Comparative environmental assessment of a gasoline and fuel cell vehicle with alternative hydrogen pathways in Iran. *Automotive Sci. Eng.*, 10, 3, 3266–3280, 2020.
3. Salkuyeh, Y.K., Saville, B.A., MacLean, H.L., Techno-economic analysis and life cycle assessment of hydrogen production from natural gas using current and emerging technologies. *Int. J. Hydrog. Energy*, 42, 30, 18894–18909, 2017.
4. El-Emam, R.S. and Özcan, H., Comprehensive review on the techno-economics of sustainable large-scale clean hydrogen production. *J. Clean. Prod.*, 220, 593–609, 2019.
5. Dincer, I. and Acar, C., Review and valuation of hydrogen production methods for better sustainability. *Int. J. Hydrog. Energy*, 40, 34, 11094–111, 2015.

6. Sadeghi, S., Ghandehariun, S., Rosen, M.A., Comparative economic and life cycle assessment of solar-based hydrogen production for oil and gas industries. *Energy*, 208, 118347, 2020.
7. Nematollahi, O., Alamdari, P., Jahangiri, M., Sedaghat, A., Alemrajabi, A.A., A techno-economical assessment of solar/wind resources and hydrogen production: A case study with GIS maps. *Energy*, 175, 914–930, 2019.
8. Ghandehariun, S., Naterer, G.F., Rosen, M.A., Wang, Z., Indirect contact heat recovery with solidification in thermochemical hydrogen production. *Energy Convers. Manag.*, 82, 212–218, 2014.
9. Steinmann, W.D., Thermal energy storage systems for concentrating solar power (CSP) technology, in: *Advances in Thermal Energy Storage Systems*, pp. 511–531, Woodhead Publishing, Cambridge, UK, 2015.
10. Nikolaidis, P. and Poullikkas, A., A comparative overview of hydrogen production processes. *Renew. Sust. Energ. Rev.*, 67, 597–611, 2017.
11. Guo, L., Chen, Y., Su, J., Liu, M., Liu, Y., Obstacles of solar-powered photocatalytic water splitting for hydrogen production: A perspective from energy flow and mass flow. *Energy*, 172, 1079–1086, 2019.
12. Chi, J. and Yu, H., Water electrolysis based on renewable energy for hydrogen production. *Chin. J. Catal.*, 39, 3, 390–394, 2018.
13. Wang, H., Li, W., Liu, T., Liu, X., Hu, X., Thermodynamic analysis and optimization of photovoltaic/thermal hybrid hydrogen generation system based on complementary combination of photovoltaic cells and proton exchange membrane electrolyzer. *Energy Convers. Manag.*, 183, 97–108, 2019.
14. Mastropasqua, L., Pecenati, I., Giostri, A., Campanari, S., Solar hydrogen production: Techno-economic analysis of a parabolic dish-supported high-temperature electrolysis system. *Appl. Energy*, 261, 114392, 2020.
15. Hoffmann, J.E., On the outlook for solar thermal hydrogen production in South Africa. *Int. J. Hydrog. Energy*, 44, 2, 629–640, 2019.
16. Steinfeld, A., Solar thermochemical production of hydrogen—A review. *Sol. Energy*, 78, 5, 603–615, 2005.
17. Çelik, D. and Yıldız, M., Investigation of hydrogen production methods in accordance with green chemistry principles. *Int. J. Hydrog. Energy*, 42, 36, 23395–23401, 2017.
18. Sadeghi, S., Ghandehariun, S., Naterer, G.F., Exergoeconomic and multi-objective optimization of a solar thermochemical hydrogen production plant with heat recovery. *Energy Convers. Manag.*, 225, 113441, 2020.
19. Sattler, C., Monnerie, N., Roeb, M., Lange, M., Solar thermal water decomposition, in: *Hydrogen Science and Engineering: Materials, Processes, Systems and Technology*, pp. 85–108, 2016.
20. Russ, B., *Sulfur Iodine Process Summary for the Hydrogen Technology Down-Selection (No. INL/EXT-12-25773)*, Idaho National Laboratory (INL) United States, 2009.

21. Perret, R., *Solar Thermochemical Hydrogen Production Research (STCH) (No. SAND2011-3622)*, Sandia National Lab. (SNL-CA, Livermore, CA (United States, 2011.
22. Lewis, M. and Masin, J., *An Assessment of the Efficiency of the Hybrid Copper-Chloride Thermochemical Cycle*, Cincinnati, OH, United States, American Institute of Chemical Engineers, New York, United States, 2005.
23. Wang, Z.L., Naterer, G.F., Gabriel, K.S., Gravelsins, R., Daggupati, V.N., Comparison of sulfur–iodine and copper–chlorine thermochemical hydrogen production cycles. *Int. J. Hydrog. Energy*, 35, 10, 4820–4830, 2010.
24. Naterer, G.F., Dincer, I., Zamfirescu, C., *Hydrogen Production from Nuclear Energy*, Springer, London, 2013.
25. Ishaq, H. and Dincer, I., A comparative evaluation of three CuCl cycles for hydrogen production. *Int. J. Hydrog. Energy*, 44, 16, 7958–7968, 2019.
26. Ozbilen, A., Dincer, I., Rosen, M.A., Development of new heat exchanger network designs for a four-step Cu–Cl cycle for hydrogen production. *Energy*, 77, 338–351, 2014.
27. Ghandehariun, S., Rosen, M.A., Naterer, G.F., Wang, Z., Pinch analysis for recycling thermal energy in the Cu-Cl cycle. *Int. J. Hydrog. Energy*, 37, 21, 16535–16541, 2012.
28. Ghandehariun, S., Wang, Z., Naterer, G. F., Rosen, M. A. Experimental investigation of molten salt droplet quenching and solidification processes of heat recovery in thermochemical hydrogen production. *Applied Energy*, 157. 267–275, 2015.
29. Ouagued, M., Khellaf, A., Loukarfi, L., Performance analyses of Cu–Cl hydrogen production integrated solar parabolic trough collector system under Algerian climate. *Int. J. Hydrog. Energy*, 43, 6, 3451–3465, 2018.
30. Ghandehariun, S., Naterer, G.F., Dincer, I., Rosen, M.A., Solar thermochemical plant analysis for hydrogen production with the copper–chlorine cycle. *Int. J. Hydrog. Energy*, 35, 16, 8511–8520, 2010.
31. Ratlamwala, T.A.H. and Dincer, I., Performance assessment of solar-based integrated Cu–Cl systems for hydrogen production. *Sol. Energy*, 95, 345–356, 2013.
32. Balta, M.T., Dincer, I., Hepbasli, A., Performance assessment of solar-driven integrated Mg-Cl cycle for hydrogen production. *Int. J. Hydrog. Energy*, 39, 35, 20652–20661, 2014.
33. Ozcan, H. and Dincer, I., Energy and exergy analyses of a solar driven Mg–Cl hybrid thermochemical cycle for co-production of power and hydrogen. *Int. J. Hydrog. Energy*, 39, 28, 15330–15341, 2014.
34. Al-Zareer, M., Dincer, I., Rosen, M.A., Development and assessment of a new solar heliostat field based system using a thermochemical water decomposition cycle integrated with hydrogen compression. *Sol. Energy*, 151, 186–201, 2017.

35. Yilmaz, F. and Selbaş, R., Thermodynamic performance assessment of solar based Sulfur-Iodine thermochemical cycle for hydrogen generation. *Energy*, 140, 520–529, 2017.
36. Sayyaadi, H. and Boroujeni, M.S., Conceptual design, process integration, and optimization of a solar CuCl thermochemical hydrogen production plant. *Int. J. Hydrog. Energy*, 42, 5, 2771–2789, 2017.
37. Ishaq, H., Dincer, I., Naterer, G.F., Development and assessment of a solar, wind and hydrogen hybrid trigeneration system. *Int. J. Hydrog. Energy*, 43, 52, 23148–23160, 2018.
38. Siddiqui, O., Ishaq, H., Dincer, I., A novel solar and geothermal-based trigeneration system for electricity generation, hydrogen production and cooling. *Energy Convers. Manag.*, 198, 111812, 2019.
39. Ishaq, H. and Dincer, I., Design and performance evaluation of a new biomass and solar based combined system with thermochemical hydrogen production. *Energy Convers. Manag.*, 196, 395–409, 2019.
40. Ishaq, H. and Dincer, I., A comparative evaluation of OTEC, solar and wind energy based systems for clean hydrogen production. *J. Clean. Prod.*, 246, 118736, 2020.
41. Sadeghi, S. and Ghandehariun, S., Thermodynamic analysis and optimization of an integrated solar thermochemical hydrogen production system. *Int. J. Hydrog. Energy*, 45, 53, 28426–28436, 2020.
42. Sadeghi, S., Ghandehariun, S., Rezaie, B., Energy and exergy analyses of a solar-based multi-generation energy plant integrated with heat recovery and thermal energy storage systems. *Appl. Therm. Eng.*, 188, 116629, 2021.
43. Temiz, M. and Dincer, I., Enhancement of solar energy use by an integrated system for five useful outputs: System assessment. *Sustain. Energy Technol. Assess.*, 43, 100952, 2021.
44. Corumlu, V. and Ozturk, M., Thermodynamic assessment of a new solar power-based multigeneration energy plant with thermochemical cycle. *Int. J. Exergy*, 34, 1, 76–102, 2021.
45. Sadeghi, S. and Ghandehariun, S., A standalone solar thermochemical water splitting hydrogen plant with high-temperature molten salt: Thermodynamic and economic analyses and multi-objective optimization. *Energy*, 240, 122723, 2022.
46. Mehrpooya, M., Ghorbani, B., Khodaverdi, M., Hydrogen production by thermochemical water splitting cycle using low-grade solar heat and phase change material energy storage system. *Int. J. Energy Res.*, 46, 7590–7609, 2022.
47. Vignarooban, K., Xu, X., Arvay, A., Hsu, K., Kannan, A.M., Heat transfer fluids for concentrating solar power systems–a review. *Appl. Energy*, 146, 383–396, 2015.
48. Ushak, S., Fernández, A.G., Prieto, C., Grageda, M., Advances in molten salt storage systems using other liquid sensible storage media for heat storage,

in: *Advances in Thermal Energy Storage Systems*, pp. 55–81, Woodhead Publishing, Cambridge, UK, 2021.
49. Moran, M.J., Bailey, M.B., Boettner, D.D., Shapiro, H.N., *Fundamentals of Engineering Thermodynamics*, Wiley, John Wiley & Sons, Hoboken, NJ, US, 2018.
50. Boles, M. and Cengel, Y., *Thermodynamics: An Engineering Approach*, 9th ed, McGraw-Hill Education, New York, 2019.
51. National Institute of Standards and Technology (NIST) National Institute of Standards and Technology, US Department of Commerce. NIST Chemistry WebBook, SRD 69. http://webbook.nist.gov/chemistry/form-ser.html; 2021 [accessed 12.01.2022].
52. Xu, C., Wang, Z., Li, X., Sun, F., Energy and exergy analysis of solar power tower plants. *Appl. Therm. Eng.*, 31, 17-18, 3904–3913, 2011.
53. Li, X., Kong, W., Wang, Z., Chang, C., Bai, F., Thermal model and thermodynamic performance of molten salt cavity receiver. *Renew. Energy*, 35, 5, 981–988, 2010.
54. An, X.H., Cheng, J.H., Su, T., Zhang, P., Determination of thermal physical properties of alkali fluoride/carbonate eutectic molten salt, in: *AIP Conference Proceedings*, vol. 1850, AIP Publishing LLC, p. 070001, 2017.
55. Bejan, A., Tsatsaronis, G., Moran, M.J., *Thermal Design and Optimization*, John Wiley & Sons, New York, NY, US, 1995.
56. Orhan, M.F., Dincer, I., Naterer, G.F., Cost analysis of a thermochemical Cu–Cl pilot plant for nuclear-based hydrogen production. *Int. J. Hydrog. Energy*, 33, 21, 6006–6020, 2008.
57. Zhuang, X., Xu, X., Liu, W., Xu, W., LCOE analysis of tower concentrating solar power plants using different molten-salts for thermal energy storage in China. *Energies*, 12, 7, 1394, 2019.
58. Deb, K., Pratap, A., Agarwal, S., Meyarivan, T., A fast and elitist multiobjective genetic algorithm: NSGA-II. *IEEE Trans. Evol. Comput.*, 6, 2, 182–197, 2002.
59. Forsberg, C.W., Peterson, P.F., Zhao, H., High-temperature liquid-fluoride-salt closed-Brayton-cycle solar power towers. *J. Sol. Energy Eng.*, 129, 141–146, 2007.
60. Naterer, G.F., *Advanced Heat Transfer*, 3rd Edition, CRC Press, Boca Raton, FL, 2022.
61. El-Emam, R.S. and Özcan, H., Comprehensive review on the techno-economics of sustainable large-scale clean hydrogen production. *J. Clean. Prod.*, 220, 593–609, 2019.
62. Agyekum, E.B., Nutakor, C., Agwa, A.M., Kamel, S., A critical review of renewable hydrogen production methods: Factors affecting their scale-up and its role in future energy generation. *Membranes*, 12, 2, 173, 2022.

4

Diversifying Solar Fuels: A Comparative Study on Solar Thermochemical Hydrogen Production Versus Solar Thermochemical Energy Storage Using Co_3O_4

Atalay Calisan and Deniz Uner*

Middle East Technical University, Faculty of Engineering, Chemical Engineering Department, Universiteler Mahallesi Dumlupinar Bulvari Cankaya, Ankara, Turkey

Abstract

A comparative analysis is provided for hydrogen production from concentrated solar energy versus storage of the solar thermal energy in the form of reduced state of cobalt oxide, which can be oxidized, whenever and wherever needed. Based on the data published in the literature so far, as well as the data generated in this study, it was demonstrated that, unless the chemical looping oxide is not reduced to the metallic state, or to a state where oxygen is permanently bound to the metal, hydrogen production is limited to a few mmole/g oxide. The residual oxygen in the oxide and hydrogen interaction is inevitable, simply because of the thermodynamic constraints of water splitting reactions. On the other hand, storage and transport of thermal energy in the form of stoichiometric or nonstoichiometric oxides is highly efficient, as demonstrated through the CoO/Co_3O_4 cycle.

Keywords: Hydrogen, thermochemical process, oxygen exchange, solar energy storage

4.1 Introduction

Storage of solar energy through thermochemical processes are very promising alternatives for several reasons. First and foremost, it is theoretically

---

Corresponding author: uner@metu.edu.tr

Nurdan Demirci Sankir and Mehmet Sankir (eds.) Solar Fuels, (137–160) © 2023 Scrivener Publishing LLC possible to utilize full spectrum of the solar radiation through thermal energy conversion processes. Second, technologies are already available to produce molecules, such as methane, hydrogen and even long chain hydrocarbons using thermal energy; hence, thermal energy provided by solar energy can be relatively easily integrated to these processes. Furthermore, as a result of these processes, chemical storage of solar energy becomes possible [1].

Presently, the most popular, and arguably the easiest process for the storage of solar energy is the production of hydrogen through water splitting. The reaction can be triggered through electrolysis, using the electrical energy generated by the photovoltaic systems, or through solar thermal means, by chemical looping [2] and by reforming reactions [3]. Thermochemical water splitting technologies are based on the thermal generation of sites depleted of oxygen atoms, such as oxygen vacancies, oxygen nonstoichiometries or a new crystalline phase with a lower oxygen to metal ratio. These sites are subsequently filled by the oxygen from water, such that hydrogen is produced as a result. The chemical processes are outlined below:

$$\text{Thermal decomposition step at } T_{high}: MO \xrightarrow{\text{heat}} MO_{1-\delta} + \frac{\delta}{2}O_2 \quad \Delta H > 0 \tag{4.1}$$

$$\text{Oxidation step at } T_{low}: MO_{1-\delta} + \delta H_2O \xrightarrow{\text{heat}} MO + \delta H_2 \quad \Delta H < 0 \tag{4.2}$$

+_____

$$\text{Net reaction (water dissociation): } H_2O \rightarrow H_2 + 0.5O_2$$

Thermal decomposition reaction is a highly endothermic process that requires thermal energy at high temperatures to remove oxygen from metal oxide. Oxidation reaction is exothermic, carried out by reacting water vapor with the reduced metal oxide at the temperatures lower than thermal decomposition step, these temperatures are dictated by the thermochemistry of these reactions.

There are two major bottlenecks in these reactions: The first bottleneck is very high temperatures needed for thermal decomposition. The root cause of this high temperature is the high energy needed to break the metal-oxygen bond [4, 5]. A conservative review of the existing literature

Table 4.1 Water splitting and thermal decomposition data for some metal/metal oxide pairs.

Metal Oxide (MO_x)	Temperature (°C)	Reported H_2 production	Normalized H_2 production (mmol H_2/mol MO_x)	Ref.
$CoFe_2O_4/Al_2O_3$	1200/1000	0.777 mmol/g	182	[6]
CeO_2	1400/1100	144 µmol/g	25	[7]
$Ce_{0.75}Zr_{0.25}O_2$		432 µmol/g	69	
$NiFe_2O_4$	1100	428 µmol/g	100	[8]
$La_{0.9}Sr_{0.1}MnO_{3-y}$	1400/800	0.91 mL/g	9	[9]
$La_{0.6}Sr_{0.4}MnO_{3-y}$		8.9 mL/g	82	
20 wt.% Fe_3O_4 in 8% YSZ	1500/1200	3 mL/g	28.9	[10]
10 wt.% $CoFe_2O_4$ in 8% YSZ		1.5 mL/g	14.6	
$Ce_{0.75}Zr_{0.25}O_2$	1400/1200	468 µmol/g	74	[11]
CeO_2	1500–750	2.12 mL/g	15.2	[12]
$Ce_{0.95}Pr_{0.05}O_{2-y}$		2.1 mL/g	15	
$Ce_{0.9}Pr_{0.1}O_{2-y}$		2.3 mL/g	16.5	

(Continued)

Table 4.1 Water splitting and thermal decomposition data for some metal/metal oxide pairs. (*Continued*)

Metal Oxide (MO$_x$)	Temperature (°C)	Reported H$_2$ production	Normalized H$_2$ production (mmol H$_2$/mol MO$_x$)	Ref.
CeO$_2$	1500/1150	10.2 mL/g	73	[13]
5–15 mol% Fe-CeO$_2$		12 mL/g	84–86	
5–15 mol% Co-CeO$_2$		11–17.2 mL/g	78–123	
5–15 mol% Ni-CeO$_2$		8.5–15.4 mL/g	61–110	
5–15 mol% Mn-CeO$_2$		10–11 mL/g	72–79	
CeO$_2$	1300–1550/400–1000	9.3 mL/g	66.5	[14]
NiFe$_2$O$_4$		11.4 mL/g	111	
NiFe$_2$O$_4$/m-ZrO$_2$		11.3 mL/g	111	
NiFe$_2$O$_4$/c-YSZ		6.4 mL/g	62.4	
Ba$_{0.5}$Sr$_{0.5}$Co$_{0.8}$Fe$_{0.2}$O$_{3-\delta}$	1000/800	83 μmol/g	18	[15]
La$_{0.65}$Sr$_{0.35}$MnO$_{3-\delta}$	1400/1050	124 μmol/g	27	
La$_{0.6}$Sr$_{0.4}$Co$_{0.2}$Fe$_{0.8}$O$_{3-\delta}$	1200/800	162 μmol/g	36	
LaSrCoO$_4$	1300/800	164 μmol/g	57	

is given in Table 4.1, summarizing the process temperatures of thermal decomposition and oxidation and the amount of water that could be produces using a particular oxide.

The data reported in Table 4.1 for hydrogen amounts reveal that there is an upper limit in the amount of hydrogen that can be produced with thermochemical cycles [6–17]. The core question under investigation in this study is to answer if there is a technological limit for solar thermochemical hydrogen production using metal/metal oxide pairs. Inspired by the literature [2], we selected CoO/Co_3O_4 to investigate this system in terms of hydrogen production efficiencies, as well as the efficiency of this material as a thermal energy storage material.

4.2 Materials and Methods

Co_3O_4 powder was used as received from Ege Ferro (99.6% metal basis).

Pt/Co_3O_4 samples were synthesized by using incipient wetness impregnation method. First, deionized (DI) water was added dropwise to 100 mg of Co_3O_4 (Ege Ferro) until a paste was formed. The added amount was noted as V_x (milliliters DI water per 100 mg Co_3O_4). Depending on the percentage of platinum on Co_3O_4, the volume of platinum source (0.035 M aqueous platinum solution in this thesis) was calculated as V_y (milliliters of platinum solution per 100 mg Co_3O_4). This calculated volume was diluted with DI water until total volume V_x was reached. This final solution was added dropwise to 100 mg Co_3O_4 until a paste was formed. The paste was mixed with a glass rod for 30 minutes. The paste was left at room temperature overnight. The mixture was dried at 80°C for 5 hours. The sample was calcined at 300°C for 3 hours.

The samples were characterized by XRD. XRD patterns were measured on a Philips model PW1840 (1729) X-ray diffractometer using Ni-filtered Cu-Kα 945 radiation at a scan rate of 0.05 degrees/s. The redox properties of the materials were evaluated by thermal gravimetric analysis using Schimadzu DTG-60H with simultaneous DTA-TG. The reference material for DTA was $α-Al_2O_3$. Unless otherwise stated, the heating rate was 10°C/min.

The reductive characteristics of the materials were performed by using Micromeritics Chemisorp 2720. The reactive gases were 10% H_2 in Ar (Linde) for TPR, 2% O_2 in He (Linde) for TPO, and 99.999% He (Linde) for TPTD. The flow rates were set to 25 cm^3/min (sccm) by built-in rotameter for all reactions. The heating rate and desired final temperature was changed depending on the reducibility performance of the material.

The composition changes in the flowing gas detected by a built in thermal conductivity detector (TCD).

Two-step thermochemical reactions tests were performed by a home built reactor system, coupled to a home built gas sensing unit. The multi-gas analyzer equipment has four different sensors for CO (maximum range, 3% with 10 ppm resolution, Gascard NG/Edinburg Instruments), CO_2 (maximum range, 10% with 0.01% resolution, Gascard NG/Edinburg Instruments), O_2 (0–1% and 0–21% ranges Model 3290/Teledyne Analytical Instruments), and H_2 (0–5% range Model 2000 XTC/Teledyne Analytical Instruments). CO and CO_2 detectors have NDIR type sensors equipped with a tungsten lamp as the infrared source. The oxygen sensor is a micro-fuel cell that follows electrochemical reactions $O_2 + 2H_2O + 4e^- \rightarrow 4OH^-$ at the cathode side and $Pb + H_2O \rightarrow PbO + 2H^+ + 2e^-$ at the anode side. The hydrogen thermal conductivity transmitter (TCT) measures the concentration of a component in a binary or pseudo-binary mixture of gases by comparing the difference in thermal conductivity.

In the reactor system, metal oxides were supported and fixed in the middle of the quartz reactor (6 mm OD) by the help of quartz wools. The gas flow rate was adjusted by a Teledyne HFC-202 Mass Flow Controller (MFC), driven by a TERRALAB mass flow control station (MFCS). The reactant gases were 99.999% H_2 (Habas) and 99.999% dry air (Oksan) while the inert gases were 99.999% Ar (Linde) and 99.999% N_2 (Oksan). The heating rate and desired final temperature was changed depending on the tested material. Temperature of the oven is measured by K-type thermocouple.

4.3 Thermodynamics of Direct Decomposition of Water

Direct decomposition of water is the most straight forward methodology to produce hydrogen from water. The process requires extremely high temperatures due to the thermodynamic constraints as shown in Figure 4.1. The products are explosive, and the separation process is very complex when the operating conditions are considered. In addition, hydrogen and oxygen molecules also split into their atomic forms as the temperature increases to 2000°C; that is, quenching the products is nearly impossible. Therefore, the overall process becomes extremely difficult to operate. Bilgen reported 2% to 3% hydrogen (nearly 50% of equilibrium conversion) in the product stream during the operation around 2000°C to 2500°C [18]. The main

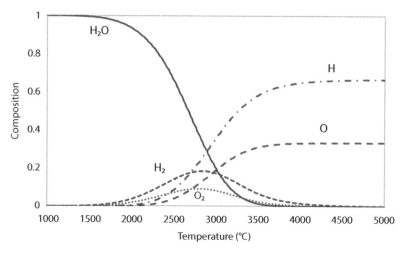

Figure 4.1 Equilibrium composition analysis of direct decomposition of water $\left(P_{H_2O} = 0.1\,\text{bar and } P_{system} = 1\,\text{bar}\right)$.

drawback of this process is the high temperature (>2000°C) needed for water decomposition. This temperature causes techno-economic difficulties such as separation of highly explosive mixture gas (H_2/O_2), radiative losses, material durability, high capital costs and scale up related problems. Therefore, hydrogen production at an industrial level by direct decomposition is impractical.

The thermodynamic estimations shown in Figure 4.1 were performed using a Gibbs free energy minimization method described in [19]. The data shown in Figure 4.1 clearly reveals that at atmospheric pressures, temperatures at the excess of 1500°C are needed for any appreciable level of molecular hydrogen production. Further increase in temperatures renders both O_2 and H_2 unstable, and they are dissociated to their atomic building blocks. Due to this instability, direct decomposition of water is not a viable alternative for hydrogen production.

4.4 A Critical Analysis of Two-Step Thermochemical Water Splitting Cycles Through the Red/Ox Properties of Co_3O_4

Thermochemical red/ox processes are based on the oxygen exchange properties of the red/ox materials. Oxygen exchange properties are related to the ability of the material to extract oxygen from water while releasing its hydrogen, followed by a facile release of oxygen from its structure. For the two-step thermochemical water splitting reactions, red/ox materials undergo a thermal decomposition reaction followed by oxidation by water as given in equations (4.1) and (4.2). The products of this reaction system, oxygen and hydrogen, are produced at different steps, hence products are separated as a consequence of the time dependent nature of the operation of this process.

The amount of hydrogen produced depends strongly on the level of oxygen removal from the material, i.e. reduction, during thermal decomposition step. The level of reduction at the thermal decomposition step is directly related to the thermochemical properties of the materials and the level of T_{high}. In other words, the net reaction (water dissociation) yield is controlled by the oxygen exchange properties of the red/ox materials and maximum achievable temperatures. On this front, the best red/ox material for two step thermochemical water splitting reaction can be defined as a red/ox material that can completely decompose, release all of its oxygen, and can be reoxidized by water to recover its initial oxidation state.

The rate of oxygen desorption and adsorption process is expected to be driven by kinetics, mass transfer, heat transfer and thermodynamics. At low temperatures, there is a steric hindrance of oxygen evolution: oxygen atoms, displaced from their lattice structures are not in close proximity to other oxygen atoms to recombine and desorb. To recuperate this problem, precious metals can be incorporated to provide active centers for these rouge oxygen atoms to be collected, recombined and desorb. As the temperature increases, the mobility and the availability of the oxygen atoms increase and the recombination and desorption can occur spontaneously, and overall rate is controlled by transport of oxygen atoms migrating on the surface. Further increase in temperature also enhances the diffusion process. At moderately high temperatures, mass transfer and heat transfer rates are similar. At even higher temperatures, thermodynamic barriers are reached, and further progress is not possible. Thermodynamic barriers can be both in terms of the chemical transformation and also in terms of a phase transition. The change in oxidation state of active sites, the structural reorganization during phase transformation create new defect sites for recombinative desorption of oxygen, and can alter the resistance of mass and heat transfer, respectively.

With this background, the limitations imposed on thermochemical hydrogen production is investigated around Co_3O_4. Co_3O_4 reduction to CoO is thermodynamically favorable, for complete transformation, temperatures of ~850°C are sufficient. However, further decomposition is thermodynamically impractical for the reasons we will discuss below.

4.4.1 Red/Ox Characteristics of Co_3O_4 Measured by Temperature-Programmed Analysis

Red/ox characteristics of Co_3O_4 was investigated by TPR, TPO, and TPtD analyses by using Micromeritics Chemisorp 2720. The temperature-programmed reduction (TPR) profile of Co_3O_4 conducted under 10% H_2-Ar is given in Figure 4.2. The reduction of Co_3O_4 took place in the temperature range of 250°C to 500°C. The total hydrogen consumption was calculated as 4 mol H_2/mol Co_3O_4. This amount is consistent with the overall reduction of the spinel oxide to metallic cobalt according to the reaction $Co_3O_4 + 4H_2 \rightarrow Co + 4 H_2O$.

A curve fitting was performed to de-convolute the shoulder at a lower temperature. Two peaks were identified in the TPR profile, indicating the existence of two events. The simulated peaks are symmetric in nature indicating a second-order process was prevailing [20]. Dissociative adsorption of hydrogen is inherently a second-order process. A symmetric Gaussian peak was used to represent the second-order behavior for peak fitting. The estimated areas as a result of fitting are summarized in Table 4.2. The results of the curve fitting compared against the two-step reduction reaction of Co_3O_4 presented in Table 4.2 are in full agreement, within the limits of the estimations, with the spinel decomposition and CoO reduction stoichiometries of the starting material.

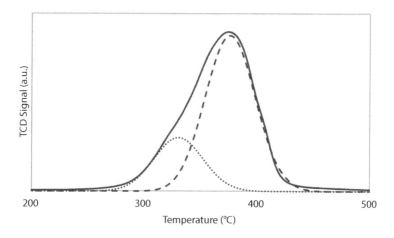

Figure 4.2 Temperature-programmed reduction profile of Co_3O_4 (solid line, dot, and dashed lines represents experimental data, reduction of $Co^{3+,4+}$ to Co^{2+} and Co^{2+} to Co^0, respectively).

Table 4.2 Estimated areas as a result of curve fitting the TPR data of Co_3O_4.

Reaction	T_{Median} (°C)	% of the total area
$Co_3O_4 + H_2 \rightarrow 3CoO + H_2O$	330	25.7
$3CoO + 3H_2 \rightarrow 3Co + 3H_2O$	375	74.3

Reduced samples were cooled to room temperature under 10% H_2-Ar flow. Temperature-programmed oxidation experiments were carried out under 2% O_2-He flow after purging the reactor with helium. Reoxidation patterns of reduced Co_3O_4 were illustrated in Figure 4.3. A broad reoxidation peak was observed in the temperature range of 250°C to 700°C. This broad peak is coupled to a sharp peak centered about 470°C. The areas under the curves were estimated and the fractional values are given in Table 4.3.

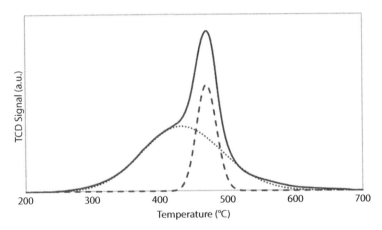

Figure 4.3 Temperature-programmed oxidation patterns of reduced Co_3O_4. Solid line represents the experimental data. Dot and dashed lines represent fitted curves for oxidation of Co^0 to Co^{2+} and Co^{2+} to $Co^{3+,4+}$, respectively.

Table 4.3 Estimated areas as a result of curve fitting of TPO data of reduced Co_4.

Reaction	T_{Median} (°C)	% of the total area
$3Co + 3/2\ O_2 \rightarrow 3CoO$	432°C	71.4
$3CoO + 1/2\ O_2 \rightarrow Co_3O_4$	470°C	28.6

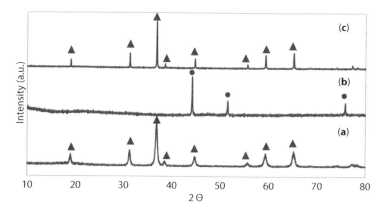

Figure 4.4 XRD patterns of Co_3O_4 (a) fresh sample; (b) after TPR; (c) after TPO following TPR (triangles label peaks due to Co_3O_4, circles represent peaks due to Co^0).

The reduction and oxidation processes were also monitored by XRD. The change in the crystal structure of Co_3O_4 during red/ox experiments were given in Figure 4.4. The fresh sample has the diffraction peaks of Co_3O_4 (JCPDS card No. 42-1467). After the TPR analysis, the sample was identified as Co (ICSD 44989) indicating complete reduction of Co_3O_4. The diffraction peaks obtained after TPO following TPR belongs to Co_3O_4 (JCPDS card No. 42-1467), demonstrating full reoxidation of sample. The broader peaks observed for fresh sample indicates the presence of smaller particles, which sinter and give rise to sharper peaks after the reduction/oxidation treatments.

Having evaluated the red/ox properties of Co_3O_4, temperature-programmed thermal decomposition experiment was performed (Figure 4.5). Oxygen evolution was detected after 700°C and gave a maximum around 870°C. Evolved oxygen was 25% of the total oxygen uptake during TPO after TPR corresponding to the reaction of $Co_3O_4 \rightarrow 3CoO + 1/2\ O_2$. The oxygen evolution peak is asymmetric indicating a first-order desorption process [20].

4.4.2 The Role of Pt as a Reduction Promoter of Co_3O_4

Having evaluated the temperature programmed reduction, oxidation and thermal decomposition profiles of Co_3O_4, the effect of a reduction promoter precious metal was also investigated. For this purpose, 0.5 wt.% Pt was impregnated to Co_3O_4, the effects were monitored by temperature programmed experiments. As expected, the reducibility of Co_3O_4 was

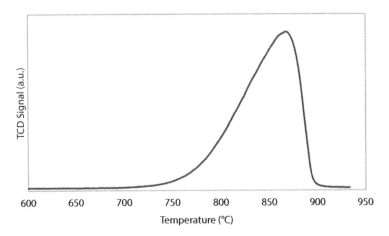

Figure 4.5 Oxygen evolution during thermal decomposition (TPtD) of Co_3O_4.

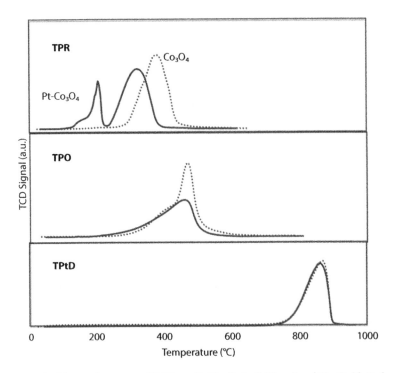

Figure 4.6 Red/ox performance of 0.5% wt. Pt/Co_3O_4 (solid lines) and Co_3O_4 (dotted lines).

THERMOCHEMICAL H2 VS SOLAR THERMOCHEMICAL HEAT   149

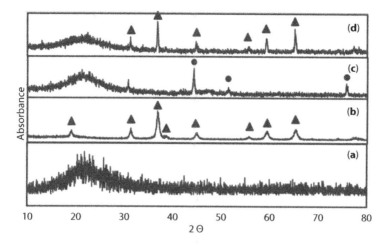

Figure 4.7 XRD patterns of 0.5% wt. Pt/Co_3O_4 (a) empty sample holder; (b) fresh sample; (c) after TPR; (d) after TPO following TPR (Triangles label peaks due to Co_3O_4, circles represent peaks due to Co^0).

enhanced in the presence of platinum, the data were reported in comparison to that of Co_3O_4 in Figure 4.6. In the presence of Pt, the reduction of Co^{3+} to Co^{2+} was completed around 240°C. Further reduction to Co^0 was completed at 420°C. Hydrogen consumption was calculated as 4 mol H_2/mol Co_3O_4 which is in good agreement with the stoichiometry. Reduction of platinum was not observed due to the low amounts.

After the reduction of 0.5% wt. Pt/Co_3O_4, residue consists of Co^0 as shown in Figure 4.7. These metals were reoxidized under 2% O_2-He flow. The oxygen uptake profile of reduced sample (Figure 4.6) becomes broader when compared to reduced form of Co_3O_4. The oxidation starts around 120°C which 100°C lower than Co_3O_4. After the completion of oxidation, the crystallography changed to its initial state, which is Co_3O_4 (Figure 4.7). Due to the limited amount of the samples obtained after reduction and oxidation, some of the characteristic peaks of Co_3O_4 were under the background noise. Finally, the oxygen desorption profile resembles that of unpromoted Co_3O_4 during the thermal decomposition.

4.4.3 A Critical Analysis of the Solar Thermochemical Cycles of Water Splitting

With the background provided so far, two step thermochemical reaction was tested in a home build reactor system. 100mg fresh Co_3O_4 was reduced under 100ccpm 10% H_2-Ar flow up to 900°C. Thereafter, reactor was swept with pure argon and cooled to 600°C. Reactant flow was changed to argon+water vapor mixture prepared by passing argon through a wash bottle containing deionized water. During the water splitting, effluent gasses were analyzed by H_2 sensitive TCD detector. Hydrogen evolution rate during isothermal water splitting experiment was given in Figure 4.8. This process was irreversible, in other words, removal of oxygen was not possible by thermal decomposition at 900 °C. These temperatures, as can be inferred from the data provided so far, are not sufficient to decompose CoO to metallic cobalt. In the presence of CoO, even though hydrogen is produced by forming spinel oxide, it is immediately consumed locally to reduce CoO to Co hence resulting in a net hydrogen production to a very low level.

It is important for redox material to completely decompose and release all of its reducible oxygen, since the residual oxygen will later consume the produced hydrogen as illustrated in Figure 4.9. Thus, it is of crucial importance to achieve required high decomposition temperatures. Although efficient solar collector designs are available, achieving high reduction temperatures required for the redox material to fully decompose and release its oxygen in the reactor are challenging and in most cases unattainable. In our previous studies, we demonstrated the application of 70cm OD parabolic dishes to solar thermal hydrogen production by thermochemical [5] and steam reforming processes [3]. The reactor temperatures easily reached 800°C and this temperature was sustainable. However, for most of the materials, including cobalt based oxide used in this study, these temperatures are not sufficient to remove all of the oxygen. Hence, hydrogen

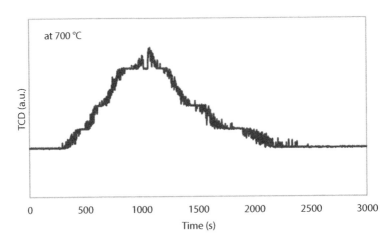

Figure 4.8 Hydrogen evolution during water splitting over reduced cobalt oxide.

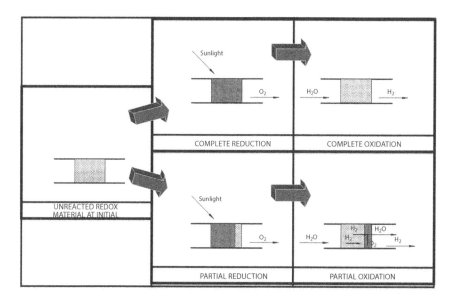

Figure 4.9 Representation of two cases as complete and partial reductions.

produced over limited number of oxygen vacancies are used by the surviving lattice oxygens and the net reaction has a very limited hydrogen production efficiency. The efficiency of produced hydrogen is constrained by the thermodynamics (Figure 4.1) and the reported variations (e.g. those in Table 4.1) are around thermodynamic values within the limits dictated by experimental conditions and experimental errors.

4.5 Cyclic Thermal Energy Storage Using Co_3O_4

Storage of thermal energy in chemical bonds is an important challenge. Solar thermochemical water splitting applications suffer from low hydrogen yield (the ratio of formed hydrogen to maximum hydrogen that can be produced from water H_2 (*formed*)/H_2 (*max*) = $4 \cdot 10^{-7}$% at 900K) during oxidation reaction [17, 21]. If not in the form of hydrogen, it is possible to use solar thermal energy to decompose Co_3O_4 to CoO at the readily available temperatures of 850 °C we have reported earlier [3]. But before reporting the cyclic nature of the decomposition, the stability of the compound, the mass and heat transfer effects during the redox processes were investigated. In the next section, the results of these studies are reported.

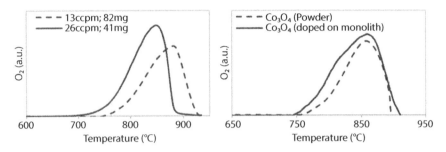

Figure 4.10 The effect of film mass transfer (top) and pore diffusion (bottom) on oxygen desorption of Co_3O_4.

4.5.1 Mass and Heat Transfer Effects During Red/Ox Processes

The results of the experiments designed to test limits of the film and pore diffusion resistances are given in Figure 4.10. The film mass transfer resistance was investigated by changing the linear velocity of the flowing gas while keeping the space time constant. This was achieved by increasing the solid material and gas flow rate by the same amounts. It was observed that at higher linear velocities, the amount of oxygen desorption was higher and started earlier.

The pore diffusion effect was studied by observing the difference between oxygen desorption rates of a bulk powder and the sample coated on a cordierite monolith. The sample coated on monolith exhibited a larger amount of oxygen desorption, when normalized with respect to the amount of Co_3O_4, since coating decreases the desorption pathway of one oxygen molecule. In addition, structured construction of monolith offers a distributed gas flow through channels and reduces pressure drop as well as overcoming film mass transfer resistances.

The thermal decomposition of Co_3O_4 is an endothermic reaction. Reaction pathway follows the $2Co_3O_4 \rightarrow 6CoO + O_2$ requires 403.5 kJ/mol O_2 at 800°C. To test the presence and the limits of heat transfer resistances, Co_3O_4 powder was mixed with SiC, a known heat transfer medium for laboratory scale catalyst tests. The limits of the external heat transfer resistances were monitored by changing the heating rate. The results revealing internal and external heat transfer related resistances were given in Figure 4.11. The external heat transfer is more dominant during the oxygen desorption step.

4.5.2 Cyclic Thermal Energy Storage Performance of Co_3O_4

Having assessed the red/ox characteristics and transport limitations behind the oxygen transport, thermal energy storage potential of Co_3O_4

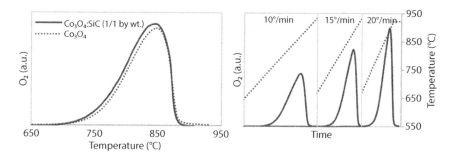

Figure 4.11 The effect of internal (left) and external (right) heat transfer on oxygen desorption rate of Co_3O_4.

was investigated by thermal gravimetric analysis. Co_3O_4 was heated up to 900°C under argon flow with a heating rate of 10°C/min for thermal decomposition. After the thermal reduction, residuals were oxidized under dry air flow during cooling, temperatures till 650°C were sufficient for full oxidation. Thermal decomposition followed by reoxidation is defined as one cycle. The cyclic performance of Co_3O_4 was given in Figure 4.12. Weight change of Co_3O_4 and temperature were indicated by solid line and dashed line, respectively.

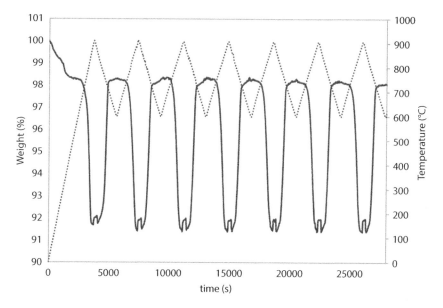

Figure 4.12 Cyclic oxygen transfer of Co_3O_4 based on weight change during heating under Ar flow up to 900°C and cooling under dry air flow to 650°C.

$$Co_3O_4 \rightarrow 3CoO + 1/2O_2 \quad \Delta H^o_{298K} = 196.2 kJ/mol$$

$$3CoO + 1/2O_2 \rightarrow Co_3O_4 \quad \Delta H^o_{298K} = -196.2 kJ/mol$$

The thermal decomposition of Co_3O_4 starts right after 750°C and reaches its maximum in the temperature range of 850 to 870°C. After the removal of the moisture (~2% by weight), the amount of oxygen exchange under alternating atmospheres of argon and dry air was found to be highly stable. A similar experiment was also conducted for 0.5 wt. %Pt/Co_3O_4 as shown in Figure 4.13. Similar thermal storage stability was observed in the presence of platinum also.

Oxygen exchange performances of Co_3O_4 and 0.5% wt. Pt/Co_3O_4 were compared by utilizing the TGA data. The oxygen desorption and adsorption data was taken from the fourth cycle presented in Figure 4.12 and Figure 4.13 for Co_3O_4 and 0.5% wt. Pt/Co_3O_4 respectively. The profiles coincided by adding offset to weight change data. Red/ox cyclic thermogram was given in Figure 4.14. Oxygen uptake and desorption profile of Pt/Co_3O_4 was broader than Co_3O_4. Oxygen desorption from Co_3O_4 shifts to higher temperature in the presence of platinum. Therefore, CoO produced

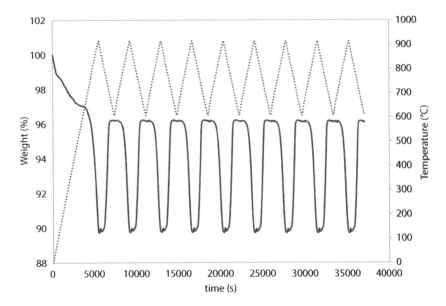

Figure 4.13 Cyclic oxygen exchange of 0.5 wt.% Pt- Co_3O_4 based on weight change during heating under Ar flow to 900°C and cooling under dry air flow to 650°C.

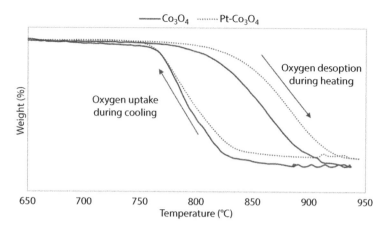

Figure 4.14 Comparison of oxygen uptake and oxygen desorption profiles of Co_3O_4 and 0.5% wt. Pt/Co_3O_4.

by solar thermochemical processes, can be used very effectively as a heat storage material [22].

Heat storage materials are important components for storage and transportation of renewable energy. In this case, using the concentrated solar energy, Co_3O_4 is reduced to CoO and stored thermal energy. When the thermal energy need arises, it is possible to release ~800 kJ/kg Co_3O_4 by exposing the material to air. This type of energy storage, using a solid compound enabling a cyclic operation can offer creative alternatives to heating by combustion, both for domestic purposes and industrial uses such as greenhouses.

The chemical potential (free energy) difference between the oxygen atom in the solid and molecular oxygen in the gas phase defines the thermodynamic barrier over the process. The thermodynamic barrier imposes a penalty over the ease with which the oxygen is regenerated: the difficulty in the oxygen release from the oxide. In other words, the exotherm in one direction has to be compensated by an endotherm of equal magnitude in the opposite direction. Therefore, a difficult reduction step is compensated by easier oxidation steps. This phenomenon is demonstrated in Figure 4.15 in the form of Ellingham diagrams. The driving force towards to oxidation of platinum is higher than cobalt. When Co and Pt form an alloy, which is the case after the reduction of $0.5\%Pt/Co_3O_4$, the transport of oxygen follows the pattern $Pt \rightarrow Co^0 \rightarrow Co^{2+} \rightarrow Co^{3+} \rightarrow Co^{4+}$ dictated by the electron energy levels in these atoms. This fact implies that platinum oxide is not likely to form unless all cobalt atoms oxidized to Co_3O_4. In other words, the presence of platinum assists the oxygen transfer from

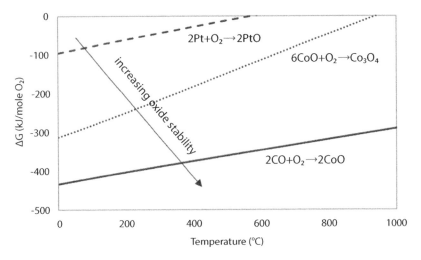

Figure 4.15 Driving force towards oxygen in the presence of cobalt and platinum.

reaction atmosphere to cobalt. This has a certain drawback during oxygen desorption since platinum behaves as another step for oxygen release from bulk structure.

Despite the issues of direct solar thermochemical splitting, solar energy integration to the methane steam reforming or dry reforming can alter

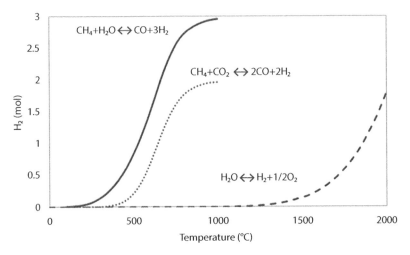

Figure 4.16 Equilibrium hydrogen production for direct water splitting, steam methane reforming and dry reforming of methane.

the reaction thermodynamics and increases the favorability of the process. Hydrogen production by steam reforming, regardless of the catalyst, has favourable equilibrium conversions at much lower temperatures, as shown in Figure 4.16. As such, integration of solar thermal energy for reforming reactions of methane or biomass followed by a syngas route to methanol or Fischer Tropsch synthesis is a very favourable solar energy integration to a sustainable fuel production route.

4.6 Conclusions

The solar thermochemical hydrogen production is inhibited by very high temperatures demanded by oxygen removal from the oxide. The residual oxygen in the structure, if reducible, further hinders the process by consuming the hydrogen produced. Steam reforming, as an alternative, provides a more favorable route, supported by both kinetics and thermodynamics. However, CoO produced by solar thermochemical processes, can be used very effectively as a heat storage material. The regeneration of the material was possible in air, the material could survive several cycles. Integration of this heat in the absence of solar irradiation toward chemical conversions is the next challenge of seamless hydrogen production by solar energy.

Acknowledgements

Partial financial support was provided by The Scientific and Technological Research Council of Turkey (TUBITAK) through 1003 program (project code: 213M006). Project coordinator, Dr. Serkan Kincal, and team member, Ezgi Yavuzyilmaz, are kindly acknowledged.

References

1. Uner, D., Storage of chemical energy and nuclear materials, in: *Energy Storage Systems, Encyclopaedia of Life Support Systems*, Y. Gogus (Ed.), pp. 119–139, UNESCO, Oxford, 2009.
2. Chueh, W.C., Falter, C. *et al.*, High-flux solar-driven thermochemical dissociation of CO_2 and H_2O using nonstoichiometric ceria. *Science*, 330, 1797, 2010.

3. Calisan, A., Ogulgonen, C.G. et al., Steam methane reforming over structured reactors under concentrated solar irradiation. *Int. J. Hydrog. Energy*, 44, 18682, 2019.
4. Gokon, N., Hara, K. et al., Thermochemical two-step water splitting cycle using perovskite oxides based on LaSrMnO3 redox system for solar H2 production. *Thermochim. Acta*, 680, 178374, 2019.
5. Calisan, A., Ogulgonen, C.G. et al., Finding the optimum between volatility and cycle temperatures in solar thermochemical hydrogen production: Pb/PbO pair. *Int. J. Hydrog. Energy*, 44, 18671, 2019.
6. Scheffe, J.R., Li, J., Weimer, A.W., A spinel ferrite/hercynite water-splitting redox cycle. *Int. J. Hydrog. Energy*, 35, 3333, 2010.
7. Le Gal, A., Abanades, S., Flamant, G., CO2 and H2O splitting for thermochemical production of solar fuels using nonstoichiometric ceria and ceria/zirconia solid solutions. *Energy Fuels*, 25, 4836, 2011.
8. Kostoglou, M. and Lorentzou, S., Improved kinetic model for water splitting thermochemical cycles using nickel ferrite. *Int. J. Hydrog. Energy*, 39, 6317, 2014.
9. Yang, C.K., Yamazaki, Y. et al., Thermodynamic and kinetic assessments of strontium-doped lanthanum manganite perovskites for two-step thermochemical water splitting. *J. Mater. Chem. A*, 2, 13612, 2014.
10. Singh, A.K., AuYeung, N.J. et al., Thermal reduction of iron oxide under reduced pressure and implications on thermal conversion efficiency for solar thermochemical fuel production. *Ind. Eng. Chem. Res.*, 54, 6793, 2015.
11. Le Gal, A. and Abanades, S., Catalytic investigation of ceria-zirconia solid solutions for solar hydrogen production. *Int. J. Hydrog. Energy*, 36, 4739, 2011.
12. Meng, Q.L., Lee, C. et al., Solar thermochemical process for hydrogen production via two-step water splitting cycle based on Ce1–xPrxO2–δ redox reaction. *Thermochim. Acta*, 532, 134, 2012.
13. Gokon, N., Suda, T., Kodama, T., Thermochemical reactivity of 5–15mol% Fe, Co, Ni, Mn-doped cerium oxides in two-step water-splitting cycle for solar hydrogen production. *Thermochim. Acta*, 617, 179, 2015.
14. Gokon, N., Sagawa, S., Kodama, T., Comparative study of activity of cerium oxide at thermal reduction temperatures of 1300–1550°C for solar thermochemical two-step water-splitting cycle. *Int. J. Hydrog. Energy*, 38, 14402, 2013.
15. Demont, A., Abanades, S., Beche, E., Investigation of perovskite structures as oxygen-exchange redox materials for hydrogen production from thermochemical two-step water-splitting cycles. *J. Phys. Chem. C*, 118, 12682, 2014.
16. Demont, A., Abanades, S., Beche, E., Investigation of perovskite structures as oxygen-exchange redox materials for hydrogen production from thermochemical two-step water-splitting cycles. *J. Phys. Chem. C*, 118, 12682, 2014.
17. Kodama, T., Kondoh, Y. et al., Thermochemical hydrogen production by a redox system of ZrO2-supported Co (II)-ferrite. *Sol. Energy*, 78, 623, 2005.

18. Bilgen, E., Ducarroir, M. et al., Use of solar energy for direct and two-step water decomposition cycles. *Int. J. Hydrog. Energy*, 2, 251, 1997.
19. Calisan, A., *Syngas Production Over Reducible Metal Oxides*, MS Thesis, Middle East Technical University, Ankara Turkey, pp. 20–22, 2013, https://avesis.metu.edu.tr/yonetilen-tez/e592b248-9507-4287-86fb-56c6eb1d5513/syngas-production-over-reducible-metal-oxides.
20. Masel, R.I., *Principles of Adsorption and Reaction on Solid Surfaces*, Wiley, New York, 1996.
21. Lundberg, M., Model calculations on some feasible two-step water splitting processes. *Int. J. Hydrog. Energy*, 18, 369, 1993.
22. Nekokar, N., Pourabdoli, M. et al., Effect of mechanical activation on thermal energy storage properties of Co_3O_4/CoO system. *Adv. Powder Technol.*, 29, 333, 2018.

Part II
ARTIFICIAL PHOTOSYNTHESIS AND SOLAR BIOFUEL PRODUCTION

5

Shedding Light on the Production of Biohydrogen from Algae

Thummala Chandrasekhar[1]* and Vankara Anuprasanna[2]

[1]Department of Environmental Science, Yogi Vemana University, Kadapa, A.P., India
[2]Department of Animal Science, Yogi Vemana University, Kadapa, A.P., India

Abstract

It is a well-known fact that fossil fuels cannot be regenerated or renewed in a short duration and therefore research on renewable energy resources, such as solar energy, wind energy, geothermal energy, bioenergy, etc., is a hot topic in recent times. Production of biomass-based hydrogen or biohydrogen (H_2) is considered as a best alternative source of energy that offer large potential benefits. Moreover, biohydrogen is a cleanest fuel with energy efficiency, and its production does not lead to any pollution. Production of hydrogen from photosynthetic algae is well established recently, but the output has to reach all the income groups of the society which is the ultimate goal in energy research. Algae are the potential candidate species for the production of biofuels due to their simple life cycle with efficient photosynthetic capacity. In certain places, these algal species are neglected and remained as waste. In this regard, mechanisms of hydrogen production from algae were emphasized in this chapter. In addition, the factors that influence the hydrogen production were also discussed. Overall, this chapter intends to summarize the developments in hydrogen production from certain algal species, which is helpful for commercial practice in the near future.

Keywords: Algae, biohydrogen, photosynthesis, photolysis, fermentation, enhancement

---

Corresponding author: tcsbiotech@gmail.com

5.1 Introduction

Energy is an essential component for both living and nonliving things to perform various activities. It is a well-known fact that the energy sources can be classified into non-renewable (conventional) and renewable (nonconventional) types based on their availability [1]. But till date, main portion of the global consumption of fuel is met by conventional fossil fuels such as coal, natural gas and petroleum products [2]. As per the information of world energy outlook (2015), fossil fuels are being used predominantly for energy requirement in different sectors providing more than 80% of global energy between 2013 and 2035 [3]. In addition, these fossil fuels are getting depleted due to enhancing human population, transportation and industrialization. Therefore, the gap between demand and supply of fossil fuel is increasing regularly, thus rising price in the recent past [4]. This crucial condition is a serious threat to modern human existence and economy, which urges the human community to focus more on alternate renewable fuels. Hence, the production of renewable energy, such as solar energy, wind energy, bioenergy, etc., from various sources is a trend now worldwide [5].

Bioenergy is a considerable form of renewable source, which was initially produced from plant species specifically using crop plants. Generally, plant materials, such as leaves, stem parts including wood, seeds, and roots, are useful for biofuel generation [6]. In certain cases, whole plant and its waste is also useful for fuel production. Later, research was continued using microorganisms and animals, including their waste and dead organisms for biofuel production. In general, four classes of biofuels exist which were developed gradually based on the demand [7]. The first-generation biofuels are produced from edible crops such as wheat, corn, maize, sugarcane, cassava, etc. But using crop plants for fuel production lead to food insecurity, which is a serious concern than energy demand. In contrast, second-generation biofuels are generated from nonedible crops and their waste. The second-generation biofuel production has set backs, such as land insufficiency, soil degradation, loss of vegetation, climate change, and global warming. The third-generation biofuels are generated from biomass of algae or diatoms through advanced technology. Fourth generation is an extension work of third-generation biofuels, i.e., production of genetically modified algae or microorganism for higher biomass in turn for improvement in biofuel production [8]. At present, the production of advanced fourth-generation fuels is in progress using nonarable land along with electrofuels and photobiological solar fuels. Advanced biofuels have advantages over first and second-generation biofuels as mentioned above which can generate in limited space [9]. In addition, advanced biofuels could capture solar energy 10 times more efficiently to produce sufficient fuel. In this technology, algae and microorganisms are cultured in bioreactors by providing necessary facilities. As they grow well in the reactors, it is easy to collect fuel from the bioreactor, and this process is not laborious when compared to first- and second-generation fuel production methods [10].

Biomass-based energy generally called as biofuels, which include bioethnol, biobutanol, biodiesel, biohydrogen, etc. [11]. The chemical composition of biomass differs among different living organisms, particularly in plant and algal species. Normally, biomass consists majorly carbohydrates, protein, and lipids, which is generally useful for production of bioethanol, biodiesel, and biohydrogen. These biofuels are also applicable for lighting, cooking, and generation of electricity. In certain cases, after biofuel production remaining waste can also be used as manure. The United States of America (USA) has announced that it would target the substitution of 20% of fossil fuels with biofuel by this year [12]. Thus, biofuels also contributes global energy supply in a gradual manner. Before that, understanding the Physico-chemical properties of biomass and fuel production mechanism from all the living organisms including algae, bacteria, diatoms, etc., is a prerequisite. Apart from advanced plants, primitive plants such as algae are the beneficial for the generation of biofuels specifically for biohydrogen [13]. In certain regions, these algal species are neglected and remained as orphan species. Moreover, using of these algal species for fuel production is considered as waste management practice [14]. In this chapter, we highlighted the production of biohydrogen from algal species using different methods. Also, certain algal species used for biohydrogen production were emphasized. This knowledge may be helpful for enhancing the production of biohydrogen and will improvise economy, life sustainibility, and environmental condition.

5.2 Hydrogen or Biohydrogen as Source of Energy

H_2 is considered as a secondary source of energy, which offers large potential benefits in terms of alternate energy sources. It is a carbon-free fuel, which burned with oxygen and is a prominent future source of energy. Although H_2 is one of the most copious elements on this planet in the form of water but its pure form exists at extremely low levels (< 1 ppm). Along with water, hydrocarbons and other organic matter are the key sources of

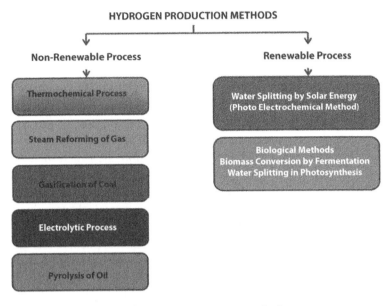

Figure 5.1 Production of pure hydrogen using various methods.

pure hydrogen [15]. Due to available in other forms, it is quite difficult to generate pure form of hydrogen. Hence, production of hydrogen from other forms is a challenging issue worldwide in recent times. At present, hydrogen is generating using several methods and can be stored easily for future usage. The most common methods today are natural gas reforming and electrolysis [16].

Hydrogen is one of the best alternative energy sources and its production specifically in biological method does not lead to any environmental pollution. Particularly, biological hydrogen is considered to be the cleanest with energy efficiency of approximately 122 kJ/g [11]. Further, it is easily changed to electricity in fuel cells and liberates a huge amount of energy per unit mass without producing any pollutants. Several reports exist on H_2 production methods, their resources and utilization (Figure 5.1). Till today, more emphasis was laid on nonbiological methods of hydrogen production [17]. But still, the output is not adequate for commercial practice. Moreover, biological hydrogen production has advantages such as low production cost and less release of greenhouse gases when compared to conventional methods [18]. Hence, it is the time to look into hydrogen production through different biological organisms including algae and their waste, which is undoubtedly an environmental friendly method. Biohydrogen production was generally done with either photosynthetic or fermentative organisms [19]. In extent, both non-governmental and governmental organizations are ready to invest on renewable fuels in number of countries. According to latest report, bioenergy including biohydrogen is contributing a considerable amount of energy to the overall energy usage in certain countries.

5.3 Hydrogen Production From Various Resources

Hydrogen majorly locked with various forms and its production is carried out using both renewable and non-renewable processes (Figure 5.1). There are various methods of H_2 production, such as thermo chemical process, reforming of natural gas/methane, gasification of coal, electrolytic process, pyrolysis of oil, and renewable processes, such as water splitting by solar energy and biological methods [10, 20, 21]. All these methods of hydrogen production use various resources and their waste [11, 19]. Most of the pure hydrogen (96%) has been generated from non-renewable fossil fuels, such as coal, natural gas, and oil and water electrolysis, contribute the remaining 4% [22, 23]. Although several scientific teams are focusing seriously on hydrogen generation chemically but productivity is not satisfactory [24]. Advanced fuels generated through recent technology using nonarable land include electrofuels and photobiological solar fuels [7, 8].

Hydrogen fuel generated through biological organisms is called biological hydrogen or biohydrogen. But still production of hydrogen carried out with chemical methods and output is not satisfactory [25]. In addition, biological hydrogen production is the easiest method due to raw material is available easily from the nature and cost-effective [26]. Most of the times, biological researchers considered the hydrogen metabolism primarily as a domain of bacteria, plants including microflora and produced regularly with photosynthetic and fermentative process. Improvement of biomass of algae, and microbial organisms, in turn enhancement of hydrogen production through recent biotechnological methods is being practiced now [14].

Biohydrogen production was generally done with either photosynthetic and/or fermentative organisms [27]. In detail, production through microbial organisms such as photosynthetic bacteria or algal species is the regular practice. Other method includes fermentation which utilizes biomass or biowaste as substrate. In addition, generation of hydrogen from waste water and industrial waste with the help of microbial organisms was also reported [28, 29]. There is a separate category of bacteria involved in fermentation process for hydrogen generation. Augmentation in H_2 generation by

Enterobacter cloacae IIT-BT 08 was achieved by Kumar and Das [30]. Koku *et al.* [31] observed the metabolism of H_2 generation in purple bacterium *Rhodobacter sphaeroides*. Hydrogen generation by *Clostridium thermocellum* 27405 using cellulosic substrates was observed by Levin *et al.* [32]. Both ethanol and hydrogen production by two thermophilic anaerobic bacteria located in Icelandic geothermal areas was reported [33]. Kobayashi *et al.* [34] proved the improvement of hydrogen production from acetate by isolated *Rhodobacter sphaeroides*. Similarly, Androga *et al.* [35] standardized the optimal light and temperature conditions for improvement of H_2 generation using *Rhodobacter capsulatus* DSM 1710 strain. Generation of hydrogen through *Clostridium* was reported by Avci *et al.* [36] and Tian *et al.* [37] using beet molasses and sugarcane bagasse. Production of H_2 from waste water by *Klebsiella oxytoca* ATCC 13182 and also under salt stress conditions through *Rhodopseudomonas palustris* were reported [38, 39]. Hydrogen generation through fermentation of cotton stalk hydrolysate by *Klebsiella* WL1316 was noticed by Li *et al.* [40]. Apart from the above mentioned examples, several bacteria, such as *Ruminococcus albus* [24], *Cladicellulosiruptor saccharolyticus* [41], *Klebsiella oxytoca* GS-4-08 [42], *Klebsiella aerogenes* or *Enterobacter aerogenes* [43], *Escherichia coli* [44], *Clostridium thermocellum* [45] and *Rhodobacter sphaeroides* [46] were widely studied using diverse species feeds, starchy and cellulosic waste, raw substrates, chemicals, etc. [4, 14]. In contrast, with certain bacteria the output is insufficient and not cost-effective. Biofuels generally involve contemporary carbon fixation, such as those that occur in microalgae or certain bacteria through the process of photosynthesis. Moreover, algae have the ability to bypass the CO_2 fixation for production of hydrogen.

5.4 Mechanism of Biological Hydrogen Production from Algae

Hydrogen production using biological samples or biowaste is one of the best, clean, and suitable methods. In extent, generation of biological hydrogen from algae is less expensive and energy intensive when compared to other species [11]. Although biohydrogen is an example of an advanced fuel, long back Gaffron and Rubin [47] observed the property of H_2 production from algae in dark anaerobic condition. Moreover, the generation of hydrogen from algae is one of the emerging trends in renewable energy area of research due to their photosynthetic capacity and short life span [48, 49]. Algae are simple photosynthetic flora categorized as micro and macro groups based on size which can be either unicellular or multicellular organisms. Based on habitat, algae are classified into terrestrial and aquatic species [50]. Among aquatic species, both marine and fresh water algal species are available in bulk number. Certain algae are also found in the tissues of other plants or on the rocks in the moist places [13]. Both sexual and asexual modes of reproduction take place in algal cells and later one occurs mostly by spore formation. Also, algae have the capacity to grow in nonpotable water and are rich macromolecule sources which may be used as food, fodder, medicine, nutraceutical, etc. Hence, they are often used as high-yield feedstock in bioenergy sector. In addition, most of the basic works related to hydrogen production have been done in model alga *Chlamydomonas reinhardtii* (*C. reinhardtii*). Apart from model alga different species, strains and mutants of both fresh and marine water algae such as *Scenedesmus, Chlorella, Chlorococcum, Nannochloropsis, Dunaliella,* and *Tetraselmis* were studied to know the hydrogen production levels [2, 11, 14, 19, 47, 50, 51].

Production of hydrogen from algae was normally done with photosynthesis and/or fermentation processes [4]. In photosynthesis, hydrogen generation majorly requires protons, electrons and O_2-sensitive hydrogenase. Photolysis of water-based hydrogen production in photosynthesis of algae is one of the best options [11, 14]. In addition, hydrogen evolution happens in the presence of water, hydrocarbons and organic matter in the process of fermentation. Several basic studies were conducted to know the mechanism of hydrogen production from algae using various physicochemical factors [17]. Almost two decades ago, Melis *et al.* [2] generated hydrogen for several days by removing sulfur in algal medium. This observation boosted research on hydrogen production from algal cells. After this innovation, several key steps and limitations for hydrogen generation were identified using different algal species. Later, different algal cells were subjected to various *in vitro* conditions to augment the hydrogen production. For instance, in model algae, such as *C. reinhardtii*, several mutants were created for enhancement of the output. In photosynthesis, solar energy is converted into chemical energy in the presence of water and carbon dioxide through a series of electron transfer mechanism. In extent, oxygenic and anoxygenic photosynthesis take place in plants including algae, green sulfur and purple nonsulfur bacteria [10, 52]. The enhancement of hydrogen generation from algal source by including copper in the algal medium was achieved by Surzycki *et al.* [27]. Facilitating anoxic stress and fermentation are the most crucial factors for the generation of hydrogen in algae.

In the process of photosynthesis in algae, photolysis of water at photosystem II (PS-II) can be considered as one of the chief sources of protons

(H⁺) and electrons (Figure 5.2). Various enzymes are involved in both PS I and II, including splitting of water at PSII and oxygen tolerance process [11]. In addition to photolysis of water, fermentative catabolism is also a source of electrons. Thus, obtained electrons reach ferredoxin (Fd) from PSI through electron transport system. Later, these electrons moved to ferredoxin-NADP⁺ reductase (FNR) from ferredoxin to produce NADPH (Figure 5.2). If the CO_2 fixation (Calvin-Benson-Bassham or CBB cycle) is not necessary in the cell in special conditions, such as unavailability or abundance of CO_2, then hydrogen generation will be activated [4, 17]. In this situation, electrons will be diverted to the enzyme hydrogenase from ferredoxin (Figure 5.2). Later, the reduction of protons (H⁺) occurred in the presence of hydrogenase using a series of electrons through photosynthetic electron transport. But the entire process is materialized under unfavorable conditions such as anoxic/hypoxic situation as hydrogenase is oxygen-sensitive [53]. Thus photosynthetically produced O_2 (PS-II by-product) irreversibly inhibits the H_2 production via hydrogenase enzyme. There are

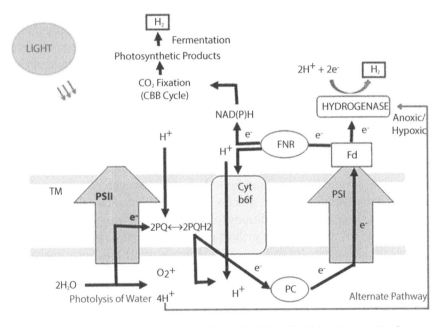

Figure 5.2 Hydrogen generation from an algal cell. CBB cycle, Calvin Benson Bassham cycle; CO_2, carbon dioxide; Cyt b6f, Cytochrome b6f; e−, electrons; Fd, Ferridoxin; FNR, ferridoxin NADP reductase; H_2, hydrogen; H+, protons; NADPH, nicotinamide adenine dinucleotide phosphate-oxidase; PC, plastocyanin; PQ, plastoquinone; PS, photosystem; TM, thylakoid membrane.

three major hydrogenases play an important role in the generation of hydrogen [54]. Figure 5.2 represents the whole schematic mechanism of biological hydrogen production in algal cells.

Glucose or photosynthetic products generated in CBB cycle may undergo fermentation in certain period, which leads to hydrogen production (Figure 5.2). In general, photosynthetic microalgae with metabolic elasticity have specific light and dark fermentative metabolism that allows them into hydrogen production. But chances of hydrogen production through the fermentation process are less when compared to photolysis of water-based method [4]. Apart from this route, several other methods include indirect photolysis was also possible in certain times. Inclusion of extra copies of hydrogenase gene in algal cells to improve hydrogen production through advanced biotechnological methods is need of the hour [55]. In addition, non-photochemical plastoquinone reduction pathway, which is another source of electrons to obtain molecular H_2 was established [25]. Apart from these methods, other mechanisms also exist in different organisms depending on source of the biomass/biowaste and raw materials [4, 18].

5.5 Production of Hydrogen from Different Algal Species

In recent times, using of algae for the production of biological hydrogen is one of the potential methods for commercial practice. Till date, many algal species were used for the hydrogen generation, and only few were highlighted here which are familiar in algal community. Table 5.1 explains about the algal species, i.e., *Scenedesmus obliquus*, *Chlorella vulgaris* and *Chlamydomonas reinhardtii* used for H_2 evolution.

5.5.1 Generation of Hydrogen in *Scenedesmus obliquus*

One of the first investigations on hydrogen production from green alga *S. obliquus* in dark anaerobic condition was reported by Gaffron and Rubin [47]. Effect of glucose and carbonyl cyanide m-chloro phenylhydrazone (CCCP) on light-dependent H_2 production was studied in similar *S. obliquus* [56]. Pow and Krasna [57] reported that hydrogenase is responsible for H_2 generation in both *S. obliquus* and *C. vulgaris* species. Enhanced hydrogen production was observed in *S. obliquus* by meta-substituted dichlorophenols biodegradation [58]. Batista *et al.* [59] produced hydrogen using *S. obliquus* and *C. vulgaris* species combined with

Table 5.1 Hydrogen production from certain algal species.

S. no.	Name of the alga	Description	Authors	Year & reference no.
1	*Scenedesmus obliquus*	It was initially used green algal species for hydrogen production using various physico-chemical factors.	Gaffron and Rubin	1942 [47]
			Kaltwasser *et al.*	1969 [56]
			Pow and Krasna	1979 [57]
			Papazi *et al.*	2012 [58]
			Batista *et al.*	2015 [59]
			Ruiz-Marrin *et al.*	2020 [60]
2	*Chlorella vulgaris*	This alga belongs to green algae group and available in both fresh and marine waters. It was also used for biological hydrogen generation using various physicochemical factors.	Kessler	1973 [61]
			Greenbaum	1977 [62]
			Pow and Krasna	1979 [57]
			Rashid *et al.*	2011 [63]
			Amutha and Murugesan	2011 [64]
			Lakaniemi *et al.*	2011 [51]
			Rashid *et al.*	2013 [65]
			Hwang *et al.*	2014 [66]
			Batista *et al.*	2015 [59]
			Lakshmikandan and Murugesan	2016 [67]
			Alalayah *et al.*	2017 [68]
			Gan *et al.*	2019 [52]
			Ruiz-Marrin *et al.*	2020 [60]
			Touloupakis *et al.*	2021 [69]

(Continued)

Table 5.1 Hydrogen production from certain algal species. (*Continued*)

S. no.	Name of the alga	Description	Authors	Year & reference no.
3	Chlamydomonas reinhardtii	This is belongs to fresh water green algae and considered as model alga. Most of the basic and applied research was carried out with this alga for hydrogen production. Number of biochemical and molecular studies were conducted with this alga to know the hydrogen production levels.	Greenbaum	1982 [70]
			Miura *et al.*	1982 [71]
			Bamberger *et al.*	1982 [72]
			Ohta *et al.*	1987 [73]
			Graves *et al.*	1988 [74]
			Ghirardi *et al.*	1997 [26]
			Melis *et al.*	2000 [2]
			Zhang *et al.*	2002 [75]
			Tsygankov *et al.*	2002 [76]
			Kosourov *et al.*	2003 [77]
			Antal *et al.*	2003 [78]
			Kruse *et al.*	2005 [79]
			Kosourov *et al.*	2005 [80]
			Mus *et al.*	2005 [81]
			Kim *et al.*	2006 [82]
			Surzycki *et al.*	2007 [27]
			Hemschemeier *et al.*	2008 [25]
			Giannelli *et al.*	2009 [83]
			Ma *et al.*	2011 [84]
			Philipps *et al.*	2012 [85]
			Scoma *et al.*	2012 [86]
			Li *et al.*	2013 [87]
			Cox *et al.*	2014 [88]
			Reifschneider-Wegner *et al.*	2014 [55]

(*Continued*)

Table 5.1 Hydrogen production from certain algal species. (*Continued*)

S. no.	Name of the alga	Description	Authors	Year & reference no.
			Volgusheva et al.	2015 [89]
			Xu et al.	2016 [49]
			Noone et al.	2017 [90]
			Wei et al.	2017 [91]
			Nagy et al.	2018 [92]
			Anandraj et al.	2019 [93]
			Kanygin et al.	2020 [18]
			Raga Sudha et al.	2021 [19]

urban wastewater in an integrated approach. Production of hydrogen was noticed by *S. obliquus* and *C. vulgaris* immobilized cells cultivated in artificial wastewater under various light conditions [60].

5.5.2 Production of Hydrogen in *Chlorella vulgaris*

Kessler [61] observed the photoproduction of hydrogen along with other components in different algal species including *C. vulgaris* and emphasized the importance of hydrogenase. Standardization of control parameters for photoproduction of hydrogen under *in situ* conditions in *C. vulgaris* was performed by Greenbaum [62]. Inclusion of glucose to the sulfur-deficient medium increased the H_2 generation by 18 times under partial light due to the external supply of carbon source [63]. The anaerobic condition was helpful in the conversion process of carbon source, which increased the hydrogen evolution, and this mechanism was proved in *C. vulgaris* MSU 01 strain isolated from pond sediment [64]. Biohydrogen and methane was generated using fresh and marine water algal species including *C. vulgaris* [51]. Fructose is the best carbon source for the improvement of hydrogen production in *C. vulgaris* when compared to glucose as proved by Rashid *et al.* [65]. In addition, sodium dithionite an oxygen scavenger was used to estimate the hydrogenase activity in *C. vulgaris* to check the hydrogen production levels [66]. Improved growth and biohydrogen generation was noticed in *C. vulgaris* utilizing aqueous extract of *Valoniopsis pachynema* as nutrient source [67]. Alalayah *et al.* [68] demonstrated that glucose concentration plays a key role in hydrogen production in *C. vulgaris*, which is quickly degraded in the metabolism. Gan *et al.* [52] proved the relationship between selenium removal efficiency and generation of hydrogen in *C. vulgaris* which indicates the involvement of various pathways in hydrogen production. Sustained photobiological hydrogen generation was noticed in *C. vulgaris* without nutrient starvation by Touloupakis *et al.* [69].

5.5.3 Generation of Hydrogen in Model Alga *Chlamydomonas reinhardtii*

Initially, quantitative studies were conducted on the production of hydrogen and oxygen using model alga *C. reinhardtii* by Greenbaum [70]. Miura *et al.* [71] proved that H_2 evolution in model alga was highly dependent on temperature and pH conditions. CO_2 was evolved through breakdown of starch with the help of the un-coupler carbonyl cyanide-p-tri fluoromethoxy phenylhydrazone (FCCP) in dark conditions and H_2 was released in the presence of light [72]. Ohta *et al.* [73] compared the fermentation products, including hydrogen in different green algal species including *C. reinhardtii*. The effect of sunlight intensity on sustained hydrogen production was studied in *C. reinhardtii* by Graves *et al.* [74]. Interference of O_2 in H_2 evolution was noticed in model alga *C. reinhardtii* [26]. Melis *et al.* [2] proved that sulfur-deprived medium improves the hydrogen generation in *C. reinhardtii* cells and this is one of the breakthroughs in biohydrogen sector. This work also emphasizes the role of hydrogenase in H_2 evolution. Morphological and biochemical characterization was conducted in hydrogen producing *C. reinhardtii* cells grown in sulfur-deprived medium [75]. Continuous light condition augmented the generation of hydrogen in sulfur-deprived, synchronous *C. reinhardtii* cultures [76]. In this model alga, the maximum rate of hydrogen production occurred when the pH was maintained at 7.7 in S-deprived medium. H_2 yield was decreased if pH was altered to 6.5 or 8.2 due to alteration of photosystem II function [77]. It was proved that PS II and oxygen consumption activities are the key in hydrogen production in sulfur-deprived model algal cells [78].

Enhancement of hydrogen production was noticed in state transition mutant 6 (*stm6*) of *C. reinhardtii* due to accumulation of large starch reserves along with low dissolved oxygen [79]. The impact of sulfur readdition on H_2 photoproduction in sulfur-deprived *C. reinhardtii* cells was studied. Readdition of low amount of sulfate leads to reactivation of PSII,

O_2 evolution temporally and also restarted the electron transport system in sulfur-deprived *C. reinhardtii* cells [80]. Hydrogen generation was also occurred in the absence of PSII through a non-photochemical pathway of PQ reduction [81]. H_2 production was carried out using a two-step process, i.e., anaerobic conversion and photosynthetic fermentation in model alga by Kim *et al.* [82]. Surzycki *et al.* [27] proved that Cu^{2+} inhibited the O_2 generation in photosystem-II, which in turn is helpful for the production of hydrogen in model alga. Interplay of electron sources and sinks leads to hydrogen generation in *C. reinhardtii* cells was proved by Hemschemeier *et al.* [25].

Effect of light intensity, chlorophyll concentration, and mixing of cultures on H_2 production was studied using in sulfur-deprived *C. reinhardtii* cells [83]. Sodium bisulfite, an oxygen scavenger, improved the photobiological hydrogen production in this model alga [84]. Production of photosynthetic H_2 was achieved under nitrogen deprivation in *C. reinhardtii* cells [85]. Hydrogen production was achieved using outside sunlight in a 50-L tubular photobioreactor with sulfur-deprived model alga [86]. Efficient hydrogen production was achieved in model alga by cocultivating with isolated bacteria [87]. Improvement in hydrogen production was observed in red light than blue light using sulfur-deprived *C. reinhardtii* cells [88]. Recently, *C. reinhardtii* have been genetically engineered with extra copies of hydrogenase gene for improvement of hydrogen production [55]. Hydrogen production in model alga under magnesium deprivation was observed by Volgusheva *et al.* [89]. Efficient hydrogen production was noticed through the cocultivation of *C. reinhardtii* with *Bradyrhizobium japonicum* [49]. Clostridial [FeFe]-hydrogenase expression augmented the hydrogen production through water splitting mechanism in this model alga [90]. Role of light intensity on hydrogen generation was demonstrated in sodium bisulphite treated *C. reinhardtii* cells [91]. Efficient hydrogen production was achieved using substrate limitation in CBB cycle through water splitting-based method [92]. Hydrogen generation was attained keeping photosystem I fluorescence as physiological indicator in model alga [93]. We achieved the augmented hydrogen production in *C. reinhardtii* through inserting an extra copy of hydrogenase gene [18]. Recently, our laboratory established the protocols for hydrogen production from *C. reinhardtii* [19]. In addition, other algal species were also used for hydrogen production.

5.6 Concluding Remarks

Hydrogen is one of the clean and best sources of energy, and investors are ready to spend more on this mission. The production of hydrogen through biological methods is considered as environmental friendly when compared to chemical methods. Algae have the capacity to generate hydrogen through photolysis of water-based and fermentation methods. In detail, PSII drives the initial stage of the process, by splitting H_2O into protons (H^+), electrons (e^-) and O_2. Later, these protons generate H_2 in alternate pathway with the help of hydrogenase. In another pathway, CBB cycle produce carbohydrates, which act as the fuel used in mitochondrial respiration and cell growth and partly divide into hydrogen in the process of fermentation. The number of algal species was used for hydrogen generation using various physico-chemical factors, and few species were emphasized in this chapter. This information will be useful for commercial practice of hydrogen using the candidate algal species.

Acknowledgments

This work is dedicated to the late Dr. Kurva Paramesh who left us due to COVID-19 pandemic.

References

1. Dincer, I., Renewable energy and sustainable development. A crucial review. *Renew. Sustain. Energy Rev.*, 4, 157–175, 2000.
2. Melis, A., Zhang, L., Forestier, M., Ghirardi, M.L., Seibert, M., Sustained photobiological hydrogen gas production upon reversible inactivation of oxygen evolution in the green alga *Chlamydomonas reinhardtii*. *Plant Physiol.*, 2122, 127–136, 2000.
3. W.E.O., World Energy Outlook. https://www.iea.org/reports/world-energy-outlook-2015.
4. Sharma, A. and Arya, S.K., Hydrogen from algal biomass: A review of production process. *Biotechnol. Rep.*, 14, 63–69, 2017.
5. Govindjee, G., Matthias Rögner (ed): Biohydrogen. *Photosynth. Res.*, 124, 337–339, 2015.

6. Varaprasad, D., Narasimham, D., Paramesh, K., Raga Sudha, N., Himabindu, Y., Keerthi Kumari, M., Improvement of production of ethanol using green alga *Chlorococcum minutum*. *Environ. Technol.*, 42, 1383–1389, 2021.
7. Saladini, F., Patrizi, N., Pulselli, F.M., Marchettini, N., Bastianoni, S., Guidelines for emergy evaluation of first, second and third generation biofuels. *Renew. Sustain. Energy Rev.*, 66, 221–227, 2016.
8. Pugazhendhi, A., Shobana, S., Nguyen, D.D., Rajesh Banu, J., Sivagurunathan, P., Chang, S.W., Ponnusamy, V.K., Kumar, G., Application of nanotechnology (nanoparticles) in dark fermentative hydrogen production. *Int. J. Hydrog. Energy*, 44, 1431–1440, 2019.
9. Oey, M., Sawyer, A.L., Ross, I.L., Hankamer, B., Challenges and opportunities for hydrogen production from microalgae. *Plant Biotechnol. J.*, 14, 1487–1499, 2016.
10. Das, D. and Veziroglu, T.N., Advances in biological hydrogen production processes. *Int. J. Hydrog. Energy*, 33, 6046–6057, 2008.
11. Paramesh, K., Lakshmana Reddy, N., Shankar, M.V., Chandrasekhar, T., Enhancement of biological hydrogen production using green alga *Chlorococcum minutum*. *Int. J. Hydrog. Energy*, 43, 3957–3966, 2018.
12. Varaprasad, D., Raghavendra, P., Raga Sudha, N., Sarma, L.S., Parveen, S.N., Sri Chandana, P., Chandra, M.S., Chandrasekhar, T., Bioethanol production in green alga *Chlorococcum minutum* through reduced graphene oxide-supported platinum-ruthenium (Pt-Ru/RGO) nanoparticles. *Bioenergy Res.*, 15, 280–288, 2022.
13. Juneja, A., Ceballos, R.M., Murthy, G.S., Effects of environmental factors and nutrient availability on the biochemical composition of algae for biofuels production-a review. *Energies*, 6, 4607–4638, 2013.
14. Paramesh, K. and Chandrasekhar, T., Improvement of photobiological hydrogen production in *Chlorococcum minutum* using various oxygen scavengers. *Int. J. Hydrogen Energy*, 13, 7641–7646, 2020.
15. Bleeker, M.F., Veringa, H.J., Kersten, S.R., Pure hydrogen production from pyrolysis oil using the steam-iron process: Effects of temperature and iron oxide conversion in the reduction. *Ind. Eng. Chem. Res.*, 49, 53–64, 2009.
16. Reddy, N.L., Emin, S., Valant, M., Venkatakrishnan, S.M., Nanostructured Bi_2O_3@TiO_2 photocatalyst for enhanced hydrogen production. *Int. J. Hydrog. Energy*, 42, 6627–6636, 2017.
17. Nagarajan, D., Lee, D.J., Kondo, A., Chang, J.S., Recent insights into biohydrogen production by microalgae-from biophotolysis to dark fermentation. *Bioresour. Technol.*, 227, 373–378, 2017.
18. Kanygin, A., Milrad, Y., Thummala, C.S., Reifschneider, K., Baker, P., Marco, P., Yacoby, I., Redding, K.E., Rewiring photosynthesis: A photosystem I-hydrogenase chimera that makes H_2 *in vivo*. *Energy Environ. Sci.*, 13, 2903–2914, 2020.
19. Raga Sudha, N., Varaprasad, D., Bramhachari, P.V., Sudhakar, P., Chandrasekhar, T., Effects of various factors on biomass, bioethanol and

biohydrogen production in green alga *Chlamydomonas reinhardtii*. *J. Appl. Biol. Biotechnol.*, 9, 152–156, 2021.
20. Wang, L., Chen, M., Wei, L., Gao, F., Lv, Z., Wang, Q., Ma, W., Treatment with moderate concentrations of $NaHSO_3$ enhances photobiological H_2 production in the cyanobacterium *Anabaena sp.* strain PCC 7120. *Int. J. Hydrogen Energy*, 35, 12777–12783, 2010.
21. Ursua, A., Gandia, L.M., Sanchis, P., Hydrogen production from water electrolysis: current status and future trends. *Proc. IEEE.*, 100, 410–426, 2012.
22. Tashie-Lewis, B.C., Somtochukwu Godfrey Nnabuife, S.G., Hydrogen production, distribution, storage and power conversion in a hydrogen economy- A technology review. *Chem. Eng. Journal Advances*, 8, 100172, 2021.
23. Patel, S.K.S., Lee, J.-K., Kalia, V.C., Nanoparticles in biological hydrogen production. An overview. *Indian J. Microbiol.*, 58, 8–18, 2018.
24. Ntaikou, I., Koutros, E., Kornaros, M., Valorisation of wastepaper using the fibrolytic/hydrogen producing bacterium *ruminococcus albus*. *Bioresour. Technol.*, 100, 5928–5933, 2009.
25. Hemschemeier, A., Fouchard, S., Cournac, L., Peltier, G., Happe, T., Hydrogen production by *Chlamydomonas reinhardtii*: An elaborate interplay of electron sources and sinks. *Planta*, 227, 397–407, 2008.
26. Ghirardi, M.L., Togasaki, R.K., Seibert, M., Oxygen sensitivity of algal H_2 production. *Appl. Biochem. Biotechnol.*, 63, 141–151, 1997.
27. Surzycki, R., Cournac, L., Peltier, G., Rochaix, J.D., Potential for hydrogen production with inducible chloroplast gene expression in *Chlamydomonas*. *Proc. Natl. Acad. Sci.*, 104, 17548–17553, 2007.
28. Ueno, Y., Otsuka, S., Morimoto, M., Hydrogen production from industrial wastewater by anaerobic microflora in chemostat culture. *J. Ferment. Bioeng.*, 82, 194–197, 1996.
29. Antoni., D., Zverlov., V.V., Schwarz, W.H., Biofuels from microbes. *Appl. Microbiol. Biotechnol.*, 77, 23–35, 2007.
30. Kumar, N. and Das, D., Enhancement of hydrogen production by *Enterobacter cloacae* IIT-BT 08. *Process. Biochem.*, 35, 589–593, 2000.
31. Koku, H., Eroğlu, I., Gündüz, U., Turker, L., Aspects of the metabolism of hydrogen production by *Rhodobacter sphaeroides*. *Int. J. Hydrog. Energy*, 27, 1315–1329, 2002.
32. Levin, D.B., Islam, R., Cicek, N., Sparling, R., Hydrogen production by *Clostridium thermocellum* 27405 from cellulosic biomass substrates. *Int. J. Hydrog. Energy*, 31, 1496–1503, 2006.
33. Koskinen, P.E.P., Beck, S.R., Örlygsson, J., Puhakka, J.A., Ethanol and hydrogen production by two thermophilic, anaerobic bacteria isolated from Icelandic geothermal areas. *Biotechnol. Bioeng.*, 101, 679–690, 2008.
34. Kobayashi, J., Yoshimune, K., Komoriya, T., Kohno, H., Efficient hydrogen production from acetate through isolated *Rhodobacter sphaeroides*. *J. Biosci. Bioeng.*, 112, 602–605, 2011.

35. Androga, D.D., Sevinç, P., Koku, H., Yucel, M., Gunduz, U., Eroglu, I., Optimization of temperature and light intensity for improved photofermentative hydrogen production using *Rhodobacter capsulatus* DSM 1710. *Int. J. Hydrog. Energy*, 39, 2472–2480, 2014.
36. Avci, A., Kilic, N.K., Dönmez, G., Donmez, S., Evaluation of hydrogen production by *Clostridium* strains on beet molasses. *Environ. Technol. (United Kingdom)*, 35, 278–285, 2014.
37. Tian, Q.Q., Liang, L., Zhu, M.J., Enhanced biohydrogen production from sugarcane bagasse by *Costridium thermocellum* supplemented with $CaCO_3$. *Bioresour. Technol.*, 197, 422–428, 2015.
38. Thakur, V., Tiwari, K.L., Jadhav, S.K., Production of biohydrogen from wastewater by *Klebsiella oxytoca* ATCC 13182. *Water Environ. Res.*, 87, 683–686, 2015.
39. Adessi, A., Concato, M., Sanchini, A., Rossi, F., De Philippis, R., Hydrogen production under salt stress conditions by a fresh water *Rhodopseudomonas palustris* strain. *Appl. Microbiol. Biotechnol.*, 100, 2917–2926, 2016.
40. Li, Y., Zhang, Q., Deng, L., Liu, Z., Jiang, H., Wang, F., Biohydrogen production from fermentation of cotton stalk hydrolysate by *Klebsiella* sp. WL1316 newly isolated from wild carp (*cyprinus carpio* L.) of the Tarim River basin. *Appl. Microbiol. Biotechnol.*, 102, 4231–4242, 2018.
41. Talluri, S., Raj, S.M., Christopher Taifor, A.F., Consolidated bioprocessing of untreated switchgrass to hydrogen by the extreme thermophile *Caldicellulosiruptor saccharolyticus* DSM 8903. *Bioresour. Technol.*, 139, 272–279, 2013.
42. Yu, L., Cao, M., Wang, P.T., Wang, S., Yue, Y.R., Yuan, W.D., Qiao, W.C., Wang, F., Song, X., Simultaneous decolorization and biohydrogen production from xylose by *Klebsiella oxytoca* GS-4-08 in the presence of azo dyes with sulfonate and carboxyl groups. *Appl. Environ. Microbiol.*, 83, 1–36, 2017.
43. Cheng, J., Liu, M., Song, W., Ding, L., Liu, J., Zhang, L., Cen, K., Enhanced hydrogen production of *Enterobacter aerogenes* mutated by nuclear irradiation. *Bioresour. Technol.*, 227, 50–55, 2017.
44. Taifor, A.F., Zakaria, M.R., Mohd Yusoff, M.Z., Toshinari, M., Hasan, M.A., Shirai, Y., Elucidating substrate utilization in biohydrogen production from palm oil mill effluent by *Escherichia coli*. *Int. J. Hydrog. Energy*, 42, 5812–5819, 2017.
45. An, Q., Wang, J.L., Wang, Y.T., Lin, Z.L., Zhu, M.J., Investigation on hydrogen production from paper sludge without inoculation and its enhancement by *Clostridium thermocellum*. *Bioresour. Technol.*, 263, 120–127, 2018.
46. Shimizu, T., Teramoto, H., Inui, M., Introduction of glyoxylate bypass increases hydrogen gas yield from acetate and l-glutamate in *Rhodobacter sphaeroides*. *Appl. Environ. Microbiol.*, 85, e01873–18, 2018.
47. Gaffron, H. and Rubin, J., Fermentative and photochemical production of hydrogen in algae. *J. Gen. Physiol.*, 26, 219–40, 1942.

48. Melis, A. and Happe, T., Hydrogen production. Green algae as a source of energy. *Plant Physiol.*, 127, 740–748, 2001.
49. Xu, L., Li, D., Wang, Q., Wu, S., Improved hydrogen production and biomass through the co-cultivation of *Chlamydomonas reinhardtii* and *Bradyrhizobium* japonicum. *Int. J. Hydrog. Energy*, 41, 9276–9283, 2016.
50. Guo, Z., Li, Y., Guo, H., Effect of light/dark regimens on hydrogen production by *Tetraselmis subcordiformis* coupled with an alkaline fuel cell system. *Appl. Biochem. Biotechnol.*, 183, 1295–1303, 2017.
51. Lakaniemi, A.M., Hulatt, C.J., Thomas, D.N., Tuovinen, O.H., Puhakka, J.A., Biogenic hydrogen and methane production from *Chlorella vulgaris* and *dunaliella tertiolecta* biomass. *Biotechnol. Biofuels*, 4, 34, 2011.
52. Gan, X., Huang, J.C., Zhou, C., He, S., Zhou,W., Relationship between selenium removal efficiency and production of lipid and hydrogen by *Chlorella vulgaris*. *Chemosphere*, 217, 825–832, 2019.
53. Meuser, J.E., Ananyev, G., Wittig, L.E., Kosourov, S., Ghirardi, M.L., Seibert, M., Dismukes, G.C., Posewitz, M.C., Phenotypic diversity of hydrogen production in chlorophycean algae reflects distinct anaerobic metabolisms. *J. Biotechnol.*, 142, 21–3, 2009.
54. Posewitz, M.C., Dubini, A., Meuser, J.E., Seibert, M., Ghirardi, M.L., Hydrogenases, hydrogen production and anoxia, in: *The Chlamydomonas Sourcebook*, 2nd Ed., E.H. Harris, D.B. Stern, G.B. Witman, (Eds.), pp. 217–255, Academic Press, Oxford, 2009.
55. Reifschneider-Wegner, K., Kanygin, A., Redding, K.E., Expression of the [FeFe] hydrogenase in the chloroplast of *Chlamydomonas reinhardtii. Int. J. Hydrog. Energy*, 39, 3657–3665, 2014.
56. Kaltwasser, H., Stuart, T.S., Gaffron, H., Light-dependent hydrogen evolution by *Scenedesmus. Planta*, 89, 309–332, 1969.
57. Pow, T. and Krasna, A.I., Photoproduction of hydrogen from water in hydrogenase-containing algae. *Arch. Biochem. Biophys.*, 194, 413–421, 1979.
58. Papazi, A., Andronis, E., Ioannidis, N.E., Chaniotakis, N., Kotzabasis, K., High yields of hydrogen production induced by meta-substituted dichlorophenols biodegradation from the green alga *Scenedesmus obliquus. PloS One*, 7, 49037, 2012.
59. Batista, A.P., Ambroson, L., Graca, S., Viegas de Sousa, C., Marques, P.A., Ribeiro, B., Botrel, E.P., Neto, P.C., Gouveia, L., Combining urban wastewater treatment with biohydrogen production-an integrated microalgae-based approach. *Bioresour. Technol.*, 184, 230–235, 2015.
60. Ruiz-Marin, A., Canedo-López, Y., Chávez-Fuentes, P., Biohydrogen production by *Chlorella vulgaris* and *Scenedesmus obliquus* immobilized cultivated in artificial wastewater under different light quality. *AMB Express*, 10, 191, 2020.
61. Kessler, E., Effect of anaerobiosis on photosynthetic reactions and nitrogen metabolism of algae with and without hydrogenase. *Arch. Mikrobiol.*, 93, 91–100, 2004.

62. Greenbaum, E., The photosynthetic unit of hydrogen evolution. *Science*, 196, 879–880, 1977.
63. Rashid, N., Lee, K., Mahmood, Q., Bio-hydrogen production by *Chlorella vulgaris* under diverse photoperiods. *Bioresour. Technol.*, 102, 2101–2104, 2011.
64. Amutha, K.B. and Murugesan, A.G., Biological hydrogen production by the algal biomass *Chlorella vulgaris* MSU 01 strain isolated from pond sediment. *Bioresour. Technol.*, 102, 194–199, 2011.
65. Rashid, N., Lee, K., Han, J.I., Gross, M., Hydrogen production by immobilized *Chlorella vulgaris*: Optimizing pH, carbon source and light. *Bioprocess Biosyst. Eng.*, 36, 867–872, 2013.
66. Hwang, J.H., Kim, H.C., Choi, J.A., Abou-Shanab, R.A.I., Dempsey, B.A., Regan, J.M., Kim, R.A., Song, H., Nam, I.H., Kim, S.N., Lee, W., Park, D., Kim, Y., Choi, J., Jim, M.K., Jung, W., Jeon, B.H., Photoautotrophic hydrogen production by eukaryotic microalgae under aerobic conditions. *Nat. Commun.*, 5, 1–6, 2014.
67. Lakshmikandan, M. and Murugesan, A.G., Enhancement of growth and biohydrogen production potential of *Chlorella vulgaris* MSU-AGM 14 by utilizing seaweed aqueous extract of *valoniopsis pachynema*. *Renew. Energy*, 96, 390–399, 2016.
68. Alalayah, W.M., Al-zahrani, A., Edris, G., Demirbas, A., Kinetics of biological hydrogen production from green microalgae *Chlorella vulgaris* using glucose as initial substrate. *Energy Sources A: Recover Util. Environ. Eff.*, 39, 1210–1215, 2017.
69. Touloupakis, E., Faraloni, C., Benavides, A.M.S., Masojidek, J., Torzillo, G., Sustained photobiological hydrogen production by *Chlorella vulgaris* without nutrient starvation. *Int. J. Hydrog. Energy*, 46, 3684–3694, 2021.
70. Greenbaum, E., Photosynthetic hydrogen and oxygen production: Kinetic studies. *Science*, 215, 291–293, 1982.
71. Miura, Y., Yagi, K., Shoga, M., Miyamoto, K., Hydrogen production by a green alga, *Chlamydomonas reinhardtii*, in an alternating light/dark cycle. *Biotechnol. Bioeng.*, 24, 1555–1563, 1982.
72. Bamberger, E.S., King, D., Erbes, D.L., Gibbs, M., H_2 and CO_2 evolution by anaerobically adapted *Chlamydomonas reinhardtii* F-60. *Plant Physiol.*, 69, 1268–1273, 1982.
73. Ohta, S., Miyamoto, K., Miura, Y., Hydrogen evolution as a consumption mode of reducing equivalents in green algal fermentation. *Plant Physiol.*, 83, 1022–1026, 1987.
74. Graves, A.D.A., Reeves, M.E., Greenbaum, E., Establishment of control parameters for *in situ*, automated screening of sustained hydrogen photoproduction by individual algal colonies. . *Plant Physiol.*, 87, 603–608, 1988.
75. Zhang, L., Happe, T., Melis, A., Biochemical and morphological characterization of sulfur-deprived and H_2-producing *Chlamydomonas reinhardtii* (green alga). *Planta*, 214, 552–561, 2002.

76. Tsygankov, A., Kosourov, S., Seibert, M., Ghirardi, M.L., Hydrogen photoproduction under continuous illumination by sulfur-deprived, synchronous *Chlamydomonas reinhardtii* cultures. *Int. J. Hydrog. Energy*, 27, 1239–1244, 2002.
77. Kosourov, S., Seibert, M., Ghirardi, M.L., Effects of extracellular pH on the metabolic pathways in sulfur-deprived, H_2 producing *Chlamydomonas reinhardtii* cultures. *Plant Cell Physiol.*, 44, 146–155, 2003.
78. Antal, T.K., Krendelava, T.E., Laurinavichene, T.V., Makarova, V.V., Ghirardi, M.L., Rubin, A.B., Tsygankov, A.A., Seibert, M., The dependence of algal H_2 production on photosystem II and O_2 consumption activities in sulfur-deprived *Chlamydomonas reinhardtii* cells. *Biochem. Biophys. Acta*, 1607, 153–160, 2003.
79. Kruse, O., Rupprecht, J., Bader, K.P., Thomas-Hall, S., Schenk, P.M., Finazzi, G., Improved photobiological H_2 production in engineered green algal cells. *J. Biol. Chem.*, 280, 34170–34177, 2005.
80. Kosourov, S., Makarova, V., Fedorov, A.S., Tsygankov, A., Seibert, M., Ghirardi, M.L., The effect of sulfur re-addition on H_2 photoproduction by sulfur-deprived green algae. *Photosynth. Res.*, 85, 295–305, 2005.
81. Mus, F., Cournac, L., Cardettini, V., Caruana, G., Peltier, G., Inhibitor studies on non-photochemical plastoquinone reduction and H_2 photoproduction in *Chlamydomonas reinhardtii*. *Biochim. Biophys. Acta.-Bioenerg.*, 1708, 322–332, 2005.
82. Kim, M.S., Baek, J.S., Yun, Y.S., Sim, S.J., Park, S., Kim, S.C., Hydrogen production from *Chlamydomonas reinhardtii* biomass using a two-step conversion process: Anaerobic conversion and photosynthetic fermentation. *Int. J. Hydrog. Energy*, 31, 812–816, 2006.
83. Giannelli, L., Scoma, A., Torzillo, G., Interplay between light intensity, chlorophyll concentration and culture mixing on the hydrogen production in sulfur-deprived *Chlamydomonas reinhardtii* cultures grown in laboratory photobioreactors. *Biotechnol. Bioeng.*, 104, 76–90, 2009.
84. Ma, W., Chen, M., Wang, L., Wei, L., Wang, Q., Treatment with $NaHSO_3$ greatly enhances photobiological H_2 production in the green alga *Chlamydomonas reinhardtii*. *Bioresour. Technol.*, 102, 8635–8638, 2011.
85. Philipps, G., Happe, T., Hemschemeier, A., Nitrogen deprivation results in photosynthetic hydrogen production in *Chlamydomonas reinhardtii*. *Planta*, 235, 729–745, 2012.
86. Scoma, A., Giannelli, L., Faraloni, C., Torzillo, G., Outdoor H_2 production in a 50-L tubular photobioreactor by means of a sulfur-deprived culture of the microalga *Chlamydomonas reinhardtii*. *J. Biotechnol.*, 157, 620–627, 2012.
87. Li, X., Huang, S., Yu, J., Wang, Q., Wu, S., Improvement of hydrogen production of *Chlamydomonas reinhardtii* by co-cultivation with isolated bacteria. *Int. J. Hydrog. Energy*, 38, 10779–10787, 2013.

88. Cox, G.M., Mccrimmon, B.M.D., Nemati, T.A., Olson, C.M., The production of hydrogen gas by *Chlamydomonas reinhardtii in* sulfur-deprived conditions under red light and white light. *Expedition*, 3, 1–1, 2014.
89. Volgusheva, A., Kukarskikh, G., Krendeleva, T., Rubin, A., Mamedov, F., Hydrogen photoproduction in green algae *Chlamydomonas reinhardtii* under magnesium deprivation. *RSC Adv.*, 5, 5633–5637, 2015.
90. Noone, S., Ratcliff, K., Davis, R.A., Subramanian, V., Meuser, J., Posewitz, M.C., King, P.W., Ghiraridi, M.L., Expression of a clostridial [FeFe]-hydrogenase in *Chlamydomonas reinhardtii* prolongs photo-production of hydrogen from water splitting. *Algal Res.*, 22, 116–121, 2017.
91. Wei, L., Yi, J., Wang, L., Huang, T., Gao, F., Wang, Q., Ma, W., Light intensity is important for hydrogen production in $NaHSO_3$ treated *Chlamydomonas reinhardtii*. *Plant Cell Physiol.*, 58, 451–457, 2017.
92. Nagy, V., Podmaniczki, A., Vidal-Meireles, A.M., Tengolics, R., Rakhely, G., Scoma, A., Water-splitting-based, sustainable and efficient H_2 production in green algae as achieved by substrate limitation of the Calvin-Benson-Bassham cycle. *Biotechnol. Biofuels*, 11, 69, 2018.
93. Anandraj, A., White, S., Mutanda, T., Photosystem I fluorescence as a physiological indicator of hydrogen production in *Chlamydomonas reinhardtii*. *Bioresour. Technol.*, 273, 313–319, 2019.

6

Photoelectrocatalysis Enables Greener Routes to Valuable Chemicals and Solar Fuels

Dipesh Shrestha[1], Kamal Dhakal[1], Tamlal Pokhrel[1], Achyut Adhikari[1], Tomas Hardwick[2,3,4], Bahareh Shirinfar[5] and Nisar Ahmed[2*]

[1]*Central Department of Chemistry, Tribhuvan University, Kirtipur Kathmandu, Nepal*
[2]*School of Chemistry, Cardiff University, Main Building, Park Place, Cardiff, UK*
[3]*National Graphene Institute, University of Manchester, Oxford Road, Manchester, UK*
[4]*Department of Materials, University of Manchester, Oxford Road, Manchester, UK*
[5]*Department of Chemistry, University of Bath, Bath, UK*

Abstract

The transformation of visible light to electrical energy for the activation and functionalization of organic compounds, under the synergistic conditions of visible light and the photoelectric current, is an atom-economical, environmentally benign, self-powered, and interfacial enabling technology. Photoelectrocatalysis (PEC) enables selective chemical transformations under mild conditions via the generation of reactive intermediates. This catalysis process is advantageous in chemicals synthesis due to its broad functional group tolerance, site-specific selectivity, biocompatibility, and operational simplicity. Photoelectrode materials are applied as useful tools to perform organic oxidations and reductions for the chemical syntheses of important compounds. In this chapter, we summarize state-of-the-art applications of PEC in the synthesis of valuable chemicals and solar fuels. Specific prominence has been given to chemical transformations concerning C–H bond activation, and C–C and C–heteroatom coupling, whereby new chemical bonds are formed via late-stage functionalization of interesting compounds or coupling of two molecular building blocks. We also

*Corresponding author: AhmedN14@cardiff.ac.uk

discuss challenges and future directions of PEC that will enable this technology to broaden its scope in chemicals synthesis.

Keywords: Photoelectrocatalysis, C–H bond activation, solar fuels, late-stage functionalization, reactor design

6.1 Introduction

Selective functionalization of the C–H bond is a preeminent step in organic synthesis since the incorporation of desired functional groups increase the value of organic compounds. Activation of the C–H bond means the process of introducing a functional group in place of the C–H bond. To synthesize the desirable product specific functional groups are required for particular chemical and physical properties [1]. Due to the unreactive and strong C–H (sp^3) and C–C (sp^3) bond, activation of these bond can be onerous [2]. C–H bonds are not commonly ascribed as functional groups, and thus the formation of new bonds in the carbon backbone require the addition of heteroatoms, such as oxygen, halogens, or unsaturated (absence of hydrogen) moieties [3]. Organic molecules contain many C–H bonds, however, their inertness means that many of them cannot be used for chemical processes. Of the few previous methods available, all of them were nonselective, resulting in a complex mix of compounds. As a result, scientists were compelled to investigate this seldom investigated area of C–H bond reactivity [4]. Nevertheless, the novel applications of C–H bonds [4–6] have been realized which has seen recent advances in selective C–H functionalization's in the form of transition-metal catalysis [7], as well as methodologies involving free radicals [2], oxidative cross coupling [8], bidentate directing group-assisted C–H bond functionalization's [9], and enzymatic functionalization [10]. Due to the burdensome nature of C–H functionalization's (time constraints, expense, environmental hazards, insufficient productivity) novel techniques are being sort out.

An electrochemical method for C–H bond functionalization was first developed in the 1840s when Kolbe studied the electrolysis of organic compounds [11]. Photochemistry [12, 13] and electrochemistry [14] in organic synthesis have been recognized as economical and environmentally benign methods with attractive efficiency. The constructive fusion of photochemistry and electrochemistry can produce a scenario resulting in a potent new tool for green chemistry where the shortcomings of each technique are perfectly compensated by their respective advantages, allowing for novel reaction pathways that would otherwise be impossible to achieve using each individual method [15]. Due to the renewable source of energy, photoelectrochemical cells have attracted the interest of many researchers. Photoelectrochemical cells convert solar radiation into electrical energy, and help to facilitate the photo-electrolysis reaction [12, 13]. Through the photovoltaic effect, photovoltaic cells convert radiation into electrical energy, typically using semiconductors. An n-type semiconductor has excess of free moving electrons while a p-type semiconductor has so with holes [16]. When a p- and n-type semiconductor come into contact, electrons and holes near the interface combine, forming a depletion region of now free charge carriers. Photons then strike this region, forming photogenerated electron-hole pairs, which contribute to the current if a load is applied to the circuit, or to a chemical reaction [17]. Performance of the photovoltaic cell is associated with the semiconductor [18]. There are predominantly six types of PEC cell: i) single light absorber, ii) heterojunction photoelectrode, iii) wired PEC tandem cell, iv) wireless PEC tandem cell, v) PV-PEC tandem cell, and vi) PV-electrolyzer cell [19]. In a simple PEC cell, energy available to catalyze the reaction from the excitation of a single visible light photon is typically insufficient. Due to many illusive problems and challenges of single photoelectrode configurations, the use of tandem systems has been increasingly rising in organic synthesis [13], [20]. In 2021, Chae *et al.* reported a photocathode for efficient photochemical production of H_2 gas, employing the semiconductor p-$CuInS_2$, which had a band gap of ~1.5 eV [21]. Currently, photoanode and photocathode materials have been extended not only to water splitting for hydrogen production but also to organic synthesis.

Photoelectrodes play a vital role in the photocatalytic cell by utilizing solar energy for chemical oxidation and reduction [22–24]. Due to the important role of the semiconductor in the photocatalytic process, it must be effective in maximizing the absorbance of appropriate light wavelengths, and in the transfer of charge between the semiconductor/electrolyte interface [23]. Typically, metal oxide and hydroxides are used as the semiconductor, since such materials have important requirements, such as possessing good stability in a chemical solution, being nontoxic and highly efficient. When a semiconductor is irradiated with a photon with energy equal to or higher than that of its band gap, an electron from valence band of the semiconductor is excited to the conduction band, generating e^-/h^+ pairs (Figure 6.1). The separated photogenerated charge carriers migrate to the surface of the catalyst with the help of a bias; the surface-migrated holes participate in the oxidation reactions of the organic compounds, while the electrons undergo reduction reactions with electron-deficient substrates or

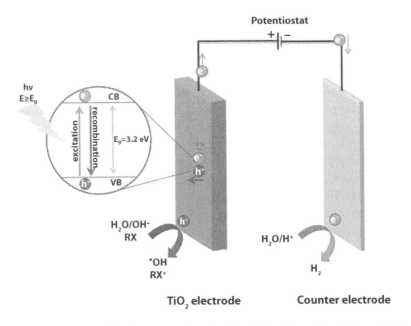

Figure 6.1 Generation of e−/h+ pairs under photoirradiation. Reproduced from ref [27].

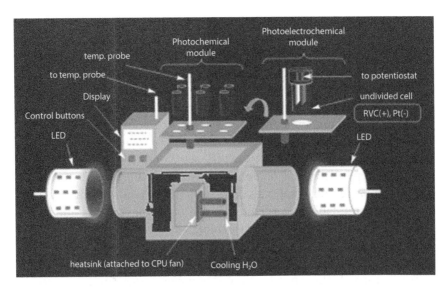

Figure 6.2 A 3D-printed reactor for photochemical and photoelectrochemical reactions. Reproduced from ref [28].

radicals. Usually, the required intermediates are generated concurrently by the electrochemical and photochemical routes over an activated catalyst. The bias can accelerate the migration of e^-/h^+ pairs, resulting in a noticeable increase in catalytic activity [25-27].

Light intensity, light wavelength, current, potential, reaction temperature, catalyst, electrode materials, and electrolyte choice influence the reaction outcome and reproducibility in photoelectrochemical reactions [28]. A recent development has been made to address this issue, a 3D-printed photoreactor equipped with thermoelectric coolers was designed for the purpose of reproducibility (Figure 6.2). The reactors can access and accurately regulate temperatures ranging from −17°C to 80°C, while reactions under high-intensity illumination were achieved with high-powered LED lamps. For large-scale reactions, the catalyst loading can be lowered to 0.1 mol% due to the high light intensity. The systems flexibility was demonstrated by integrating IKA's ElectraSyn 2.0 with the photoreactor to achieve reactions in a repeatable manner [28, 29].

6.2 C–H Functionalization in Complex Organic Synthesis

In a wide sense, the conversion of carbon–hydrogen bonds into carbon–carbon or carbon–heteroatom bonds are referred to as "C–H functionalization" [6]. The efficient and selective activation/functionalization (or transformation) of carbon–hydrogen [$C(sp^3)$–H, $C(sp^2)$–H, and/or $C(sp)$–H] bonds in saturated, unsaturated, and aromatic hydrocarbons, as well as other organic compounds, to produce C–X bonds has remained a major target in organic chemistry over the years. Due to the high stability of a $C(sp^3)$–H bond, strong reaction conditions are usually necessary to replace a hydrogen atom with a functional group, such as in alkanes [30]. Photoelectrochemical cells use light, including sunlight, to generate electrical energy. One or two semiconducting photoelectrodes, as well as auxiliary metal and reference electrodes, are immersed in an electrolyte in each cell [31]. A photocathode is typically used to reduce water to give hydrogen, while a photoanode is used to oxidize water to produce oxygen. PEC cells are well adapted to catalyze the redox transformation of organic molecules to provide high added-value compounds because they can generate a high reducing or oxidizing power under mild conditions under light illumination [32].

6.3 Examples of Photoelectrochemical-Induced C–H Activation

Xu and co-workers used primary, secondary, and tertiary alkyltrifluoroborates **2** (or 1,4-dihydropyridine) to create a variety of functionalized heteroarenes **3** with good regioselectivity and chemoselectivity (Scheme 6.1) [32].

Furthermore, Hu *et al.* found that the abundant and robust haematite photoanode may be used in the extremely orthoselective, nondirected arene C–H amination process in hexafluoroisopropanol **14** with a wide spectrum of azoles **15** as shown in Figure 6.3 [33].

A Br^+/Br^- mediator was used to achieve photoelectrochemical dimethoxylation of furan (Figure 6.4) at low applied bias [34]. The reaction was carried out in methanol with a Faradaic efficiency of up to 99% at +0.1 V vs. standard hydrogen electrode (SHE) using Et_4NBr and Et_4NBF_4 as a cosupporting electrolyte. Using solar energy, the applied potential could be decreased as compared to the electrochemical method.

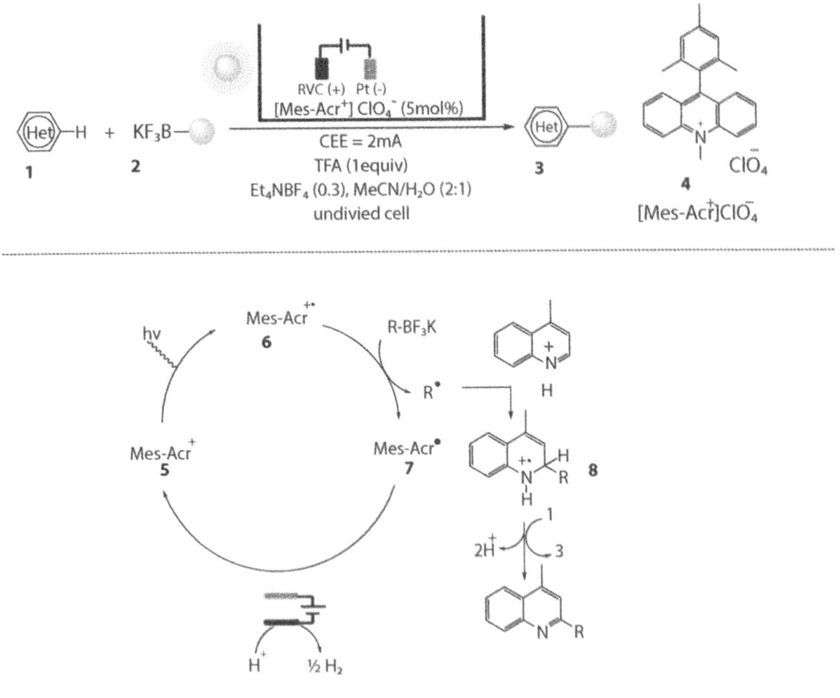

Scheme 6.1 Photoelectrochemical C–H alkylation of heteroarenes with organotrifluoroborates [32].

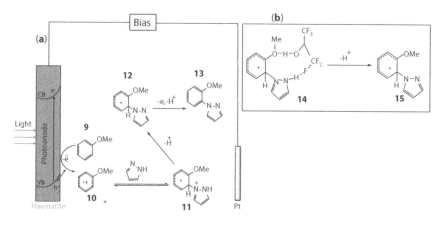

Figure 6.3 Mechanistic hypothesis. (a) Proposed mechanism of C–N bond formation. (b) Proposed hydrogen bonding among anisole, HFIP, and pyrazole [33].

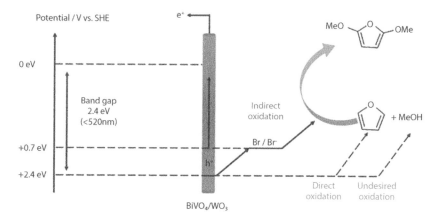

Figure 6.4 Photoelectrochemical dimethoxylation of furan. Reproduced from ref [34].

Using amides as nitrogen donors, Pandey and Laha used photoredox catalysis to comply with oxidative activation of benzylic C–H bonds toward C–N bond synthesis in 2015. The photocatalytic cycle begins with the singlet state excitation of the 9,10-dicyanoanthracene (DCA) photocatalyst, as indicated in Scheme 6.2. This is involved in the production of radical **20**, which can activate the benzyl position via the hydrogen atom transfer (HAT) reaction step [35].

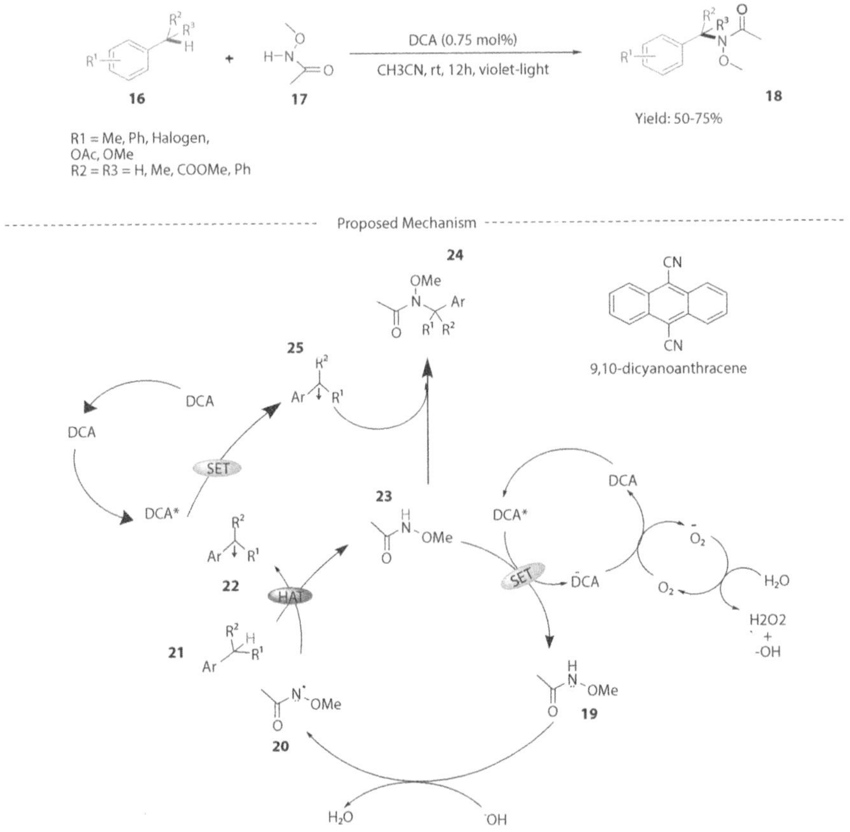

Scheme 6.2 Direct benzylic amination reaction by DCA.

6.4 C–C Functionalization

Solar energy is a sustainable, cheap and environment friendly source of energy. Plants utilize solar energy to convert CO_2 into organic compounds, a process which is used as inspiration for PEC cells [36, 37]. It is difficult to achieve the goal of reducing CO_2 in photocatalytic system using only a photocathode or photoanode. The combination of the photocathode for CO_2 reduction and photoanode for H_2O oxidation may allow the reduction of CO_2 without an external electric bias. An example occurred when InP/[MCE2-A + MCE4] was used as the photocathode (MCE2-A = [Ru(4,4′-diphosphate ethyl-2,2′-bipyridine) $(CO)_2 Cl_2$], MCE4 = [Ru{4,4′-di(1Hpyrrolyl-3-propyl carbonate-2,2′-bipyridine}$(CO)_2$ (MeCN)Cl_2) and Pt/TiO_2 was used as the photoanode (Figure 6.5) [38].

PHOTOELECTROCATALYSIS AS ENABLING TECHNOLOGY   193

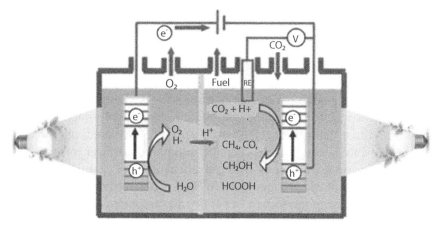

Figure 6.5 Photoelectrocatalytic reduction of CO_2 with H_2O. Reproduced from ref [38].

C/N-doped heterojunctions of Zn_x:Co_y@Cu were fabricated by Wang et al. [39] to generate the favourable heterojunction for the CO_2 reduction by mimicking the Calvin cycle of plants (Figure 6.6). During the irradiation of light electrons and holes were generated and quickly separated by

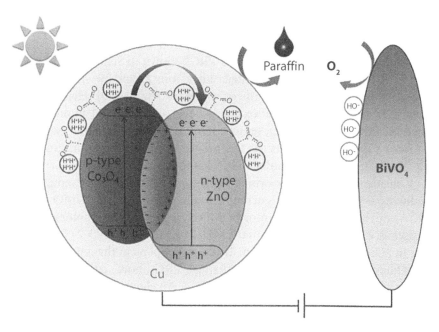

Figure 6.6 Proposed mechanism for paraffin synthesis from CO_2 reduction. Reproduced from ref [39].

the built in-electric field, resulting in a higher electron mobility. Protons were converted into active hydrogen by capturing electrons, the active hydrogen of which then reduced CO_2 into paraffin.

In organic synthesis, photoredox catalysis (PRC) and synthetic organic electrochemistry (SOE) are often regarded as competing approaches. Their fusion has mostly gone unnoticed up until now. We have considered the state-of-the-art synthetic organic photoelectrochemistry, grouping examples into three categories: 1) electrochemically mediated photoredox catalysis (e-PRC), where electrochemical and photochemical components have an interdependent role of synergistic benefit in the chemical process; 2) decoupled photoelectrochemistry (dPEC), where electrochemical and photochemical components have separate and discrete roles; and 3) interfacial photoelectrochemistry (iPEC), where reactions occur at the photoelectrode surface [40].

6.5 Electrochemically Mediated Photoredox Catalysis (e-PRC)

Electrochemically mediated photoredox catalysis (e-PRC) was created to characterize a type of PEC that involves an intimate interaction between electrochemical and photochemical processes, as well as subsequent phases within the same catalytic cycle. There are two subcategories in this section: "radical ion e-PRC" and "recycling e-PRC," shown in Scheme 6.3. Radical ion e-PRC involves electrogenerated radical ions **27** and **29** which are photoexcited to yield super-oxidants or super-reductants [29]. Here, electrochemistry provides a basal redox energy level (e.g., a radical anion), which then, in a transitory manner, redox energy is delivered by photoexcitation to form super-redox agents. As the mediator regenerates, it accumulates electrons and photons in order to surpass the activation energy barrier.

Consider the molecular orbital transition of DCA as an example of a reducing e-PRC. Cathodic current initially populates the LUMO (ψ_2) of DCA with an electron, becoming SOMO-2 (ψ_2) of DCA$^{•-}$. Photoexcitation causes an electron to move from the MO-1 (ψ_1) to the SOMO-2 (ψ_2), resulting in SOMO–HOMO inversion. This also happens in the analogous case with phenothiazine (PTZ) as an oxidizing e-PRC; anodic current removing an electron converts HOMO-4 to SOMO-4. An electron is then promoted from MO-1 to SOMO-4 by 514 nm light. In either situation, the e-PRC achieves a doublet excited state (Figure 6.7) [40].

PHOTOELECTROCATALYSIS AS ENABLING TECHNOLOGY   195

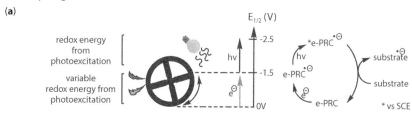

Scheme 6.3 Radical ion electroactivated photoredox catalysis and photocatalyst electrorecycling.

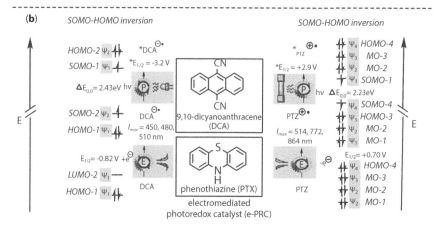

Figure 6.7 (a) Conceptual redox energy level diagram for the photoexcitation of electrochemically generated ions in e-PRC. (b) SOMO-HOMO inversion concept for two electromediated photoredox catalysts (e-PRCs). Reproduced from ref [40].

As an example, a new PEC approach was proposed to access extremely reduced intermediates using simultaneous cathodic reduction and photoexcitation. Cathodic reduction of **33** (DCA; $E_{1/2}$ = 0.82 V) produced the radical anion **34** that absorbed visible light and emited a high fluorescence wavelength (excitation energy $E_{0,0}$ = 2.38 eV). The reduction potential of the photoexcited DCA **35** is reported to be −3.2 V vs SCE, which time-dependent density-functional theory (TD-DFT) calculations show emerges from a SOMO-HOMO level inversion exhibiting an extremely unstable electronic structure with a half-filled bonding orbital (ψ_1) and a filled antibonding orbital (ψ_2). This PEC method has also been applied to the reductive functionalization of aryl halides **36** with reduction potentials (E_{red}) as low as −2.94 V (Scheme 6.4) [41].

On the other hand, recycling e-PRC entails the turnover of a photocatalyst that is a recognized photoredox catalyst (PRCat) in PRC, and is a colourful species in its ground state [29]. Lambert described S_NAR reactions of unactivated aryl fluorides in the presence of e-PRC. Photoexcited 2,3-dichloro-5,6-dicyanoquinone (DDQ) was oxidizing enough

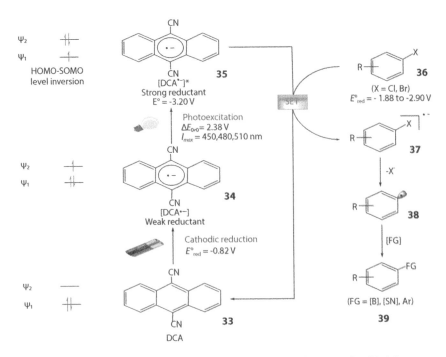

Scheme 6.4 Reductive photoelectrocatalysis for the functionalization of aryl halides. Reproduced from ref [41].

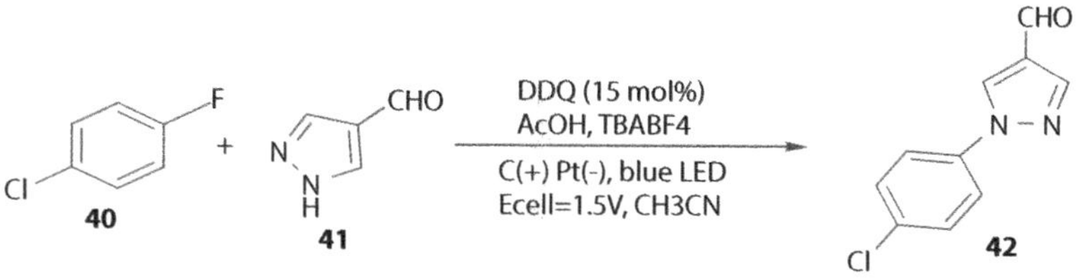

Scheme 6.5 SNAr reactions of unactivated aryl fluorides at ambient temperature and without a base under e-PRC. Reproduced from ref [42].

(E_{red} = +3.18 V vs. SCE) to engage chlorofluoroarenes **40** in a single electron transfer (SET) oxidation (Scheme 6.5) [42].

6.6 Interfacial Photoelectrochemistry (iPEC)

In interfacial photoelectrochemistry, one of the electrodes is coated with a photoresponsive material, such as semiconductor whose band gap is appropriate for the reaction and to the incoming energy of the visible-light photons. When irradiated, an applied potential is employed to promote charge carrier separation. In photoanodes, an electron is promoted from the valence band to the conductive band by an applied potential followed by irradiation, resulting in a hole that is employed for oxidation chemistry (Figure 6.8) [40, 43].

According to Hu, Grätzel, and colleagues, Haematite was employed as a strong photoanode for nondirected arene C–H amination. Photogenerated holes in haematite oxidize electron-rich arenes, such as anisole **43** to radical

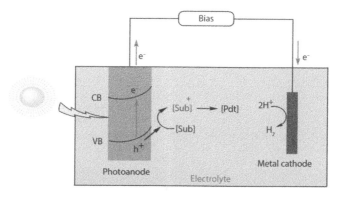

Figure 6.8 Photoelectrochemical cell for oxidative transformations of organic substrates. CB, conduction band; VB, valence band. Modified from ref [33].

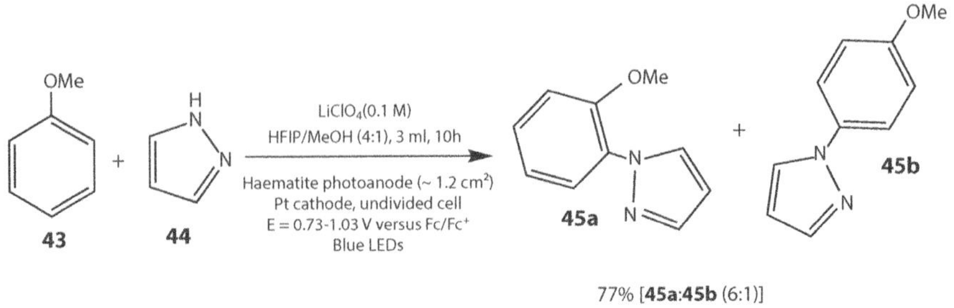

Scheme 6.6 iPEC C–H amination of electron-rich arenes with a hematite photoanode.

cations, which then react with azoles **44** to produce nitrogen heterocycles **45** (Scheme 6.6) [33].

Moreover, the TiO_2 photocatalyst has demonstrated its utility in a variety of applications [44], including the formation of C–N bonds using alcohols as mild and green alkylating reagents for amine **51** *N*-alkylation and the *de novo* synthesis of five- and six-membered *N*-heterocycles. The intense photogenerated electron/holes on TiO_2 nanoparticle surfaces efficiently activate inert C–H, C–C, and C–X bonds of common organic compounds to form useful C–N molecules. The C–N compounds **47–50** produced by these TiO_2-based photocatalysts are summarized in Scheme 6.7 below [45].

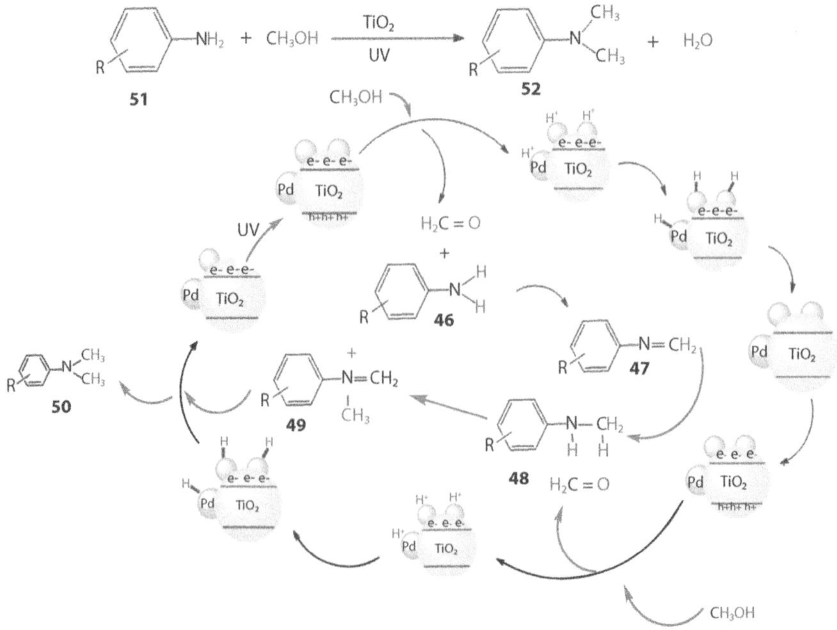

Scheme 6.7 C–N bond construction by TiO_2 photocatalysis.

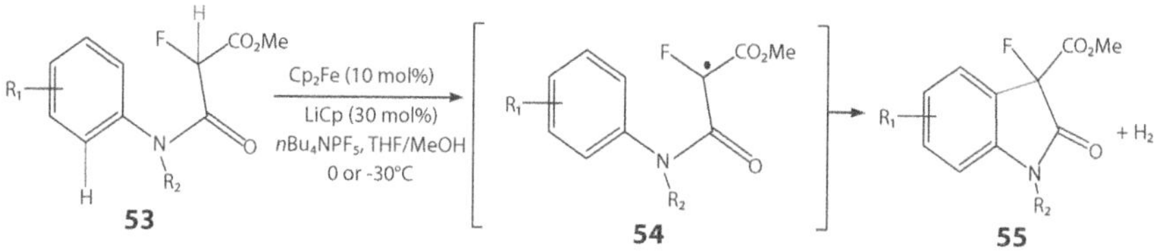

Scheme 6.8 Synthesis of 3-fluorooxindole.

In terms of activation and modification of organic substrates, it is clear that photocatalysis and electrochemistry have many similarities. The combination of a homogeneous photocatalytic system and an electrochemical cell opens new unexplored synthesis pathways and enables net-oxidative photocatalytic processes in the absence of a chemical oxidant, such as the C–H alkylation of heteroarenes and the coupling of azoles with arenes in the presence of an electrogenerated photocatalyst [46, 47].

6.7 Reagent-Free Cross Dehydrogenative Coupling

Through cross-coupling of C(sp^3)–H and C(sp^2)–H bonds, a novel method was proposed for producing functionalized monofluoroalkyl radicals **54** by electrochemical activation of C–H bonds, as well as its use in the synthesis of 3-fluorooxindoles **55**. As a redox catalyst, the low-cost organometallic compound ferrocene (Cp$_2$Fe) was used for the electrochemical cyclization of substrate **53**. The reaction conditions were specified as electrolyzing the substrate **53** at 0°C in a mixed solvent of MeOH/THF (1:2) with 10 mol% of Cp$_2$Fe as the catalyst and 30 mol% of LiCp as the additive. The constant current electrolysis was carried out in a single cell that was outfitted with a reticulated vitreous carbon (RVC) anode and a Pt plate cathode. After using 2.5 F of charge, the required 3-fluorooxindole **55** was extracted in 82% yield under these moderate conditions (Scheme 6.8) [48].

6.8 Conclusion

With the advancement of PEC research, a vast amount of information and experience has been gained, resulting in the discovery of a variety of active materials. In the valence band (VB), electrons might be activated and enter the conduction band (CB), resulting in the generation of photogenerated electrons (e$^-$) and holes (h$^+$); the photogenerated charge carriers (electrons and holes) could then travel to the surface of the semiconductor. Suitable semiconducting photoelectrode design is big challenge to increase the efficiency of photoelectrochemical cells. It is necessary to fabricate suitable electrode for PEC to increase absorption of light with suitable band gap. Moreover, fabricated photoelectrodes can increase the fast transport and efficient separation of electrons and holes to increase the productivity of PEC. Photoelectrocatalytic cells for the production of valuable chemicals and solar fuels are a potentially important technology for ensuring a long-term energy future. This is something that can be developed faster by focusing on more crucial features, such as cost effectiveness, operation at temperatures greater than room temperature, and efficient recovery of reaction products.

References

1. Hartwig, J.F. and Larsen, M.A., Undirected, homogeneous C–H bond functionalization: Challenges and opportunities. *ACS Cent. Sci.*, 2, 5, 281–292, May 2016.
2. Ren, Z., Mo, F., Dong, G., Catalytic functionalization of unactivated sp³ C–H bonds via exo-directing groups: Synthesis of chemically differentiated 1, 2-diols. *J. Am. Chem. Soc.*, 134, 41, 16991–16994, Oct. 2012.
3. Godula, K., Sames, D., C-H Bond functionalization in complex organic synthesis. *Science.*, 312, 5770, 67–72, April 2006.
4. Holland, H.L., C-H activation. *Curr. Opin. Chem. Biol.*, 3, 1, 22–27, 1999.
5. Nicolaou, K.C., Bulger, P.G., Sarlah, D., Palladium-catalyzed cross-coupling reactions in total synthesis. *Angew. Chem.-Int. Ed.*, 44, 29, 4442–4489, 2005.
6. Yamaguchi, J., Yamaguchi, A.D., Itami, K., C-H bond functionalization: Emerging synthetic tools for natural products and pharmaceuticals. *Angew. Chem.-Int. Ed.*, 51, 36, 8960–9009, 2012
7. Kim, D.-S., Park, W.-J., Jun, C.-H., Metal–organic cooperative catalysis in C–H and C–C bond activation. *Chem. Rev.*, 117, 13, 8977–9015, Jul. 2017.
8. Tang, S., Liu, Y., Lei, A., Electrochemical oxidative cross-coupling with hydrogen evolution: A green and sustainable way for bond formation. *Chem*, 4, 1, 27–45, 2018.
9. Rej, S., Ano, Y., Chatani, N., Bidentate directing groups: An efficient tool in C-H bond functionalization chemistry for the expedient construction of C-C bonds. *Chem. Rev.*, 120, 3, 1788–1887, 2020.
10. Lewis, J.C., Coelho, P.S., Arnold, F.H., Enzymatic functionalization of carbon–hydrogen bonds. *Chem. Soc. Rev.*, 40, 4, 2003–2021, Mar. 2011.
11. Kolbe, H., Untersuchungen über die elektrolyse organischer verbindungen. *Justus Liebigs Ann. Chem.*, 69, 3, 257–294, 1849.

12. Liu, J., Lu, L., Wood, D., Lin, S., New redox strategies in organic synthesis by means of electrochemistry and photochemistry. *ACS Cent. Sci.*, 6, 8, 1317–1340, Aug. 2020.
13. Hardwick, T., Qurashi, A., Shirinfar, B., Ahmed, N., Interfacial photoelectrochemical catalysis: Solar-induced green synthesis of organic molecules. *ChemSusChem*, 13, 8, 1967–1973, 2020.
14. Pokhrel, T., Bijaya, B.K., Giri, R., Adhikari, A., Ahmed, N., C–H bond functionalization under electrochemical flow conditions. *Chem. Rec.*, 22, 6, 2022, e202100338.
15. Yu, Y., Guo, P., Zhong, J.S., Yuan, Y., Ye, K.Y., Merging photochemistry with electrochemistry in organic synthesis. *Org. Chem. Front.*, 7, 1, 131–135, 2019.
16. Warner, J., How a Solar Cell Works. A Solar Future. *Chem. Matters*, American Chemical Society, 9–11, April/May 2014. https://teachchemistry.org/chemmatters/april-2014/a-solar-futurehttps://www.acs.org/content/acs/en/education/resources/highschool/chemmatters/past-issues/archive-2013-2014/how-a-solar-cell-works.html (accessed Feb. 12, 2022).
17. Luceño-Sánchez, J.A., Díez-Pascual, A.M., Capilla, R.P., Materials for photovoltaics: State of art and recent developments. *Int. J. Mol. Sci.*, 20, 4, Article number 976, Feb. 2019.
18. Wrighton, M.S., Photoelectrochemical conversion of optical energy to electricity and fuels. *Acc. Chem. Res.*, 12, 9, 303–310, Sep. 1979.
19. Khalid, S., Malik, M.A., Ahmed, E., Khan, Y., Ahmed, W., Chapter 13- Synthesis of transition metal sulfide nanostructures for water splitting, in: *Emerging Nanotechnologies for Renewable Energy*, W. Ahmed, M. Booth, E. Nourafkan (Eds.), pp. 311–341, Elsevier, 2021.
20. Chen, Q., Fan, G., Fu, H., Li, Z., Zou, Z., Tandem photoelectrochemical cells for solar water splitting. *Adv. Phys. X*, 3, 1, 1487267, Jan. 2018.
21. Chae, S.Y., Kim, Y., Park, E.D., Im, S.H., Joo, O.-S., CuInS2 photocathodes with atomic gradation-controlled (Ta, Mo) x (O, S) y passivation layers for efficient photoelectrochemical H2 production. *ACS Appl. Mater. Interfaces*, 13, 49, 58447–58457, Dec. 2021.
22. Miao, Y. and Shao, M., Photoelectrocatalysis for high-value-added chemicals production. *Chin. J. Catal.*, 43, 3, 595–610, Mar. 2022.
23. Monllor-Satoca, D., Díez-García, M., II, Lana-Villarreal, T., Gómez, R., Photoelectrocatalytic production of solar fuels with semiconductor oxides: Materials, activity and modeling. *Chem. Commun.*, 56, 82, 12272–12289, Oct. 2020.
24. Wu, Y.-C., Song, R.-J., Li, J.-H., Recent advances in photoelectrochemical cells (PECs) for organic synthesis. *Org. Chem. Front.*, 7, 14, 1895–1902, 2020.
25. Li, P., Zhang, T., Mushtaq, M.A., Wu, S., Xiang, X., Yan, D., Research progress in organic synthesis by means of photoelectrocatalysis. *Chem. Rec.*, 21, 4, 841–857, 2021.

26. Cui, Y., Ge, P., Chen, M., Xu, L., Research progress in semiconductor materials with application in the photocatalytic reduction of CO_2. *Catalysts*, 12, 4, 4, Article number 372, Apr. 2022.
27. Bessegato, G.G., Guaraldo, T.T., Zanoni, M.V.B., *Enhancement of photoelectrocatalysis efficiency by using nanostructured electrodes*, IntechOpen, 2014.
28. a) Schiel, F. *et al.*, A 3D-printed open access photoreactor designed for versatile applications in photoredox- and photoelectrochemical synthesis. *ChemPhotoChem*, 5, 431–437, 2021. b) Wu, S., Kaur, J., Karl, T.A., Tian, X., Barham, J.P., Synthetic molecular photoelectrochemistry: New frontiers in synthetic applications, mechanistic insights and scalability. *Angew. Chem. Int. Ed.*, 61, 12, Mar. 2022.
29. Wu, S., Kaur, J., Karl, T.A., Tian, X., Barham, J.P., Synthetic molecular photoelectrochemistry: New Frontiers in synthetic applications, mechanistic insights and scalability. *Angew. Chem. Int. Ed.*, 61, 2022, e202107811.
30. Raţ, C., II, Soran, A., Varga, R.A., Silvestru, C., *C–H Bond Activation Mediated by Inorganic and Organometallic Compounds of Main Group Metals*, 1st ed., vol. 70, Elsevier Inc, 2018.
31. Decker, F. and Cattarin, S., Photoelectrochemical cells | Overview, in: *Encyclopedia of Electrochemical Power Sources*, pp. 1–9, 2009.
32. Yan, H., Hour, Z.-W., Xu, H.-C., Photoelectrochemical C–H alkylation of heteroarenes with organotrifluoroborates. *Angew. Chem. Int. Ed.*, 131, 4640–4643, 2019.
33. Zhang, L., Liardet, L., Luo, J., Ren, D., Grätzel, M., Hu, X., Photoelectrocatalytic arene C–H amination. *Nat. Catal.*, 2, 4, 366–373, 2019.
34. Tateno, H., Miseki, Y., Sayama, K., Photoelectrochemical dimethoxylation of furan via a bromide redox mediator using a BiVO4/WO3 photoanode. *Chem. Commun.*, 53, 31, 4378–4381, 2017.
35. Pandey, G. and Laha, R., Visible-light-catalyzed direct benzylic c (sp^3)–H amination reaction by cross-dehydrogenative coupling. *Angew. Chem. Int. Ed.*, 54, 14875–15092, 2015.
36. Cheng, J., Zhang, M., Wu, G., Wang, X., Zhou, J., Cen, K., Photoelectrocatalytic reduction of CO_2 into chemicals using Pt-modified reduced graphene oxide combined with Pt-modified TiO_2 nanotubes. *Environ. Sci. Technol.*, 48, 12, 7076–7084, Jun. 2014.
37. Ješić, D., Lašič Jurković, D., Pohar, A., Suhadolnik, L., Likozar, B., Engineering photocatalytic and photoelectrocatalytic CO_2 reduction reactions: Mechanisms, intrinsic kinetics, mass transfer resistances, reactors and multi-scale modelling simulations. *Chem. Eng. J.*, 407, 126799, Mar. 2021.
38. Xie, S., Zhang, Q., Liu, G., Wang, Y., Photocatalytic and photoelectrocatalytic reduction of CO^2 using heterogeneous catalysts with controlled nanostructures. *Chem. Commun.*, 52, 1, 35–59, Dec. 2015.
39. Wang, J. *et al.*, Photoelectrocatalytic reduction of CO_2 to paraffin using p-n heterojunctions. *iScience*, 23, 1, 100768, Jan. 2020.

40. Barham, J.P. and König, B., Synthetic photoelectrochemistry. *Angew. Chem. Int. Ed.*, 59, 29, 11732–11747, Jul. 2020.
41. Kim, H., Kim, H., Lambert, T.H., Lin, S., Reductive electrophotocatalysis: Merging electricity and light to achieve extreme reduction potentials. *J. Am. Chem. Soc.*, 142, 5, 2087–2092, Feb. 2020.
42. Huang, H. and Lambert, T.H., Electrophotocatalytic $S_N Ar$ reactions of unactivated aryl fluorides at ambient temperature and without base. *Angew. Chem. Int. Ed.*, 59, 2, 658–662, Jan. 2020.
43. Hardwick, T. and Ahmed, N., C–H functionalization via electrophotocatalysis and photoelectrochemistry: Complementary synthetic approach. *ACS Sustain. Chem. Eng.*, 9, 12, 4324–4340, Mar. 2021.
44. Kazuya, N., Akira, F., TiO_2 photocatalysis: Design and applications. *J. Photochem. Photobiol. C: Photochem. Rev.*, 13, 3, 169–189, 2012, https://doi.org/10.1016/j.jphotochemrev.2012.06.001.(https://www.sciencedirect.com/science/article/pii/S1389556712000421)
45. Ma, D., Zhai, S., Wang, Y., Liu, A., Chen, C., Synthetic approaches for C-N bonds by TiO_2 photocatalysis. *Front. Chem.*, 7, 635, Sep. 2019.
46. Capaldo, L., Quadri, L.L., Ravelli, D., Merging photocatalysis with electrochemistry: The dawn of a new alliance in organic synthesis. *Angew. Chem. Int. Ed.*, 58, 49, 17508–17510, Dec. 2019.
47. Perathoner, S., Centi, G., Su, D., Turning perspective in photoelectrocatalytic cells for solar fuels. *ChemSusChem*, 9, 4, 345–357, Feb. 2016.
48. Holmes-Gentle, I., Alhersh, F., Bedoya-Lora, F., Hellgardt, K., Photoelectrochemical reaction engineering for solar fuels production, in: *Photoelectrochemical Solar Cells*, N.D. Sankir and M. Sankir (Eds.), pp. 1–41, John Wiley & Sons, Inc., Hoboken, NJ, USA, 2018.

Part III

PHOTOCATALYTIC CO_2 REDUCTION TO FUELS

7
Graphene-Based Catalysts for Solar Fuels

Zhou Zhang, Maocong Hu* and Zhenhua Yao[†]

Department of Chemical Engineering, Jianghan University, Wuhan, China

---

Abstract

Graphene, discovered in 2004 by A. K. Geim and K. S. Novoselov, is considered one of the most promising materials in a wide range of processes because of its outstanding electronic, optical, thermal, and mechanical properties. The growing global energy demand and environmental impact of fossil energy resources have led to energy security and global climate change issues while catalytic processes demonstrated great potential in solving both energy crisis and environmental pollution. One of the excellent examples is the application of graphene-based catalysts in the production of solar fuels including hydrocarbons (eg. formic acid, methanol, etc.) from the reduction of carbon dioxide and hydrogen from solar-driven water-splitting with photocatalysis process. In this chapter, recent progress in common synthetic methods to prepare graphene-based catalysts was first reviewed while critical comments and possible solutions were also provided. The second section introduced a summary of the characterization techniques, which were closely relevant to catalysis fields. The performance of graphene-based catalysts in typical processes, such as the conversion of carbon dioxide to value-added hydrocarbons and clean hydrogen by water-splitting were further discussed. Finally, we presented the challenges and opportunities for the future development of graphene-based catalysts in solar fuels.

Keywords: Graphene, catalyst, photocatalysis, CO2 conversion, hydrogen production

---

*Corresponding author: maocong.hu@jhun.edu.cn
[†]Corresponding author: zhenhua.yao@jhun.edu.cn

7.1 Introduction

With the growth of global energy demand and the serious destruction of the ecological environment, the production of sustainable clean energy is an alternative way for future development [1]. Production of solar fuels including hydrocarbons (eg. formic acid, methanol, etc) from the reduction of carbon dioxide and hydrogen from solar-driven water-splitting with photocatalysis process is an effective and sustainable process [2–5]. To efficiently convert solar energy with catalytic processes, photocatalytic nanostructured materials need to have suitable band gaps, exhibit strong visible light response, possess unique properties such as high photocatalytic performance, easy preparation, nontoxicity, and good thermal and chemical stability. Semiconductors are widely used in photocatalysis due to their low price, high stability, non-toxicity, and suitable energy band positions [6, 7]. Especially, visible-light-induced semiconductor photocatalysts can efficiently utilize light energy and enhance photocatalytic activity. Unfortunately, the photocatalytic activity of semiconductors has not yet reached the level required for industrial production applications. Over the past decades, explorers have proposed many techniques to improve the photocatalytic performance of semiconductors, including semiconductor recombination, structural modification, elemental doping, and surface modification [8, 9].

Graphene, a single-layer carbon sheet with a honeycomb lattice structure, was discovered in 2004, which attracted the attention of researchers due to its excellent properties including high electron mobility, large specific surface area, enhanced light absorption, high stability, strong corrosion-resistance, excellent electrical conductivity, and high adsorption capacity [10, 11]. It shows great potential in the field of photocatalysis [12–15], which plays a key role in improving the photocatalytic performance of semiconductor catalysts as an additive, support, electron acceptor, and transporter, a cocatalyst, and a photocatalyst [16–18]. For example, graphene-based catalysts are employed in photocatalytic CO_2 reduction, converting harmful greenhouse gases into valuable energy-rich chemicals such as CH_4 and CH_3OH by using solar energy [19, 20]. Graphene is usually involved in water reduction or hydrogen evolution reaction (HER) and water oxidation or oxygen evolution reaction (OER) with the assistance of sunlight (i.e. photocatalysis), where it severs as different roles [21, 22].

This chapter will introduce the preparation, characterization, and performance of graphene-based catalysts for solar fuels production, focusing on two rapidly developing areas: photocatalytic CO_2 reduction and water splitting. The major challenges and opportunities for future development are also discussed. This chapter provides key information to our community to design and fabricate graphene-based catalysts with great performance for solar fuels production.

7.2 Preparation of Graphene and Its Composites

7.2.1 Preparation of Graphene (Oxide)

Graphene was first peeled off manually and mechanically using Scotch tape by Geim *et al.* while different preparation technologies were continuously developed and improved in recent years [23, 24]. Generally, the preparation methods of graphene can be divided into two categories: bottom-up method with atoms or molecules as origination via chemical reactions, and top-down method using techniques such as exfoliation and electrostatic deposition.

Epitaxial growth and chemical vapor deposition (CVD) are typical bottom-up methods for synthesizing graphene flakes, where carbon atoms can be reconfigured upon the substrate to form graphene flakes. For example, free-standing graphene can be induced to grow on SiC substrates, which is commonly used for epitaxial growth [25]. Plasma-enhanced, hot-wall, and cold-wall CVD also showed great promise in the preparation of high-quality large-area graphene, which can be deposited at relatively low temperatures using plasma-enhanced chemical vapor deposition [26]. Usually, graphene harvested by epitaxial growth and chemical vapor deposition led to high-quality while it is difficult to be modified by semiconductors. In contrast, graphene with structural defects and oxygen-containing functional groups can be well dispersed in solvents, which can be further easily compounded with semiconductors to prepare graphene-based catalysts.

Redox exfoliation is a common synthesis method for preparing large amounts of graphene, namely the Hummers method as shown in Figure 7.1. Using strong oxidants (sulfuric acid, nitric acid, potassium permanganate, etc.) to oxidize graphite, introducing oxygen-containing functional groups (such as hydroxyl groups, ketone groups, ether bonds, etc.) to prepare graphene oxide (GO), and obtain GO by mechanical exfoliation, while various inorganic impurities are removed with post-purification. Graphene oxide is reduced by chemical, microwave, photothermal, and other reduction methods to obtain incompletely reduced graphene. The introduction of oxygen-containing functional groups during the oxidation process can interact with metal cations and provide active sites for

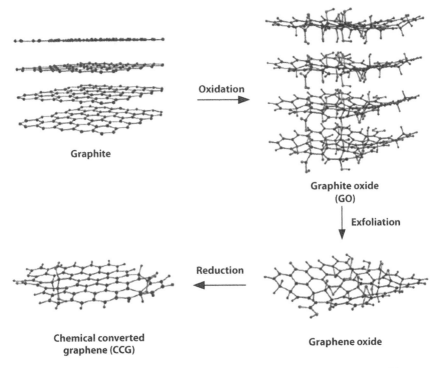

Figure 7.1 Graphene oxide prepared by hummers method and graphene prepared by chemical reduction. Reprinted with permission from [27], Copyright 2011, Wiley.

nanoparticle nucleation and growth during the preparation of composite materials. It plays an important role and benefits the preparation of graphene-based photocatalysts.

7.2.2 Preparation of Graphene-Based Photocatalysts

Commonly used semiconductors for the preparation of graphene-based photocatalysts include metal oxides (such as TiO_2 [28], ZnO [29], SnO_2 [30]), multi-component metal oxides (such as $BaTiO_3$ [31]), metal sulfides (such as ZnS [32]), non-metallic polymers (such as g-C_3N_4 [33]), etc. The preparation method has a direct impact on the morphology, structure, and size of the final graphene-based photocatalysts. Different methods lead to different combinations of graphene and semiconductors, which affects the activity of the graphene-based photocatalysts. Typical preparation methods including hydrothermal/solvothermal method, solution mixing method, and *in-situ* growth method are introduced below.

7.2.2.1 Hydrothermal/Solvothermal Method

The hydrothermal/solvothermal method refers to the use of water or organic matter as a solvent in a closed hydrothermal reaction vessel to react under high temperature and high-pressure reaction conditions, where semiconductors and graphene can be closely combined to prepare composites [34–36]. The metal contained compounds and graphene oxide, which are easily soluble in water, or organic solvents are usually used as precursors in this method. It led to the production of metal oxide/graphene composites under hydrothermal reaction. For example, Zhong *et al.* [37] prepared $NaTaO_3$/reduced graphene oxide (rGO) composite photocatalyst by hydrothermal method (Figure 7.2). By using different contents of graphene oxide (GO), $NaTaO_3$/reduced graphene oxide (rGO) composite photocatalyst can be synthesized by hydrothermal reaction in an aqueous solution. $NaTaO_3$ prepared by the hydrothermal method has high purity. The composite can improve the absorption of visible light and ultraviolet light to a certain extent. In another work, to obtain a homogeneous composite of graphene and TiO_2 photocatalyst, Lin *et al.* [38] synthesized

Figure 7.2 SEM results of (a) $NaTaO_3$ and (b, c) $NaTaO_3$/rGO. Reprinted with permission from [37], Copyright 2019, Elsevier.

reduced graphene oxide (rGO)-TiO$_2$ nanocomposites by a simple solvothermal method in an acidic solution. The electrostatic effect generated by the solvothermal reaction promotes the assembly of graphene oxide and titania. The two-dimensional wrinkled structure of reduced graphene oxide sheets encapsulates anatase TiO$_2$ in the prepared nanocomposite. To develop an inexpensive and efficient oxygen evolution (OER) catalyst, Roy et al. [39] used a hydrothermal method to prepare graphene oxide-wrapped cobalt phosphate nanotube metal electrocatalysts.

7.2.2.2 Sol-Gel Method

The sol-gel method is widely used in the preparation of graphene-based photocatalysts. Compared with the hydrothermal method, the conditions of this method are mild. For example, Pei et al. [40] prepared graphene/TiO$_2$ nanocomposite by sol-gel assisted electrospray method to prepare a photocatalyst that can directly utilize solar energy for surface self-cleaning to remove pollutants diffused in the air. To enhance the photoreduction of nitrobenzene to prepare aniline under visible light conditions, Tashkandi et al. [41] prepared MnCo$_2$O$_4$ nanoparticles (NPs) by a simple sol-gel method. The MnCo$_2$O$_4$ nanoparticles (NPs) were uniformly dispersed on reduced graphene oxide (rGO). Trusova et al. [42] utilized N, N-dimethyl octylamine to promote the formation and stabilization of metal sols as well as stabilization of graphene suspensions, with the assistance of a combination of sonochemical and sol-gel methods. Zadmehr et al. [43] explored the preparation of magnetic titania-graphene oxide (GO) composites by sol-gel method and hydrothermal method, respectively. Moreover, the TiO$_2$-Fe$_3$O$_4$ core-shell was decorated on graphene oxide flakes (Figure 7.3).

7.2.2.3 In Situ Growth Method

In situ growth method is also widely used to prepare graphene-based catalysts while metal salts are commonly used as precursors. For example, Liu et al. [44] used triethanolamine to modify graphene oxide and prepared highly dispersed nano-TiO$_2$ *in situ* on the modified functional graphene (Figure 7.4). Wang et al. [45] obtained monodisperse Cu$_3$P nanoplates/reduced graphene oxide (rGO) composites by *in situ* growth of monodisperse Cu$_3$P nanoplates on reduced graphene oxide via hydrophobic interaction in a hydrophobic interaction colloidal solution. Ji et al. [46] prepared rGO/CeO$_2$ nanocomposites by dispersing CeO$_2$ nanoparticles onto reduced graphene oxide (rGO) nanosheets using *in situ* growth and self-assembly methods. Wang et al. [47] used an *in-situ* deposition method

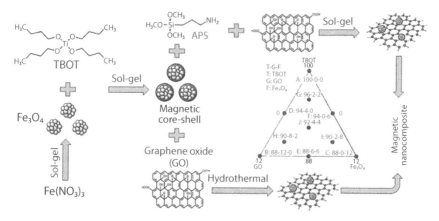

Figure 7.3 TBOT-GO-Fe$_3$O$_4$ nanocomposites were prepared by sol-gel method and hydrothermal method. Reprinted with permission from [43], Copyright 2021, Elsevier.

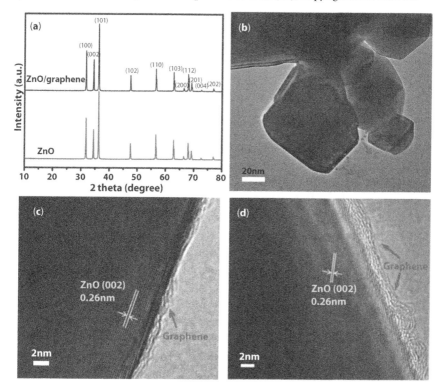

Figure 7.4 (a) XRD patterns and (b) TEM image of samples. (c) and (d) HRTEM images of ZnO/graphene, corresponding to the red dotted circles of c and d in Figure 7.3(b), respectively. (For interpretation of the references to color in this figure legend, the reader is referred to the web version of this article.) Reprinted with permission from [47], Copyright 2021, Elsevier.

to deposit a small amount of graphene as a co-catalyst on the surface of ZnO, and tightly combined graphene with oxidative properties to form a Schottky-junction graphene-based composite (Figure 7.4).

7.3 Graphene-Based Catalyst Characterization Techniques

Graphene-based catalysts have unique structures, and their catalytic characterization techniques are different from those of traditional catalysts. Commonly used characterization techniques for graphene-based catalysts include scanning electron microscopy (SEM), transmission electron microscopy (TEM), high-resolution transmission electron microscopy (HRTEM), x-ray photoelectron spectroscopy (XPS), x-ray diffraction (XRD), x-ray Ray Absorption Near Edge Structure (XANES), X-ray Absorption Fine Structure (XAFS), Atomic Force Microscopy (AFM), Fourier Transform Infrared Spectroscopy (FTIR), etc. Appropriate characterization techniques can be selected according to the specific purpose of the study.

7.3.1 SEM, TEM, and HRTEM

Scanning electron microscopy (SEM), transmission electron microscopy (TEM), and high-resolution transmission electron microscopy (HRTEM) were used to characterize the surface morphology and internal structure of graphene-based catalysts. Having a general understanding of the catalyst can also help us make a reasonable explanation of the performance of the catalyst. A visualization tool for graphene-based catalysts can measure the number of graphene layers and determine the significant impact of catalyst surface area on catalytic performance [48]. Graphene-based catalytic materials are prone to layer folding while TEM images can measure the thickness of the flakes. HRTEM observes the edges of graphene-based materials which can accurately calculate the number of layers of graphene-based catalysts [49]. For example, Ma *et al.* [50] prepared a g-C_3N_4/rGO/Bi_2MoO_6 composite catalyst, and it can be seen from Figure 7.5 that rGO is located between g-C_3N_4 and Bi_2MoO_6 to form a sandwich structure. The reduced graphene oxide acts as a good electronic medium in the z-type system, which contributes to the enhancement of photocatalytic efficiency. Pu *et al.* [51] used an *in-situ* growth method to coat TiO_2 NPs on reduced

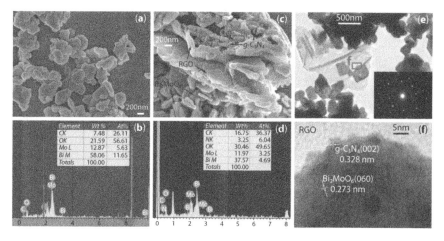

Figure 7.5 SEM images of (a) Bi_2MoO_6 and (c) MGN100 composite, and the accompanying EDX spectra (the marked area b and d) of (b) Bi_2MoO_6 and (d) MGN100 composite. TEM images of MGN100 composite (e) and the accompanying HRTEM image (f), and the inset of e is the SAED pattern of the MGN100 composite. Reprinted with permission from [50], Copyright 2017, Elsevier.

graphene oxide (TiO_2@rGO) under UV photoreduction conditions. Small TiO_2@rGO NPs with interconnected networks and polycrystalline features are prepared, and the thickness of the material can be estimated by statistical calculations.

7.3.2 X-Ray Techniques: XPS, XRD, XANES, XAFS, and EXAFS

XPS is a powerful spectroscopic technique capable of qualitative, quantitative, or semi-quantitative valence analysis of elemental composition. XPS is often used to characterize the elemental composition and content, chemical state, chemical bond, and molecular structure of materials [52–55]. By measuring the kinetic energy/binding energy of graphene-based composites under x-ray beam irradiation and the number of electrons escaping from this material, an XPS spectrum can be obtained. For example, Ansari *et al.* [56] used the sol-gel method to synthesize cobalt hexaferrite nanoparticles using natural reducing agents for the first time and prepared carbon-based nanocomposites including graphene and carbon nanotubes while XPS was employed (Figure 7.6).

X-ray diffraction (XRD) and X-ray absorption near-edge spectroscopy (XANES) were used to investigate important parameters such as the crystal

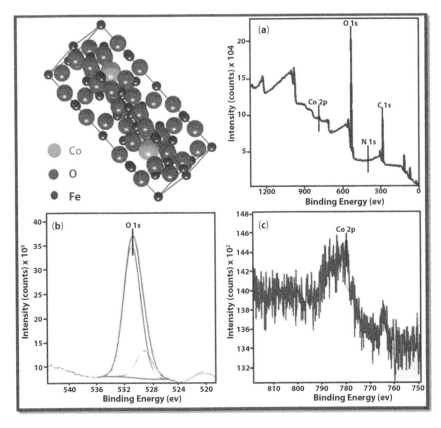

Figure 7.6 XPS spectra of CoFe$_{12}$O$_{19}$ nanoparticles (a), Co 2p (b) and O 1s (c). Reprinted with permission from [56], Copyright 2019, Elsevier.

structure and oxidation state of the catalyst [57–60]. X-ray absorption fine structure (EXAFS) is an effective characterization method for bond length and coordination. X-ray technology also has certain applications for evaluating the dispersion of catalytic sites, which helps analyze catalyst structure and performance [61, 62]. For example, Yang *et al.* [63] synthesized CoSe$_2$ nanoparticles by hydrothermal method using ascorbic acid and further immobilized them on graphene oxide (GO) nanosheets. The successful preparation of CoSe$_2$/GO nanocomposites was confirmed by the analysis of XRD patterns (Figure 7.7). Sayadi *et al.* [64] first prepared TiO$_2$/graphene (G) and then doped Bi and SnO$_2$ nanoparticles by hydrothermal method while the crystal phases of G, Bi, and Bi/SnO$_2$ components were accurately determined by XRD.

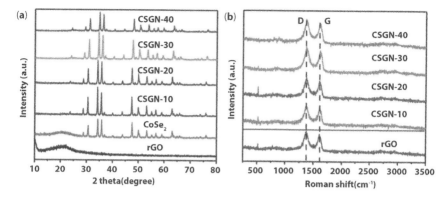

Figure 7.7 (a) XRD patterns of rGO, CoSe2, CSGN-10, CSGN-20, CSGN-30 and CSGN-40; (b) Raman spectra of CoSe2 and rGO. Reprinted with permission from [63], Copyright 2022, Elsevier.

7.3.3 Atomic Force Microscopy (AFM)

The atomic force microscope is a new type of high-resolution instrument, which has been widely used in the morphological characterization of various materials since its invention. AFM is easy to operate and the data has high confidence, and it is considered an important method for thickness measurement and layer number calculation. During the measurement, the surface of the graphene-based material is scanned by the probe on the cantilever, and the surface topography and three-dimensional topography image of the material can be obtained, and the thickness of the graphene film can be accurately measured through the image [65–70]. For example, Shen *et al.* [71] prepared silver chemically converted graphene (CCG) nanocomposites using an *in situ* chemical synthesis method. The reduction of graphene oxide flakes is accompanied by the formation of silver nanoparticles. The structure and composition of the nanocomposites were determined by transmission electron microscopy (TEM), atomic force microscopy (AFM), and X-ray diffraction. Zhong *et al.* [72] developed a novel two-dimensional sandwich-type imidazolate zeolite framework (ZIF)-derived graphene-based nitrogen-doped porous carbon sheets (GNPCSs) by *in situ* growth of ZIF on graphene oxide (GO). AFM analysis also revealed the layered structure of GNPCSs-800 with a thickness of about 20 nm. Sumathi *et al.* [73] explored the effect of hydrothermal reaction temperature on the morphology of graphene nanoparticles, and used AFM to characterize the morphology of graphene nanoparticles. The morphology of graphene nanoparticles at different hydrothermal temperatures is shown in Figure 7.8. The hydrothermal temperature had a significant

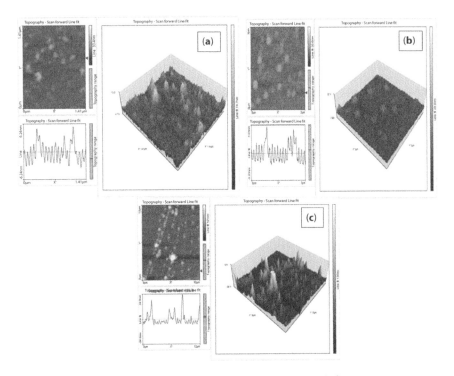

Figure 7.8 shows the AFM images of Graphene nanoparticles at different temperatures, (a) 140 °C (b) 160 °C (c) 180 °C. Reprinted with permission from [73], Copyright 2021, Elsevier.

effect on both the growth of graphene nanoparticles and the surface roughness of the material, and the agglomeration of graphene nanoparticles was reduced at 180 °C compared to other temperatures.

7.3.4 Fourier Transform Infrared Spectroscopy (FTIR)

Fourier transform infrared spectroscopy is often used to detect chemical bonds in molecules. During the preparation process of graphene oxide, graphite is exfoliated using oxidation. During this process, a series of functional groups are often introduced, and a large number of residues will remain in the subsequent reduction. Functional groups have a huge impact on the performance of catalysts. Thus, it is crucial to evaluate the removal of functional groups for graphene-based catalysts. It is necessary to use Fourier transform infrared spectroscopy (FTIR) to detect chemical bonds in the composites [74–76]. Fan *et al.* [77] used FT-IR to study the

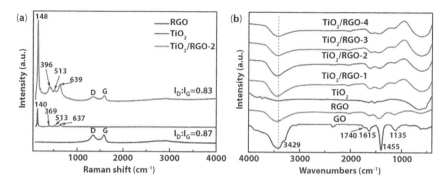

Figure 7.9 (a) Raman spectra of RGO, TiO_2, and TiO_2/RGO-2; (b) FT-IR spectra of GO, RGO, TiO_2, and TiO_2/RGO nanohybrids. Reprinted with permission from [77], Copyright 2021, Elsevier.

functional groups of graphene and titania dioxide. The absorption peaks at 740 cm^{-1}, 1615 cm^{-1}, 1455 cm^{-1} and 1135 cm^{-1} of graphene oxide in Figure 7.9 are C=O, C=C, C–OH and C–O–C characteristic absorption peaks, which also confirm the existence of various oxygen-containing functional groups in graphene. The peak of RGO at 1455 cm^{-1} is much weaker than that of graphene oxide, but there is still an absorption peak of oxygen-containing functional groups, indicating that graphene is only partially reduced during the redox process. Kocijan *et al.* [78] prepared TiO_2@rGO nanomaterials by a combination of hydrothermal method and calcination method. By comparing the FT-IR spectra of GO, rGO, and TiO_2@rGO nanomaterials, the peak intensities of oxygen-containing functional groups (O-H, C=O, C-O in rGO) are weakened. The peaks of TiO_2@rGO nanocomposites are located at 450-900 cm^{-1}, which are the stretching vibration peaks of Ti-O-Ti and Ti-O-C bonds, indicating the existence of a bond between Ti and C due to effective interaction [79, 80].

7.3.5 Other Technologies

In addition to characterization techniques mentioned above, Raman Spectroscopy, Inductively Coupled Plasma Mass Spectrometry (ICP), Ultraviolet-Visible Spectroscopy (UV-vis), X-ray Fluorescence (XRF), Thermogravimetric Analysis (TGA), Brunel-Angstrom Techniques such as Better-Taylor (BET), and Scanning Tunneling Microscopy (STM) have also been instrumental in the characterization of graphene-based catalysts, which was systematically reviewed in our previous work [60, 81].

7.4 Graphene-Based Catalyst Performance

CO_2 reduction to produce hydrocarbon fuels and water splitting for the production of hydrogen with photocatalysis are considered desirable and promising strategies to solve energy and environmental crisis [82–88]. Excellent electron transport of graphene is beneficial to the separation of photogenerated carriers, which can reduce the recombination of electrons and holes and further improve the efficiency of photocatalysis [89–91]. In this section, the applications of graphene-based composite catalysts for photocatalytic CO_2 reduction to produce hydrocarbons and water splitting to produce hydrogen are summarized (Table 7.1). The catalytic performance was provided and the positive effect of graphene on the performance was further discussed.

Table 7.1 Preparation, characterization, and application of graphene-based catalysts for solar fuels production.

Catalyst	Preparation method	Characterization techniques	Performance	Ref.
$CsPbBr_3$/GO	Ligand-assisted reprecipitation (LARP) method	XRD, TEM, HR-AEM, UV-vis, XPS, TRPL, TGA	Methane, 6.9 $\mu mol \cdot g^{-1} \cdot h^{-1}$	[92]
$AgCuInS_2$-G-TiO_2	Hydrothermal	XRD, SEM, TEM, XPS, UV-vis, Raman, FTIR	A total methanol amount of 15.21%.	[93]
TiO_2/NG	CVD	FESEM, TEM, XPS, EDS, XRD, TGA, Raman	The sum yield of CO, CH_3OH, and CH_4 of 18.11 $\mu mol \cdot g^{-1} \cdot h^{-1}$	[94]
$LaYAgO_4$-G-TiO_2	Hydrothermal	XRD, SEM-EDX, HRTEM, XPS, Raman, DRS	After 48 hours of reaction, the total methanol increased by 12.27%	[95]
G-$Zn_{0.5}Cd_{0.5}S$	Hydrothermal	XRD, FESEM, TEM, HRTEM, BET, XPS, UV-vis, Raman	Methanol, 1.96 $\mu mol \cdot g^{-1} \cdot h^{-1}$	[96]

(Continued)

Table 7.1 Preparation, characterization, and application of graphene-based catalysts for solar fuels production. (*Continued*)

Catalyst	Preparation method	Characterization techniques	Performance	Ref.
Au-TiO$_2$-NG	Photodeposition	XRD, HRSTEM, HRTEM, UV-Vis, DRS, XPS, Raman	The highest release value of 742.39µmol·g^{-1}·h^{-1}	[97]
TiO$_2$-rGO	Hydrothermal	XRD, TEM, FESEM, FTIR, BET, UV-Vis	12.26µmol/g MeOH, 9.46µmol/g CH$_4$ and 1.83µmol/g EtOH (4h)	[98]
G/Pd@PtCu	Hydrothermal	TEM, HRTEM, STEM, EDS, TEM, ICP-MS, XPS, XRD, FTIR	Methane, 129.7µmol·g^{-1}·h^{-1}	[99]
Ag/rGO	Wet assisted laser ablation	XRD, SEM, UV-Vis	Carbon monoxide, 133.1µmol·g^{-1}·h^{-1}	[100]
Cu(II)/ING/CeO$_2$	Hydrothermal	XRD, TEM, FTIR, UV-vis, XPS, Raman	Methanol, 385.8µmol·g^{-1}·h^{-1}	[101]
NG/CdS	CVD	FESEM, STEM, ISI-XPS, DRIFTS	The yields of CO and CH$_4$ reached 2.59 and 0.33 µmol, respectively	[102]
hTS-G	Hydrothermal	SEM, EDS, TEM, BET, XRD, UV-DRS, FTIR	Carbon monoxide, 187.9µmol·g^{-1}·h^{-1}	[103]
GTG	Step-by-step loading method	XRD, TEM, TG, FESEM, XPS, UV-vis, BET	Carbon monoxide, 3.4µmol·g^{-1}·h^{-1}	[104]
rGO/TiO$_2$	Electrochemical anodization	FESEM, TEM, XPS	Carbon monoxide, 760µmol·g^{-1} in 2 hrs	[105]

(*Continued*)

Table 7.1 Preparation, characterization, and application of graphene-based catalysts for solar fuels production. (*Continued*)

Catalyst	Preparation method	Characterization techniques	Performance	Ref.
G/ZnS	Hydrothermal	XRD, EDS, SEM, TEM, UV-vis, XPS	Hydrogen yield, 0.049μmol/cm^2/min	[106]
Pt-G/TiO$_2$	Photodeposition	UV-vis/DRS, XRD, TEM	Hydrogen yield, 4.71mmol·h^{-1}·g^{-1}	[107]
WO$_3$/TiO$_2$/rGO	Hydrothermal	XRD, FESEM, SEM, TEM, XPS, UV-vis, ICP, Raman	Hydrogen yield, 245.8μmol·h^{-1}·g^{-1}	[108]
Au@ZnO-G	Nanoparticle solution mixing	HRTEM, XRD, UV-vis, Raman, XPS	Hydrogen yield, 709μmol·h^{-1}·g^{-1}	[109]
ZnO-MoS$_2$-G	Solid dispersion method	SEM, EDS, XRD, XPS, Raman	Hydrogen yield, 5.4mmol·h^{-1}·g^{-1}	[110]
Ni/GO/CdS	Photodeposition	BET, ATR-FTIR, DRS, HRTEM, XRD, SEM, UV-vis, Raman	Hydrogen yield, 8866μmol·h^{-1}·g^{-1}	[111]
rGO/SrTiO$_3$/PAN	Electrospinning technology	TGA, Raman, SEM, TEM, XRD, UV-vis	Hydrogen yield, 170mmol·h^{-1}·g^{-1}	[112]
rGO-LaNiO$_3$	Self-assembled photocatalytic reduction	XRD, SEM, TEM, BET, FTIR, XPS, Raman	Hydrogen yield, 3.22mmol·h^{-1}·g^{-1}	[113]
Co@CoO/NG	Calcined ZIF-67 and *in-situ* preparation process	TGA, XRD, XPS, FESEM, HRTEM, UV-vis	Hydrogen yield, 330μmol·h^{-1}·g^{-1}	[114]
S, N-GQD/P25	Hydrothermal	TEM, HRTEM, XRD, FTIR, UV-vis, Raman	Highest hydrogen production rate, 5.7μmol·h^{-1}	[115]

7.4.1 Photocatalytic CO_2 Reduction

With the continuous development of modern technology and the continuous improvement of the quality of life, the use of fossil energy has become crazier and crazier, and the use of various energy sources has increased year by year, followed by environmental pollution hidden behind the bustling scene. Among them, the greenhouse effect and energy shortage have the greatest impact on our lives. When we enjoy the convenience brought by energy, we have to consider how to solve the worries of global warming. We try to eliminate the effects of excess CO_2 in many ways, and try to solve this problem by capturing storage, absorption, and other ways. Photocatalytic CO_2 reduction converts carbon dioxide into hydrocarbon fuels such as methanol and methane. This method has been widely recognized by people. Through the action of light everywhere in nature, under the action of catalysts, the troublesome CO_2 could be converted into valuable products. Unfortunately, the photocatalyst's photogenerated electron and hole recombination lead to the problem of low efficiency. It is well known that graphene can reduce the recombination of electrons and holes through efficient electron transport, which has attracted several researchers to use graphene to develop multi-modal photocatalysts to improve the conversion efficiency of CO_2 reduction.

Nano heterostructures (NHS) were widely considered to promote light absorption and charge separation. Chen et al. [92] reported $CsPbBr_3$ GO NHSs and $CsPbBr_3$ conductive few-layer graphene (FLG) NHSs visible light catalytic reduction of CO_2 to methane. Using different graphene materials, the conductive few-layer graphene composite material has a low potential for photo-excited electrons, which cannot complete the conversion from CO_2 to CH_4 under visible light. Different classes of $CsPbBr_3$ graphene-based NHSs have different energy band structures, resulting in different reduction potentials and different CO_2 reduction performances (Figure 7.10). Otgonbayar et al. [93] reported a novel micro/nano-structured $AgCuInS_2$-graphene-TiO_2 ternary composite for photocatalytic reduction of CO_2 to prepare methanol fuel. The research proves that $AgCuInS_2$ and TiO_2 are irregularly dispersed on the graphene surface in the graphene-based catalyst prepared by the hydrothermal method. There is close contact and chemical bond interaction between the components. The ternary composite material is compared with the binary composite material. The material shows good performance in catalyzing CO_2 reduction.

Wang et al. [94] propose an effective strategy for the in-situ construction of graphene-based catalysts. TiO_2/N-doped graphene hollow spheres were prepared by chemical vapor deposition for enhanced photocatalytic CO_2

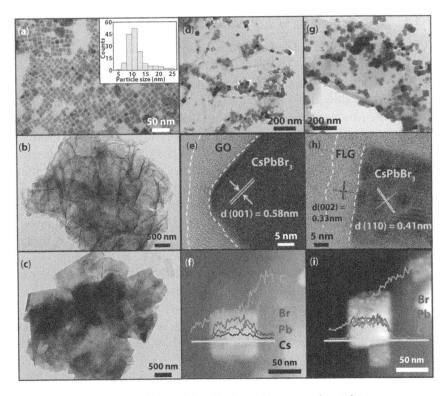

Figure 7.10 TEM images of (a) $CsPbBr_3$ NPs (inset: histogram of particle size distribution), (b) GO and (c) FLG. (d) TEM, (e) HR-TEM images and (f) EDS line-scan of $CsPbBr_3$-GO NHS. (g) TEM, (h) HR-TEM images and (i) EDS line-scan of $CsPbBr_3$-FLG NHS. Reprinted with permission from [92], Copyright 2021, Elsevier.

reduction performance. The schematic diagram of the preparation processes and the morphological characterization images of TiO_2/NG HS are shown in Figure 7.11. Pyridine nitrogen sites on the composites and the interface promote the separation and transport of photogenerated carriers, which effectively improves the photocatalytic CO_2 reduction performance. Otgonbayar et al. [95] synthesized graphene-based $LaYAgO_4$-graphene-TiO_2 nanocomposites for photocatalytic CO_2 reduction of methanol by a simple hydrothermal method. $LaYAgO_4$-graphene-TiO_2 exhibited high stability during CO_2 conversion, through the close combination of the three materials, leading to the improvement of the efficiency of CO_2 conversion.

Madhusudan et al. [96] prepared graphene-$Zn_{0.5}Cd_{0.5}S$ nanocomposites by a direct double-pot hydrothermal method. The FESEM and TEM images of the ZCS samples and the FESEM images of 2G-ZCS are shown in Figure 7.12. Through the images, we can see that the ZCS hollow nanospheres

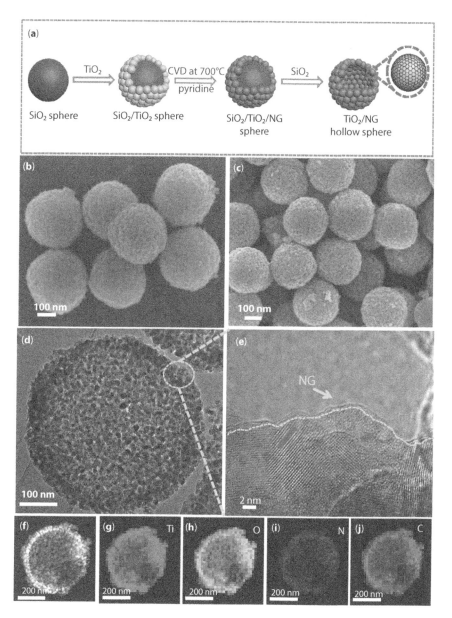

Figure 7.11 (a) Schematic illustration of TiO_2/NG HS preparation processes; (b, c) FESEM images of SiO_2/TiO_2 spheres and TiO_2/NG HS; TEM (d) and its corresponding HRTEM (e) image of TiO_2/NG HS; (f) HAADF image of TiO_2/NG HS; (g–j) EDS mapping images of Ti, O, N and C elements. Reprinted with permission from [94], Copyright 2021, Elsevier.

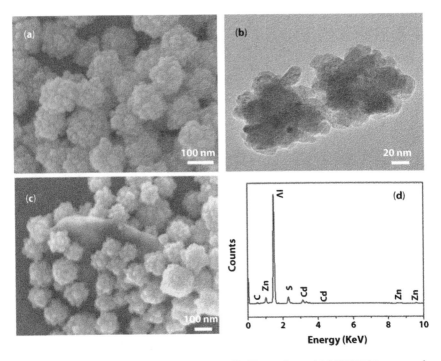

Figure 7.12 (a) FESEM and (b) TEM images of ZCS samples and (c) FESEM images and (c) EDS patterns of 2G-ZCS samples. Reprinted with permission from [96], Copyright 2020, Elsevier.

are attached to the graphene nanosheets. In the figure, the authors proposed that the folds of the graphene flakes due to the oxidation process would enhance the activity of visible light photocatalytic CO_2 reduction. The methanol yield of graphene-$Zn_{0.5}Cd_{0.5}S$ nanocomposites is 98 times higher than that of $Zn_{0.5}Cd_{0.5}S$ nanospheres under visible light illumination. Graphene plays the role of receiving and transporting electrons in the catalytic system, significantly improving the ZCS Photocatalytic activity of nanospheres for CO_2 reduction. Graphene has a unique conjugated structure, which can be used as an excellent carrier with a high specific surface area in composite catalysts. It finally promotes the uniform distribution of photoactive components and eases the agglomeration of active components. Kamal *et al.* [97] used a multi-step process to deposit plasmonic gold nanoparticles on TiO_2-decorated N-graphene heterostructure catalysts, which showed high activity for CO_2 reduction and high selectivity for methane generation. Compared with the traditional binary Au-TiO_2 photocatalyst, the optical properties and electron mobility of the catalyst are improved. N-graphene plays a key role in the composite, which reduces the activation energy for CO_2 reduction and improves the charge transfer. The coupling of the highly active Au-TiO_2 with the high electron transport of doped graphene has a synergistic effect on promoting photocatalytic CO_2 reduction. The activity of the composite catalyst is 60 times higher than that of the Au-TiO_2 photocatalyst.

Liu et al. [98] used a hydrothermal method to prepare nanostructured rGO aerogels while $TiCl_4$ was employed as precursors to prepare rod-like TiO_2-rGO composite aerogels. The conversion rate of TiO_2-rGO composite aerogel is 15.7 times higher than that of pure TiO_2. In addition, Xi et al. [99] prepared stacked graphene/Pd@PtCu composites for the efficient reduction of CO_2 into methane, and the PtCu shell in the composites had high selectivity for CO_2 reduction. The authors used atomically thin PtCu to polarize the charge at the Pd-PtCu interface, promote charge separation and migration, which improves the photocatalytic performance through interface regulation. Zhou et al. [100] used a synthetic strategy of polyvinylpyrrolidone (PVP)-assisted laser ablation to prepare silver nanocubes/reduced graphene oxide (rGO) composites, which enabled efficient photocatalytic reduction of CO_2. Images of silver nanoparticles produced by laser ablation at different PVP concentrations are shown in Figure 7.13. This wet-assisted laser ablation method can controllably prepare metal nanoparticles of different shapes and sizes. The cubic silver will be distributed on the graphene nanoflakes under the action of PVP as a surfactant. The photocatalytic CO_2 reduction conversion rate of this graphene-based catalyst

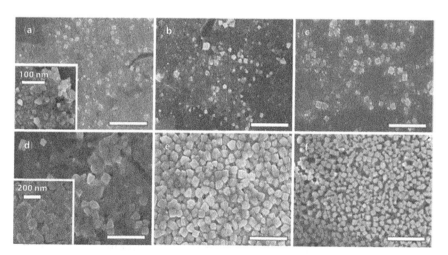

Figure 7.13 Silver nanoparticles produced by laser ablation at different PVP concentrations. (a) 4.5 mm; (b) 9.0 mm; (c) 18.0 mm; (d) 45.0 mm; (e) 90.1 mm; (f) 180.2 mm. All scales are 1μm. Reprinted with permission from [100], Copyright 2019, Elsevier.

is 133.1 $\mu mol \cdot g^{-1} \cdot h^{-1}$, and its conversion efficiency is 2-3 times higher than that of graphene oxide or silver nanoparticles. The cubic-structured silver nanoparticles can generate a strong local electromagnetic field at the edges, which can reduce CO_2 more efficiently to prepare hydrocarbon fuels than spherical silver nanoparticles under the same visible light conditions.

Cocatalysts often play an important role in photocatalysis in composites. Zhou et al. [101] used imidazole to introduce nitrogen atoms into nitrogen-doped graphene (ING) and Cu (II) as cocatalysts to promote CO_2 reduction for methanol production. The optimal ratio of 3.6% was used to prepare Cu (II)/ING/CeO_2 catalyst. The yield of CH_3OH over Cu (II)/ING/CeO_2 catalyst with adsorbed Cu (II) ions was as high as 385.8 $\mu mol \cdot g^{-1} \cdot h^{-1}$, while that of ING/$CeO_2$ catalyst without Cu (II) ions is only 3.57 $\mu mol \cdot g^{-1} \cdot h^{-1}$. The authors concluded that Cu (II) could capture electrons and play a positive role in the separation of electrons and holes during the photocatalytic process, which significantly improved the catalytic performance of CO_2 reduction to methanol.

The use of sacrificial templates to prepare hollow spherical photocatalysts also has many applications in the preparation of graphene-based composites, which is one of the effective strategies to synthesize hollow spherical high-efficiency catalysts. Bie et al. [102] synthesized CdS hollow spheres using SiO_2 microspheres as a sacrificial agent. The authors used chemical vapor deposition to *in situ* grow monolayer N-doped monolayer graphene on the CdS hollow spheres to prepare NG/CdS photocatalytic composites. Compared with pure cadmium sulfide hollow spheres, the photocatalytic production of CO and CH_4 by NG/CdS composite hollow spheres increased by 4-5 times. Chung et al. [103] developed hollow TiO_2 microspheres-graphene composites by hydrothermal method to optimize the catalytic performance of anatase TiO_2. They prepared hollow TiO_2 microspheres with SiO_2 microspheres as sacrificial templates, which were further mixed with graphene to prepare graphene-based catalytic material through hydrothermal reaction. The authors also investigated the effect of hollow microsphere cavity size, hydrothermal pH value, and the ratio of the two materials on catalytic CO_2 reduction. In addition, Yang et al. [104] modified the inner and outer surfaces of the TiO_2 spherical shell with graphene and used a double-sided modification to increase the contact area to separate electrons from the inner and outer sides at the same time, which provided a new way to design graphene-based catalysts (Figure 7.14). Compared with single-sided modified TiO_2 photocatalysts, graphene double-sided modified TiO_2 photocatalysts are more conducive to catalyzing CO_2 reduction. Rambabu et al. [105] prepared a GO/rGO-wrapped TiO_2 nanotube composite with a unique structure for

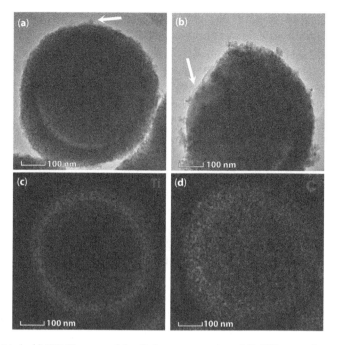

Figure 7.14 (a, b) TEM image and (c, d) element mapping of $G_{3\%}TG_{2\%}$ sample. Reprinted with permission from [104], Copyright 2021, Elsevier.

catalyzing the reduction of CO_2 to CO. The composite photocatalyst with this wrapped structure enables the charge to be better separated, which leads to the formation rate of CO being an order of magnitude higher than that of graphene-supported TiO_2 nanotubes.

7.4.2 Hydrogen Production by Water Splitting

Hydrogen is a kind of clean energy with high calorific value without the production of any harmful effluent after combustion. Many technologies were developed to boost its production while water splitting with photocatalysis is one of them. Since its use for hydrogen production via photocatalytic water splitting, graphene-based photocatalysts have attracted extensive attention due to their excellent structural properties [116–121]. In this section, the performance of different kinds of graphene-based photocatalysts was introduced and compared while the promoting mechanism was briefly summarized.

Various semiconductor materials, such as TiO_2, ZnO, WO_3, ZnS, and Fe_2O_3 are used to catalyze hydrogen evolution while ZnS showed good stability. Hassan *et al.* [106] used a hydrothermal method to prepare graphene/ZnS composites as photoanodes for water splitting. The hydrogen production was 3.7 times higher than that of ZnS. The authors concluded that the incorporation of low content of graphene enhanced the capture and separation of charges changed catalyst structure, and optimized the bandgap. It enabled light absorption in a wider wavelength range, which converted more absorbed solar energy into chemical energy and finally promoted hydrogen evolution performance of hydrogen production. Yang *et al.* [122] proposed low-cost single-atom Fe/G high-efficiency composite catalysts. They used the density functional theory (DFT) method to analyze and calculate four graphene composite iron single-atom catalysts with different structures. Iron single atoms can effectively be dispersed on the graphene substrate and exhibit unique electronic properties. The hybridization of iron single atoms and oxygen atomic orbitals enabled the Fe/G composite catalyst to strongly adsorb water on the catalyst surface. The author proposed that the interaction played a crucial role in the production of hydrogen from water splitting.

Titanium dioxide is the most commonly used photocatalyst. Titanium dioxide has stable properties and high photocatalytic performance in the ultraviolet region. Due to the limited energy in the ultraviolet region, many strategies were proposed to expand the light absorption range of titanium dioxide including semiconductor recombination, structural change, noble metal loading, element doping, sensitization, and other methods while graphene played a key role in these methods. For example, Nguyen *et al.* [107] used platinum noble metal and graphene non-metallic modified TiO_2 to prepare a Pt-GN/TiO_2 composite photocatalyst and used it for photocatalytic hydrogen production. Ternary Pt-GN/TiO_2 catalyst was prepared by photodeposition. The hydrogen production rate was as high as 4.71 mmol·h^{-1}·g^{-1}, which is 1.4 times that of Pt-TiO_2 and 2.2 times that of GN/TiO_2. The authors concluded that graphene had high electron mobility, which promoted charge capture and separation in Pt-GN/TiO_2 ternary catalyst.

In addition, the construction of heterojunction is also an effective strategy to improve the catalytic activity of TiO_2 photocatalysts, where graphene served different roles. He *et al.* [108] used a simple hydrothermal method to synthesize the WO_3/TiO_2/rGO compound in one step, where WO_3/TiO_2 dispersed on the surface of graphene sheets through chemical bonding (Figure 7.15). Graphene provided sufficient adsorption sites and active sites, leading to the hydrogen evolution rate of the ternary composite (245.8 µmol·g^{-1}·h^{-1}) being 3.5 times that of titanium dioxide. The formation of Schottky heterojunction between TiO_2 and reduced graphene oxide

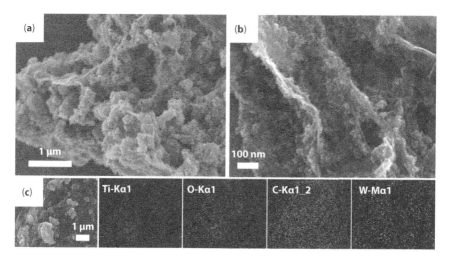

Figure 7.15 (a, b) SEM images of WTG samples; (c) EDS element maps of Ti, O, C, and W in WTG samples. Reprinted with permission from [108], Copyright 2020, Elsevier.

facilitated the separation of charges on TiO_2 and the transfer of electrons. The heterojunction within the ternary composite photocatalysts jointly suppressed the recombination separation and transfer of electrons.

Zinc oxide is also a common metal-semiconductor, which is widely used in photocatalytic degradation and other fields. For the preparation of solar fuels by photocatalysis, the rapid recombination of electrons and holes in the process of zinc oxide catalyzing water splitting is the key to a major constraint for catalytic hydrogen production. Machin et al. [109] synthesized Au@ZnO-graphene composite photocatalysts. The incorporation of graphene and Au nanoparticles improves photocatalytic hydrogen production efficiency. The highest hydrogen yield is 8.8 times higher than that of ZnO nanoparticles. The doping of gold nanoparticles and graphene has a synergistic effect on improving the catalytic water splitting of zinc oxide for hydrogen production. As a co-catalyst, graphene not only provided support for Au@ZnO-graphene composite photocatalyst but also promoted the trapping and separation of charges. It enhanced the absorption of visible light by the catalyst, which further improved the photocatalytic activity.

The zinc oxide/graphene structure can improve the electron transport rate and inhibit the recombination of electrons and holes, but its response to visible light is low. The performance of catalyzing water splitting for hydrogen production is unsatisfactory. As an n-type semiconductor with a suitable bandgap, MoS_2 is often used in photocatalysts to enhance the absorption of visible light. Dong et al. [110] reported that the heterostructured ZnO-MoS$_2$-graphene composite could significantly enhance the photocatalytic hydrogen evolution activity of ZnO nanoparticles. They obtained high-performance layered graphene and molybdenum disulfide materials through electrochemical-assisted liquid-phase exfoliation. The enhanced hydrogen evolution efficiency originated from the enhanced solar light absorption via heterojunction of the catalysts and the high electron mobility of graphene. Moreover, graphene plays a role in increasing the specific surface area of the catalyst and improving the mobility of electrons in the construction of the composite catalyst. Quiroz-Cardoso et al. [111] first prepared CdS nanofibers, and then sequentially modified CdS with GO and nickel nanoparticles to synthesize CdS nanocomposite photocatalyst for photocatalytic hydrogen evolution (Figure 7.16). CdS nanofibers were prepared by the *in-situ* method and photodeposition method, respectively. Different proportions of graphene oxide were used to modify CdS nanofibers, and nickel was used as a co-catalyst to improve photoactivity. *In-situ* synthesized CdS fibers were used. The close contact with graphene is more favorable, while the electron capture and transfer rates are enhanced.

Daulbayev et al. [112] used electrospinning to synthesize graphene flakes from biological waste for the preparation of graphene/SrTiO$_3$/PAN composite photocatalysts, which were further used for hydrogen production by splitting water (Figure 7.17). Few-layer graphene prepared from rice husks was considered a cocatalyst. Compared with the SrTiO$_3$ catalyst, the composite photocatalyst had higher hydrogen evolution efficiency. The variable metal valence and oxygen vacancies of perovskites are beneficial

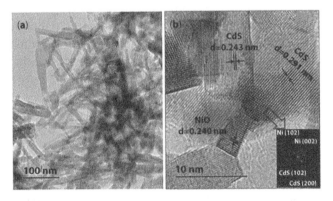

Figure 7.16 HRTEM images of photocatalysts. (a) The interplanar spacing of Ni/GO CdS at 1 wt% and (b) the diffraction pattern of Ni/GO CdS at 1 wt% of the sample. Reprinted with permission from [111], Copyright 2019, Elsevier.

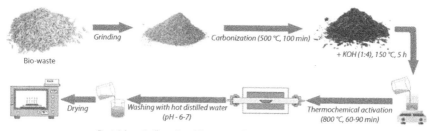

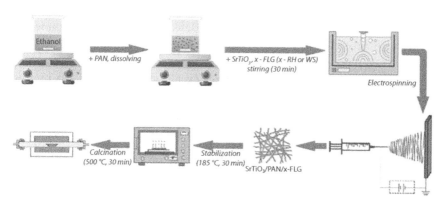

Figure 7.17 Schematic diagrams of the process of obtaining c-SrTiO$_3$/PAN/x-FLG by electrospinning. Reprinted with permission from [112], Copyright 2021, Elsevier.

to promote photocatalytic efficiency. The doping of noble metals (such as Au, Pd, etc.) can accelerate charge separation by effectively regulating oxygen vacancies. As an ideal co-catalyst, graphene with high electron transport is considered as a substitute for noble metals. Lv *et al.* [113] prepared a noble metal-free reduced graphene oxide-LaNiO$_3$ wrinkled nanocomposite photocatalyst by a self-assembly photocatalytic reduction method. The electron transfer under the photocatalytic action was intensified, which promoted the reduction of GO to RGO. LaNiO$_3$ nanoparticles were encapsulated in the hole-structured graphene. Ni-C bond strengthens the interaction of the two raw materials, through the anchoring of oxygen vacancies and the bonding of reduced graphene oxide-LaNiO$_3$ to form an encapsulated nanoreactor. The hydrogen production efficiency of the catalyst is 12 times that of LaNiO$_3$ with high stability and photocatalytic activity.

The activation energy of the water-splitting reaction is high. Moreover, it is difficult to produce oxygen and hydrogen at the same time. Many water splitting experiments are half-reactions. So far, there have been many reports on total hydrolysis where graphene can be used as a carrier or co-catalyst in perhydrolysis catalysts. Graphene has excellent electron-transport properties and a large specific surface area, which plays a key role in improving the performance of composite catalysts for perhydrolysis. For example, to achieve efficient total hydrolysis of water, Qiao *et al.* [114] developed a multifunctional composite catalyst capable of simultaneously producing hydrogen and oxygen. The images of the core-shell structured Co@CoO are shown in Figure 7.18. Under calcination at different heat treatment temperatures, core-shell cobalt particles with different properties can be obtained, which are then composited with nitrogen-doped graphene. It has more abundant crystal defects and a better framework, which can provide more active sites than other composites. The oxygen and hydrogen production rate of the catalyst were 543 198 $\mu mol \cdot g^{-1} \cdot h^{-1}$ and 330 $\mu mol \cdot g^{-1} \cdot h^{-1}$, respectively. After three cycles of experiments, the hydrogen production performance decreased by less than 20%. Nitrogen-doped graphene plays an important role in the catalysis of Co@CoO/NG-7. Characterization results show that the addition of graphene improves the absorption of light and promotes the electron pairing of Co@CoO/NG-7. The capture and separation of ions greatly reduce the recombination of electrons and holes, which improves the catalytic activity and stability of the catalyst.

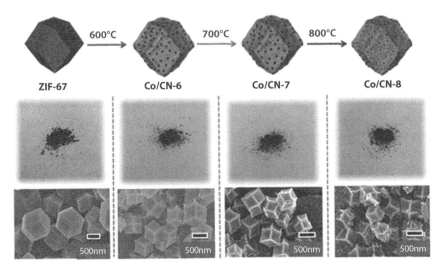

Figure 7.18 Synthesized images of Co@CoO/heat-treated at different temperatures and their SEM images analyzed. Reprinted with permission from [114], Copyright 2020, Elsevier.

In addition to being used as a catalyst carrier and co-catalyst, graphene can also be used as a catalyst for catalyzing hydrogen production. Graphene with different degrees of oxidation has different semiconductor properties. Therefore, graphene can be used for photocatalysis by changing the degree of oxidation. The bandwidth of nitrogen-doped graphene oxide quantum dots is 2.2 eV, which can be used for the photocatalytic production of hydrogen. Xie *et al.* [115] added sulfur-based nitrogen-doped graphene oxide quantum dots and combined them with TiO_2. S, N co-doped graphene quantum dots (GQDs)/TiO_2 composites were synthesized by hydrothermal method. The catalysts showed an excellent light absorption range and catalytic water-separation hydrogen activity. Comparing S, N-GQD/P25 with TiO_2, under the action of visible light, the photocatalytic activity and hydrogen production efficiency are improved. The rate of hydrogen production is up to 3.6 times that of TiO_2. The authors concluded that graphene quantum dots played a significant role in improving the hydrogen evolution rate. The addition of S, N-GQD changed the bandgap of the catalyst, which can enhance the absorption of more sunlight energy. At the same time, it greatly promoted the transfer of electrons and reduces the recombination of electrons and holes, which comprehensively improved the photocatalytic hydrogen production performance of N-GQD/P25.

7.5 Conclusion and Future Opportunities

Graphene is widely used in the development of photocatalysts and the production of solar fuels due to its excellent properties. This chapter summarizes the preparation methods of graphene and its composites, the characterization techniques for graphene-based photocatalysts, and the catalytic performance of graphene-based catalysts in catalytic carbon dioxide reduction and water splitting for hydrogen production. Graphene-modified semiconductor materials can significantly improve the efficiency of catalyzing the conversion of carbon dioxide into hydrocarbon fuels and catalyzing hydrogen evolution. The high-efficiency catalysis of graphene-based photocatalysts is closely related to the unique structure and high electron mobility of graphene. Therefore, the incorporation of graphene can reduce the recombination of electrons and holes as well as improve the efficiency of photocatalytic solar energy conversion. In addition, graphene may substitute noble metals and other co-catalysts for carbon dioxide reduction and hydrogen production. Moreover, graphene with different degrees of oxidation can be used as catalysts for the photocatalytic production of solar fuels directly.

Graphene-based photocatalysts for solar fuels have made great progress in recent years. It should be noted that the use of graphene-based catalysts for solar fuel production is only at its very early stage. Therefore, it is still a long road ahead before their commercial applications. In our opinion, the following directions may be paid more attention to in future research. First, the catalytic mechanism of graphene-based catalysts requires further study. The graphene-based photocatalytic mechanism is of great significance for the production of solar fuels and the design of photocatalysts. For example, in graphene-based photocatalytic carbon dioxide reduction, different C1 products can be selectively produced by manipulating the potential of charge carriers. The clear mechanism of photocatalysis helps us to design highly active graphene-based photocatalysts according to specific needs. The determination of the catalytic mechanism can help us leap in the field of graphene-based photocatalysts. In the future, we should pay more attention to catalysis mechanism research. Second, although the performance of graphene photocatalysts has been improved to a certain extent with the continuous development of graphene photocatalysts, the efficiency is still far from commercial application. The energy of visible light accounts for a large part of the energy of sunlight. However, the absorption of visible light by many semiconductor materials is not satisfactory. In the past few decades, researchers have modified photocatalysts to obtain high-performance graphene-based photocatalysts with suitable band gaps, but the utilization rate of sunlight by graphene-based photocatalysts is still limited while a large amount of energy is wasted. It finally leads to a low efficiency for solar fuels production. Finally, the large-scale production of graphene-based photocatalysts is another unsolved challenge. The structure and composition of graphene-based catalysts have a great influence on the performance of the catalyst. It is necessary to ensure that the structure and composition of graphene-based catalysts are completely consistent while it's a tough target. On the one hand, the preparation process of graphene is often prepared by the redox exfoliation method. The oxidation degree of graphene caused great trouble for the standardization of graphene preparation. Moreover, it is difficult to keep the structure and composition of graphene-based photocatalysts consistent during the preparation process, especially in the preparation and production of graphene-based photocatalysts with many components and complex structures. Therefore, it's essential to develop new methods or fabricate new devices to tune the structure and composition of graphene-based photocatalysts for real applications. In summary, graphene-based photocatalysts have achieved fruitful results in the production of solar fuels in the past few decades. However, on the road to its real application, it still faces huge challenges, which require more effort from our community.

Acknowledgments

This work was partially supported by the Natural Science Foundation of China (52102256), Jianghan University Science Research Project (2021-95), and Jianghan University Startup Package Project (2019-042).

References

1. Michel, A., Panagiotis, P., Chafic-Touma, S., Technologies and materials for renewable energy, environment and sustainability. *Energy Procedia*, 119, 1–2, 2017.
2. Wan, Z.Y., Wang, J., Wang, K., Hu, M.C., Wang, X.T., Photocatalytic reduction of CO_2 with H_2O vapor into solar fuels over Ni modified porous In_2O_3 nanosheets. *Catal. Today*, 374, 44–52, 2021.
3. Wan, Z.Y., Hu, M.C., Hu, B.B., Yan, T., Wang, K., Wang, X.T., Vacancy induced photocatalytic activity of La doped In(OH)(3) for CO_2 reduction with water vapor. *Catal. Sci. Technol.*, 10, 2893–2904, 2020.
4. Yan, T., Wan, Z.Y., Wang, K., Hu, M.C., Wang, X.T., A 3D carbon foam derived from phenol resin via CsCl soft-templating approach for high-performance supercapacitor. *Energy Technol.*, 8, 1901301, 2020.
5. Hu, B.B., Hu, M.C., Guo, Q., Wang, K., Wang, X.T., In-vacancy engineered plate-like In(OH)(3) for effective photocatalytic reduction of CO_2 with H_2O vapor. *Appl. Catal. B-Environ.*, 253, 77–87, 2019.
6. Meng, X., Zhuang, Y., Tang, H., Lu, C., Hierarchical structured $ZnFe_2O_4$@SiO_2@TiO_2 composites for enhanced visible-light photocatalytic activity. *J. Alloys Compd.*, 761, 15–23, 2018.
7. Zhang, Y., Chen, J., Tang, H., Xiao, Y., Qiu, S., Li, S., Cao, S., Hierarchically-structured SiO_2-Ag@TiO_2 hollow spheres with excellent photocatalytic activity and recyclability. *J. Hazard. Mater.*, 354, 17–26, 2018.
8. Byun, J., Huang, W., Wang, D., Li, R., Zhang, K.A.I., CO_2-triggered switchable hydrophilicity of a heterogeneous conjugated polymer photocatalyst for enhanced catalytic activity in water. *Angew. Chem. Int. Ed.*, 57, 2967–2971, 2018.
9. Lee, W.H., Liao, C.H., Tsai, M.F., Huang, C.W., Wu, J.C.S., A novel twin reactor for CO_2 photoreduction to mimic artificial photosynthesis. *Appl. Catal. B-Environ.*, 132, 445–451, 2013.
10. Hu, M.C., Yao, Z.H., Wang, X.Q., Graphene-based nanomaterials for catalysis. *Ind. Eng. Chem. Res.*, 56, 3477–3502, 2017.

11. Yao, Z., Hu, M., Wang, X., Characterization techniques for graphene-based materials in catalysis. *AIMS Mater. Sci.*, 4, 755–788, 2017.
12. Li, X., Yu, J., Wageh, S., Al-Ghamdi, A.A., Xie, J., Graphene in photocatalysis: A review. *Small*, 12, 6640–6696, 2016.
13. Yao, Z., Yan, T., Hu, M., Comparison of undergraduate chemical engineering curricula between China and America universities based on statistical analysis. *Educ. Chem. Eng.*, 38, 55–59, 2022.
14. Zhang, W.H., Mohamed, A.R., Ong, W.J., Z-scheme photocatalytic systems for carbon dioxide reduction: Where are we now? *Angew. Chem. Int. Ed.*, 59, 22894–22915, 2020.
15. Lin, L.H., Yu, Z.Y., Wang, X.C., Crystalline carbon nitride semiconductors for photocatalytic water splitting. *Angew. Chem. Int. Ed.*, 58, 6164–6175, 2019.
16. Bilisik, K. and Akter, M., Graphene nanocomposites: A review on processes, properties, and applications. *J. Ind. Text.*, 51, 3718S–3766S, 2021.
17. Malik, J.A., Madani, A., Pieber, B., Seeberger, P.H., Evidence for photocatalyst involvement in oxidative additions of nickel-catalyzed carboxylate o-arylations. *J. Am. Chem. Soc.*, 142, 11042–11049, 2020.
18. Chang, D.W. and Baek, J.B., Nitrogen-doped graphene for photocatalytic hydrogen generation. *Chem. Asian J.*, 11, 1125–1137, 2016.
19. Bie, C.B., Yu, H.G., Cheng, B., Ho, W., Fan, J.J., Yu, J.G., Design, fabrication, and mechanism of nitrogen-doped graphene-based photocatalyst. *Adv. Mater.*, 33, 2003521, 2021.
20. Remiro-Buenamanana, S. and Garcia, H., Photoassisted CO_2 conversion to fuels. *Chemcatchem*, 11, 342–356, 2019.
21. Marschall, R., 50 years of materials research for photocatalytic water splitting. *Eur. J. Inorg. Chem.*, 2021, 2435–2441, 2021.
22. Nasir, J.A., Munir, A., Ahmad, N., ul Haq, T., Khan, Z., Rehman, Z., Photocatalytic z-scheme overall water splitting: Recent advances in theory and experiments. *Adv. Mater.*, 33, 2105195, 2021.
23. Singla, S., Sharma, S., Basu, S., Shetti, N.P., Aminabhavi, T.M., Photocatalytic water splitting hydrogen production via environmental benign carbon-based nanomaterials. *Int. J. Hydrog. Energy*, 46, 33696–33717, 2021.
24. Darkwah, W.K., Teye, G.K., Ao, Y., Graphene nanocrystals in CO_2 photoreduction with H_2O for fuel production. *Nanoscale Adv.*, 2, 991–1006, 2020.
25. Liu, Z., Su, Z., Li, Q., Sun, L., Zhang, X., Yang, Z., Liu, X., Li, Y., Li, Y., Yu, F., Zhao, X., Induced growth of quasi-free-standing graphene on SiC substrates. *RSC Adv.*, 9, 32226–32231, 2019.
26. Khalid, A., Mohamed, M.A., Umar, A.A., Graphene growth at low temperatures using RF-plasma enhanced chemical vapour deposition. *Sains Malays.*, 46, 1111–1117, 2017.
27. Bai, H., Li, C., Shi, G., Functional composite materials based on chemically converted grapheme. *Adv. Mater.*, 23, 1089–1115, 2011.

28. Alamelu, K. and J. Ali, B.M., Au nanoparticles decorated sulfonated graphene-TiO$_2$ nanocomposite for sunlight driven photocatalytic degradation of the recalcitrant compound. *Sol. Energy*, 211, 1194–1205, 2020.
29. Ahmed, S.N. and Haider, W., Enhanced photocatalytic activity of ZnO-graphene oxide nanocomposite by electron scavenging. *Catalysts*, 11, 187, 2021.
30. Choudhari, A., Bhanvase, B.A., Saharan, V.K., Salame, P.H., Hunge, Y., Sonochemical preparation and characterization of rGO/SnO$_2$ nanocomposite: Electrochemical and gas sensing performance. *Ceram. Int.*, 46, 11290–11296, 2020.
31. Zhao, Y., Zhang, X., Liu, J., Wang, C., Li, J., Jin, H., Graphene oxide modified nano-sized BaTiO$_3$ as a photocatalyst. *Ceram. Int.*, 44, 15929–15934, 2018.
32. Chang, C.-J., Lin, Y.-G., Weng, H.-T., Wei, Y.-H., Photocatalytic hydrogen production from glycerol solution at room temperature by ZnO-ZnS/graphene photocatalysts. *Appl. Surf. Sci.*, 451, 198–206, 2018.
33. Ren, Y., Zeng, D., Ong, W.-J., Interfacial engineering of graphitic carbon nitride (g-C$_3$N$_4$)-based metal sulfide heterojunction photocatalysts for energy conversion: A review. *Chin. J. Catal.*, 40, 289–319, 2019.
34. Hu, M.C., Yao, Z.H., Li, L.L., Tsou, Y.H., Kuang, L.Y., Xu, X.Y., Zhang, W., Wang, X.Q., Boron-doped graphene nanosheet-supported Pt: a highly active and selective catalyst for low-temperature H$_2$-SCR. *Nanoscale*, 10, 10203–10212, 2018.
35. Hu, M.C., Yao, Z.H., Hui, K.N., Hui, K.S., Novel mechanistic view of catalytic ozonation of gaseous toluene by dual-site kinetic modeling. *Chem. Eng. J.*, 308, 710–718, 2017.
36. Hu, M.C., Hui, K.S., Hui, K.N., Role of graphene in MnO$_2$/graphene composite for catalytic ozonation of gaseous toluene. *Chem. Eng. J.*, 254, 237–244, 2014.
37. Zhongtian, F., Song, Z., Zhongxue, F., Hydrothermal preparation of NaTaO$_3$/rGO composite photocatalyst to enhance UV photocatalytic activity. *Results Phys.*, 15, 102669, 2019.
38. Lin, W., Xie, X., Wang, X., Wang, Y., Segets, D., Sun, J., Efficient adsorption and sustainable degradation of gaseous acetaldehyde and o-xylene using rGO-TiO$_2$ photocatalyst. *Chem. Eng. J.*, 349, 708–718, 2018.
39. Roy, O., Jana, A., Pratihar, B., Saha, D.S., De, S., Graphene oxide wrapped Mix-valent cobalt phosphate hollow nanotubes as oxygen evolution catalyst with low overpotential. *J. Colloid Interface Sci.*, 610, 592–600, 2022.
40. Pei, C., Zhu, J.-H., Xing, F., Photocatalytic property of cement mortars coated with graphene/TiO$_2$ nanocomposites synthesized via sol-gel assisted electrospray method. *J. Clean. Prod.*, 279, 123590, 2021.
41. Tashkandi, N.Y. and Mohamed, R.M., Sol-gel assembled MnCo$_2$O$_4$/rGO photocatalyst for enhanced production of aniline from photoreduction of nitrobenzene under visible light. *Ceram. Int.*, 48, 13216–13228, 2022.

42. Trusova, E.A., Kotsareva, K.V., Kirichenko, A.N., Abramchuk, S.S., Ashmarin, A.A., Perezhogin, I.A., Synthesis of graphene-based nanostructures by the combined method comprising sol-gel and sonochemistry techniques. *Diam. Relat. Mater.*, 85, 23–36, 2018.
43. Zadmehr, L. and Salem, S., Sol-gel and hydrothermal technical ability in the synthesis of magnetic anatase-graphene oxide nanocomposite with excellent photoactivity. *Mater. Sci. Eng. B*, 268, 115122, 2021.
44. Liu, G., Wang, R., Liu, H., Han, K., Cui, H., Ye, H., Highly dispersive nano-TiO_2 in situ growing on functional graphene with high photocatalytic activity. *J. Nanopart. Res.*, 18, 1–8, 2016.
45. Wang, X., Pan, Q., Xie, Y., Wang, L., Liu, J., Wang, S., Pan, K., Monodisperse Cu_3P nanoplates in situ grown on reduced graphene oxide via hydrophobic interaction for water splitting. *Mater. Lett.*, 306, 130947, 2022.
46. Ji, Z., Shen, X., Li, M., Zhou, H., Zhu, G., Chen, K., Synthesis of reduced graphene oxide/CeO_2 nanocomposites and their photocatalytic properties. *Nanotechnology*, 24, 115603, 2013.
47. Wang, L., Tan, H., Zhang, L., Cheng, B., Yu, J., In-situ growth of few-layer graphene on ZnO with intimate interfacial contact for enhanced photocatalytic CO_2 reduction activity. *Chem. Eng. J.*, 411, 128501, 2021.
48. Lin, C., Wei, W., Hu, Y.H., Catalytic behavior of graphene oxide for cement hydration process. *J. Phys. Chem. Solids*, 89, 128–133, 2016.
49. Kuila, T., Bose, S., Mishra, A.K., Khanra, P., Kim, N.H., Lee, J.H., Chemical functionalization of graphene and its applications. *Prog. Mater. Sci.*, 57, 1061–1105, 2012.
50. Ma, D., Wu, J., Gao, M., Xin, Y., Sun, Y., Ma, T., Hydrothermal synthesis of an artificial Z-scheme visible light photocatalytic system using reduced graphene oxide as the electron mediator. *Chem. Eng. J.*, 313, 1567–1576, 2017.
51. Pu, S., Zhu, R., Ma, H., Deng, D., Pei, X., Qi, F., Chu, W., Facile in-situ design strategy to disperse TiO_2 nanoparticles on graphene for the enhanced photocatalytic degradation of rhodamine 6G. *Appl. Catal. B Environ.*, 218, 208–219, 2017.
52. Yeh, T.-F., Chen, S.-J., Teng, H., Synergistic effect of oxygen and nitrogen functionalities for graphene-based quantum dots used in photocatalytic H_2 production from water decomposition. *Nano Energy*, 12, 476–485, 2015.
53. Ma, J., Shi, L., Hou, L., Yao, L., Lu, C., Geng, Z., Fabrication of graphene/$Bi_{12}O_{17}Cl_2$ as an effective visible-light photocatalyst. *Mater. Res. Bull.*, 122, 110690, 2020.
54. Pal, D.B., Rathoure, A.K., Singh, A., Investigation of surface interaction in rGO-CdS photocatalyst for hydrogen production: An insight from XPS studies. *Int. J. Hydrog. Energy*, 46, 26757–26769, 2021.
55. Sadeghi Rad, T., Khataee, A., Arefi-Oskoui, S., Sadeghi Rad, S., Orooji, Y., Gengec, E., Kobya, M., Graphene-based ZnCr layered double hydroxide nanocomposites as bactericidal agents with high sonophotocatalytic

performances for degradation of rifampicin. *Chemosphere*, 286, 131740, 2022.
56. Ansari, F., Soofivand, F., Salavati-Niasari, M., Eco-friendly synthesis of cobalt hexaferrite and improvement of photocatalytic activity by preparation of carbonic-based nanocomposites for waste-water treatment. *Compos. B Eng.*, 165, 500–509, 2019.
57. Sibul, R., Kibena-Põldsepp, E., Ratso, S., Kook, M., Sougrati, M.T., Käärik, M., Merisalu, M., Aruväli, J., Paiste, P., Treshchalov, A., Leis, J., Kisand, V., Sammelselg, V., Holdcroft, S., Jaouen, F., Tammeveski, K., Iron- and nitrogen-doped graphene-based catalysts for fuel cell applications. *ChemElectroChem*, 7, 1739–1747, 2020.
58. Yan, Q., Li, J., Zhang, X., Zhang, J., Cai, Z., Synthetic bio-graphene based nanomaterials through different iron catalysts. *Nanomaterials (Basel)*, 8, 840, 2018.
59. Hasani, A., Teklagne, M.A., Do, H.H., Hong, S.H., Van Le, Q., Ahn, S.H., Kim, S.Y., Graphene-based catalysts for electrochemical carbon dioxide reduction. *Carbon Energy*, 2, 158–175, 2020.
60. Hu, M., Yao, Z., Wang, X., Graphene-based nanomaterials for catalysis. *Ind. Eng. Chem. Res.*, 56, 3477–3502, 2017.
61. Byon, H.R., Suntivich, J., Shao-Horn, Y., Graphene-based non-noble-metal catalysts for oxygen reduction reaction in acid. *Chem. Mater.*, 23, 3421–3428, 2011.
62. Youn, D.H., Jang, J.W., Kim, J.Y., Jang, J.S., Choi, S.H., Lee, J.S., Fabrication of graphene-based electrode in less than a minute through hybrid microwave annealing. *Sci. Rep.*, 4, 5492, 2014.
63. Yang, B., Huang, Z., Wu, H., Hu, H., Lin, H., Nie, M., Li, Q., Sea urchin-like $CoSe_2$ nanoparticles modified graphene oxide as an efficient and stable hydrogen evolution catalyst. *J. Electroanal. Chem.*, 907, 116037, 2022.
64. Sayadi, M.H., Homaeigohar, S., Rezaei, A., Shekari, H., $Bi/SnO_2/TiO_2$-graphene nanocomposite photocatalyst for solar visible-light-induced photodegradation of pentachlorophenol. *Environ. Sci. Pollut. Res. Int.*, 28, 15236–15247, 2021.
65. Ibrahim, A., Klopocinska, A., Horvat, K., Hamid, Z.A., Graphene-based nanocomposites: Synthesis, mechanical properties, and characterizations. *Polym. (Basel)*, 13, 2869, 2021.
66. Zhuang, Y., Liu, Q., Kong, Y., Shen, C., Hao, H., Dionysiou, D.D., Shi, B., Enhanced antibiotic removal through a dual-reaction-center Fenton-like process in 3D graphene-based hydrogels. *Environ. Sci. Nano*, 6, 388–398, 2019.
67. Yin, W., Huang, Y., Lu, M., Tang, Y., Zhang, G., Li, D., Controllable direct growth and patterning of graphene-based transparent and conductive films on insulating substrates via Cu nanoparticles assisted-catalysis method. *Diam. Relat. Mater.*, 123, 108868, 2022.

68. Fraga, T.J.M., Carvalho, M.N., Ghislandi, M.G., da Motta Sobrinho, M.A., Functionalized graphene-based materials as innovative adsorbents of organic pollutants: A concise overview. *Braz. J. Chem. Eng.*, 36, 1–31, 2019.
69. Chikh, B., Ghiat, I., Saadi, A., Boudjemaa, A., Graphene-based nanomaterials as catalysts in solar water splitting, in: *Graphene-Based Nanomaterial Catalysis*, p. 152, 2022.
70. Weng, Z., Dixon, S.C., Lee, L.Y., Humphreys, C.J., Guiney, I., Fenwick, O., Gillin, W.P., Wafer-scale graphene anodes replace indium tin oxide in organic light-emitting diodes. *Adv. Opt. Mater.*, 10, 2101675, 2022.
71. Shen, J., Shi, M., Li, N., Yan, B., Ma, H., Hu, Y., Ye, M., Facile synthesis and application of Ag-chemically converted graphene nanocomposite. *Nano Res.*, 3, 339–349, 2010.
72. Zhong, H.X., Wang, J., Zhang, Y.W., Xu, W.L., Xing, W., Xu, D., Zhang, Y.F., Zhang, X.B., ZIF-8 derived graphene-based nitrogen-doped porous carbon sheets as highly efficient and durable oxygen reduction electrocatalysts. *Angew. Chem. Int. Ed. Engl.*, 53, 14235–14239, 2014.
73. Sumathi, N., Dhanemozhi, A.C., Thangaraju, D., Adewinbi, S.A., Mohanraj, K., Marnadu, R., Shkir, M., Hydrothermal synthesis of self-assembled potassium-doped graphene semiconducting nanoparticles for p-Si/n-graphene junction diode applications. *Surf. Interfaces*, 26, 101408, 2021.
74. Ojha, A. and Thareja, P., Graphene-based nanostructures for enhanced photocatalytic degradation of industrial dyes. *Emergent Mater.*, 3, 169–180, 2020.
75. Iftikhar, A., Yousaf, S., Ali, F.A.A., Haider, S., Khan, S.U.-D., Shakir, I., Iqbal, F., Warsi, M.F., Erbium-substituted $Ni_{0.4}Co_{0.6}Fe_2O_4$ ferrite nanoparticles and their hybrids with reduced graphene oxide as magnetically separable powder photocatalyst. *Ceram. Int.*, 46, 1203–1210, 2020.
76. Kumar, A., Sadanandhan, A.M., Jain, S.L., Silver doped reduced graphene oxide as a promising plasmonic photocatalyst for oxidative coupling of benzylamines under visible light irradiation. *New J. Chem.*, 43, 9116–9122, 2019.
77. Fan, H., Yi, G., Zhang, Z., Zhang, X., Li, P., Zhang, C., Chen, L., Zhang, Y., Sun, Q., Binary TiO_2/RGO photocatalyst for enhanced degradation of phenol and its application in underground coal gasification wastewater treatment. *Opt. Mater.*, 120, 111482, 2021.
78. Kocijan, M., Ćurković, L., Ljubas, D., Mužina, K., Bačić, I., Radošević, T., Podlogar, M., Bdikin, I., Otero-Irurueta, G., Hortigüela, M.J., Graphene-based TiO_2 nanocomposite for photocatalytic degradation of dyes in aqueous solution under solar-like radiation. *Appl. Sci.*, 11, 3966, 2021.
79. Harikrishnan, M., Athira, S., Sykam, N., Rao, G.M., Mathew, A., Preparation of rGO-TiO_2 composite and study of its dye adsorption properties. *Mater. Today: Proc.*, 9, 61–69, 2019.
80. Zhang, H., Wang, X., Li, N., Xia, J., Meng, Q., Ding, J., Lu, J., Synthesis and characterization of TiO_2/graphene oxide nanocomposites for photoreduction of heavy metal ions in reverse osmosis concentrate. *RSC Adv.*, 8, 34241–34251, 2018.

81. Hu, M., Yao, Z., Wang, X., Characterization techniques for graphene-based materials in catalysis. *AIMS Mater. Sci.*, 4, 755–788, 2017.
82. Wang, Y., Wang, S., Zhang, S.L., Lou, X.W., Formation of hierarchical FeCoS$_2$–CoS$_2$ double-shelled nanotubes with enhanced performance for photocatalytic reduction of CO$_2$. *Angew. Chem.*, 132, 12016–12020, 2020.
83. Vu, N.N., Kaliaguine, S., Do, T.O., Critical aspects and recent advances in structural engineering of photocatalysts for sunlight-driven photocatalytic reduction of CO$_2$ into fuels. *Adv. Funct. Mater.*, 29, 1901825, 2019.
84. Ren, X., Gao, M., Zhang, Y., Zhang, Z., Cao, X., Wang, B., Wang, X., Photocatalytic reduction of CO$_2$ on BiO$_X$: Effect of halogen element type and surface oxygen vacancy mediated mechanism. *Appl. Catal. B Environ.*, 274, 119063, 2020.
85. Wang, X., Wang, Y., Gao, M., Shen, J., Pu, X., Zhang, Z., Lin, H., Wang, X., BiVO$_4$/Bi$_4$Ti$_3$O$_{12}$ heterojunction enabling efficient photocatalytic reduction of CO$_2$ with H$_2$O to CH$_3$OH and CO. *Appl. Catal. B Environ.*, 270, 118876, 2020.
86. Fajrina, N. and Tahir, M., A critical review in strategies to improve photocatalytic water splitting towards hydrogen production. *Int. J. Hydrog. Energy*, 44, 540–577, 2019.
87. Pan, J., Wang, P., Wang, P., Yu, Q., Wang, J., Song, C., Zheng, Y., Li, C., The photocatalytic overall water splitting hydrogen production of g-C$_3$N$_4$/CdS hollow core-shell heterojunction via the HER/OER matching of Pt/MnOx. *Chem. Eng. J.*, 405, 126622, 2021.
88. Ou, W., Pan, J., Liu, Y., Li, S., Li, H., Zhao, W., Wang, J., Song, C., Zheng, Y., Li, C., Two-dimensional ultrathin MoS$_2$-modified black Ti^{3+}–TiO$_2$ nanotubes for enhanced photocatalytic water splitting hydrogen production. *J. Energy Chem.*, 43, 188–194, 2020.
89. Nastasi, G., Camiola, V.D., Romano, V., Direct simulation of charge transport in graphene nanoribbons. *Commun. Comput. Phys.*, 31, 449–494, 2022.
90. Mahmoudi, T., Wang, Y., Hahn, Y.B., SrTiO$_3$/Al$_2$O$_3$-graphene electron transport layer for highly stable and efficient composites-based perovskite solar cells with 20.6% efficiency. *Adv. Energy Mater.*, 10, 1903369, 2020.
91. Shen, D., Zhang, W., Xie, F., Li, Y., Abate, A., Wei, M., Graphene quantum dots decorated TiO$_2$ mesoporous film as an efficient electron transport layer for high-performance perovskite solar cells. *J. Power Sources*, 402, 320–326, 2018.
92. Chen, Y.-H., Ye, J.-K., Chang, Y.-J., Liu, T.-W., Chuang, Y.-H., Liu, W.-R., Liu, S.-H., Pu, Y.-C., Mechanisms behind photocatalytic CO$_2$ reduction by CsPbBr$_3$ perovskite-graphene-based nanoheterostructures. *Appl. Catal. B Environ.*, 284, 119751, 2021.
93. Otgonbayar, Z., Cho, K.Y., Oh, W.C., Novel micro and nanostructure of an AgCuInS$_2$-graphene-TiO$_2$ ternary composite for photocatalytic CO$_2$ reduction for methanol fuel. *ACS Omega*, 5, 26389–26401, 2020.

94. Wang, L., Zhu, B., Cheng, B., Zhang, J., Zhang, L., Yu, J., *In-situ* preparation of TiO_2/N-doped graphene hollow sphere photocatalyst with enhanced photocatalytic CO_2 reduction performance. *Chin. J. Catal.*, 42, 1648–1658, 2021.
95. Otgonbayar, Z., Liu, Y., Cho, K.Y., Jung, C.-H., Oh, W.-C., A novel ternary composite of $LaYAgO_4$ and TiO_2 united with graphene and its complement: Photocatalytic performance of CO_2 reduction into methanol. *Mater. Sci. Semicond. Process.*, 121, 105456, 2021.
96. Madhusudan, P., Wageh, S., Al-Ghamdi, A.A., Zhang, J., Cheng, B., Yu, Y., Graphene-$Zn_{0.5}Cd_{0.5}S$ nanocomposite with enhanced visible-light photocatalytic CO_2 reduction activity. *Appl. Surf. Sci.*, 506, 144683, 2020.
97. Kamal, K.M., Narayan, R., Chandran, N., Popović, S., Nazrulla, M.A., Kovač, J., Vrtovec, N., Bele, M., Hodnik, N., Kržmanc, M.M., Likozar, B., Synergistic enhancement of photocatalytic CO_2 reduction by plasmonic Au nanoparticles on TiO_2 decorated N-graphene heterostructure catalyst for high selectivity methane production. *Appl. Catal. B Environ.*, 307, 121181, 2022.
98. Liu, S., Jiang, T., Fan, M., Tan, G., Cui, S., Shen, X., Nanostructure rod-like TiO_2-reduced graphene oxide composite aerogels for highly-efficient visible-light photocatalytic CO_2 reduction. *J. Alloys Compd.*, 861, 158598, 2021.
99. Xi, Y., Zhang, Y., Cai, X., Fan, Z., Wang, K., Dong, W., Shen, Y., Zhong, S., Yang, L., Bai, S., PtCu thickness-modulated interfacial charge transfer and surface reactivity in stacked graphene/Pd@PtCu heterostructures for highly efficient visible-light reduction of CO_2 to CH_4. *Appl. Catal. B Environ.*, 305, 121069, 2022.
100. Zhou, R., Yin, Y., Long, D., Cui, J., Yan, H., Liu, W., Pan, J.H., PVP-assisted laser ablation growth of Ag nanocubes anchored on reduced graphene oxide (rGO) for efficient photocatalytic CO_2 reduction. *Prog. Nat. Sci. Mater. Int.*, 29, 660–666, 2019.
101. Zhou, S.S. and Liu, S.Q., Photocatalytic reduction of CO_2 based on a CeO_2 photocatalyst loaded with imidazole fabricated N-doped graphene and Cu(ii) as cocatalysts. *Photochem. Photobiol. Sci.*, 16, 1563–1569, 2017.
102. Bie, C., Zhu, B., Xu, F., Zhang, L., Yu, J., In situ grown monolayer n-doped graphene on CdS hollow spheres with seamless contact for photocatalytic CO_2 reduction. *Adv. Mater.*, 31, e1902868, 2019.
103. Chung, Y.-C., Xie, P.-J., Lai, Y.-W., Lo, A.-Y., Hollow TiO_2 microsphere/graphene composite photocatalyst for CO_2 photoreduction. *Catalysts*, 11, 1532, 2021.
104. Yang, Y., Liu, M., Han, S., Xi, H., Xu, C., Yuan, R., Long, J., Li, Z., Double-sided modification of TiO_2 spherical shell by graphene sheets with enhanced photocatalytic activity for CO_2 reduction. *Appl. Surf. Sci.*, 537, 147991, 2021.
105. Rambabu, Y., Kumar, U., Singhal, N., Kaushal, M., Jaiswal, M., Jain, S.L., Roy, S.C., Photocatalytic reduction of carbon dioxide using graphene oxide wrapped TiO_2 nanotubes. *Appl. Surf. Sci.*, 485, 48–55, 2019.

106. Hassan, A., Liaquat, R., Iqbal, N., Ali, G., Fan, X., Hu, Z., Anwar, M., Ahmad, A., Photo-electrochemical water splitting through graphene-based ZnS composites for H_2 production. *J. Electroanal. Chem.*, 889, 115223, 2021.
107. Nguyen, N.T., Zheng, D.D., Chen, S.S., Chang, C.T., Preparation and Photocatalytic Hydrogen production of Pt-Graphene/TiO_2 composites from water splitting. *J. Nanosci. Nanotechnol.*, 18, 48–55, 2018.
108. He, F., Meng, A., Cheng, B., Ho, W., Yu, J., Enhanced photocatalytic H_2-production activity of WO_3/TiO_2 step-scheme heterojunction by graphene modification. *Chin. J. Catal.*, 41, 9–20, 2020.
109. Machin, A., Arango, J.C., Fontanez, K., Cotto, M., Duconge, J., Soto-Vazquez, L., Resto, E., Petrescu, F.I.T., Morant, C., Marquez, F., Biomimetic catalysts based on Au@ZnO-graphene composites for the generation of hydrogen by water splitting. *Biomimetics (Basel)*, 5, 39, 2020.
110. Dong, H., Li, J., Chen, M., Wang, H., Jiang, X., Xiao, Y., Tian, B., Zhang, X., High-throughput production of ZnO-MoS_2-graphene heterostructures for highly efficient photocatalytic hydrogen evolution. *Mater. (Basel)*, 12, 2233, 2019.
111. Quiroz-Cardoso, O., Oros-Ruiz, S., Solís-Gómez, A., López, R., Gómez, R., Enhanced photocatalytic hydrogen production by CdS nanofibers modified with graphene oxide and nickel nanoparticles under visible light. *Fuel*, 237, 227–235, 2019.
112. Daulbayev, C., Sultanov, F., Korobeinyk, A.V., Yeleuov, M., Azat, S., Bakbolat, B., Umirzakov, A., Mansurov, Z., Bio-waste-derived few-layered graphene/$SrTiO_3$/PAN as an efficient photocatalytic system for water splitting. *Appl. Surf. Sci.*, 549, 149176, 2021.
113. Lv, T., Wu, M., Guo, M., Liu, Q., Jia, L., Self-assembly photocatalytic reduction synthesis of graphene-encapsulated $LaNiO_3$ nanoreactor with high efficiency and stability for photocatalytic water splitting to hydrogen. *Chem. Eng. J.*, 356, 580–591, 2019.
114. Qiao, S., Guo, J., Wang, D., Zhang, L., Hassan, A., Chen, T., Feng, C., Zhang, Y., Wang, J., Core-shell cobalt particles Co@CoO loaded on nitrogen-doped graphene for photocatalytic water-splitting. *Int. J. Hydrog. Energy*, 45, 1629–1639, 2020.
115. Xie, H., Hou, C., Wang, H., Zhang, Q., Li, Y., S, n Co-doped graphene quantum Dot/TiO_2 composites for efficient photocatalytic hydrogen generation. *Nanoscale Res. Lett.*, 12, 400, 2017.
116. N. R. A. M, Sham,, R. M, Yunus,, N. N., Rosman,, W. Y., Wong,, K., Arifin,, L. J., Minggu,, Current progress on 3D graphene-based photocatalysts: From synthesis to photocatalytic hydrogen production. *Int. J. Hydrog. Energy*, 46, 9324–9340, 2021.
117. Albero, J., Mateo, D., García, H., Graphene-based materials as efficient photocatalysts for water splitting. *Molecules*, 24, 906, 2019.

118. Nemiwal, M., Zhang, T.C., Kumar, D., Graphene-based electrocatalysts: Hydrogen evolution reactions and overall water splitting. *Int. J. Hydrog. Energy*, 46, 21401–21418, 2021.
119. Tiwari, J.N., Singh, A.N., Sultan, S., Kim, K.S., Recent advancement of p-and d-block elements, single atoms, and graphene-based photoelectrochemical electrodes for water splitting. *Adv. Energy Mater.*, 10, 2000280, 2020.
120. Kuang, P., Sayed, M., Fan, J., Cheng, B., Yu, J., 3D graphene-based H_2-production photocatalyst and electrocatalyst. *Adv. Energy Mater.*, 10, 1903802, 2020.
121. Ali, A. and Shen, P.K., Recent progress in graphene-based nanostructured electrocatalysts for overall water splitting. *Electrochem. Energy Rev.*, 3, 370–394, 2020.
122. Yang, J., Fan, Y., Liu, P.-F., Theoretical insights into heterogeneous single-atom Fe_1 catalysts supported by graphene-based substrates for water splitting. *Appl. Surf. Sci.*, 540, 148245, 2021.

8

Advances in Design and Scale-Up of Solar Fuel Systems

Ashween Virdee and John Andresen*

Research Centre for Carbon Solutions (RCCS), Heriot-Watt University, Edinburgh, UK

Abstract

Production of solar fuels through CO_2 reduction with water and sunlight is a key process towards a net-zero future. Significant progress has been made in the development of novel materials and catalysts. However, solar fuel technologies still face obstacles towards scale-up. This chapter reviews the current state-of-the-art technologies developed and the effect of operating conditions on system performance. The most promising technologies are photocatalytic (PC), photovoltaic powered electrochemical (PV+EC) and photoelectrochemical (PEC) solar fuel production systems. PC systems have simple designs and are used where the driving force of the reduction reaction is only dependent on the solar energy/incident light. Electrochemical (EC) systems utilize catalytic conversion of CO_2 to other chemicals and fuels in an electrolytic cell powered by electrical energy. The integration of an electrochemical reactor with a light source and a photocatalyst obtains a photoelectrochemical (PEC) reactor. The three PEC system configurations used are photocathode – anode, cathode – photoanode and photocathode – photoanode systems. Of the various reactors, the three main operating conditions, i.e., irradiation, temperature, and pressure of the system, were studied to understand the effect they have on the reactor performances. Finally, key considerations were highlighted that need to be considered when designing and scaling up solar fuel systems.

Keywords: Photoreduction, electrochemical, photo-electrochemical, solar fuels, CO_2 reduction, scale-up

---

*Corresponding author: j.andresen@hw.ac.uk

8.1 Introduction

Scale-up of solar fuel systems can be key for the global transition from fossil fuels to mitigate climate change while ensuring a sustainable and affordable shift to a net zero future [1-3]. Solar energy is vastly underutilized where an estimated 62,000 PWh of energy is striking the earth's surface every year while in comparison the global energy consumption is less than 0.3% of this at about 170 PWh [4]. To date, solar energy conversion into electrical energy has progressed at speed, but the global need for electricity is only 27PWh while the remaining 84% is largely utilized as carbon-based fuels. Hence, the conversion of carbon dioxide through photoreduction to produce solar fuels is vital for a just transition to a net zero future [5, 6]. Solar fuel systems have gained significant advances over the past decade on the fundamental level and state-of-the-art scale-up considerations are being discussed in this chapter.

Scale-up of solar fuel production through CO_2 reduction in the presence of water and sunlight to produce chemicals have generally been targeting C1 molecules such as carbon monoxide, methane, and methanol, although some longer-chain hydrocarbons have also been produced [7, 8]. Although significant progress has been made to develop novel materials and design reactors, the complexity of the technology still imposes several obstacles for large-scale applications. The general factors that affect the feasibility of the process are reactor design configurations and physiochemical properties of the novel materials [7]. In particular, advancements are required for the synthesis of the novel catalysts that offer high reaction selectivity, stability, and improved reaction kinetics. Currently, knowledge on the atomic processes occurring at the catalyst surface and design of highly active and selective catalysts is still in early stages [5]. Once these novel catalysts have been developed, a detailed understanding of the reactions and optimized process conditions need to be determined for scale-up. Hence, there has been limited success in the development of suitable scalable reactor designs.

8.2 Strategies for Solar Photoreactor Design

Strategies for scale-up of solar fuels has been considered ever since Giacomo Ciamician in 1916 proposed a process to store solar energy using a photochemical reaction that can significantly reduce CO_2 concentrations in the air [9, 10]. Since then various production methods have been developed to carry out sunlight assisted CO_2 reduction and produce value-added chemicals [11]. These methods can be categorized into three

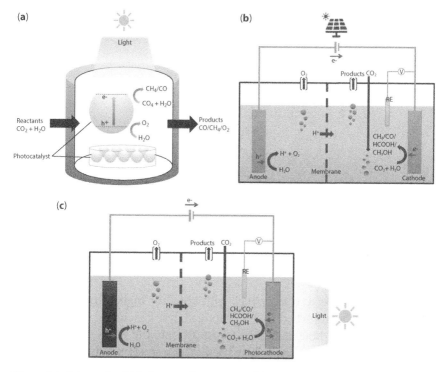

Figure 8.1 Schematic of (a) photocatalytic system, (b) photovoltaic + electrochemical system, (c) photoelectrochemical system.

general groups: photocatalytic (PC) or photosynthetic (PS) systems, photo-electrochemical systems (PEC) and photovoltaic powered electrochemical systems (PV+EC) as outlined in Figure 8.1. The three systems mentioned above have been receiving significant attention compared to other methods, such as bio-photosynthetic and photothermal systems as they have the capability of carrying out the reduction reaction at relatively milder conditions, i.e., at lower temperatures and around ambient pressures [12]. Additionally, they use greener and safer catalysts in comparison to dangerous heavy metal catalysts used in thermal catalytic systems [13].

8.2.1 Photocatalytic Systems

In photocatalytic systems, the CO_2 photoreduction reaction can be generalized to three steps, i.e., solar/light energy harvesting, separation and transportation of the charges, and surface reactions [6]. Figure 8.2 outlines the CO_2 photocatalytic reduction steps with H_2O on a heterogenous catalyst particle. During the first step, the photons from the light source that have

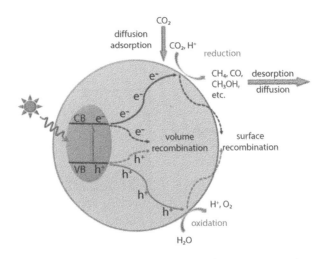

Figure 8.2 CO_2 photocatalytic reduction with H_2O on a heterogenous catalyst particle [14].

the required band-gap energy (i.e., with equal or higher energy than the bandgap) will be absorbed by the photocatalyst generating electron–hole pairs. The electron-hole pairs then transport onto the surface of the photocatalyst where the electrons reduce the CO_2 into fuels, while the holes oxidize H_2O into O_2 [6]. It should be noted, that during the transportation of the charges a large fraction of the photogenerated electron-hole pairs recombine and do not participate in the reactions. Moreover, the CO_2 photoreduction reaction displays similarities to solar water splitting. In a H_2O rich environment, the proton produced from the H_2O oxidation reaction competes with CO_2 reduction. Thus, CO_2 photoreduction is a complex process involving multiple proton-coupled electron transfers resulting in the formations of a wide range of products [6].

During photocatalytic CO_2 reduction, several reactions take place on the catalyst as a result the product rate and selectivity is complex to control. Table 8.1 outlines reaction mechanisms identified during photocatalytic CO_2 conversion and the issue during scale-up is to maintain control. The product selectivity varies depending on a range of factors, for example the reaction conditions and type of catalyst used. When the reaction takes place in a gaseous environment, the common products produced are carbon monoxide and methane. While in aqueous environments, the common products produced are alcohols such as methanol and ethanol [15]. Looking at the photocatalysts, a wide range of metal and non-metal catalysts have been studied e.g., TiO_2, CuO, ZnO, ZrO_2, ZnS, graphitic carbon nitride (g-C3N4), and carbon nanotubes. TiO_2 based catalysts are the

Table 8.1 Summary of the reaction mechanism studies for photocatalytic CO_2 reduction [17].

Mechanism 1	Mechanism 2
$2CO_2 \rightarrow 2CO + O_2$	$2CO_2 + 4H^* \rightarrow 2HCOOH + O_2$
$2CO \rightarrow 2C^* + O_2$	$HCOOH + 2H^* \rightarrow HCOH + H_2O$
$C^* + H^* \rightarrow CH_3OH$	$HCOH + 2H^* \rightarrow CH_3OH$
$CH^* + H^* \rightarrow CH_2^*$	$CH_3OH + H^* \rightarrow CH_3^*$
$CH_2^* + H^* \rightarrow CH_3^*$	$CH_3^* + H^* \rightarrow CH_4$
$CH_3^* + H^* \rightarrow CH_4$	$CH_3^* + CH_3^* \rightarrow C_2H_6$
$CH_3^* + OH \rightarrow CH_3OH$	

benchmark as they are the most extensively studied because of its high chemical stability, availability, non-toxicity, and low costs of the materials [15]. TiO_2 displays good activity under UV light radiations; however, the efficiency of the reactions remains low due to the rapid recombination of the electrons-hole pairs generated, has a mild reducing strength and limited visible light absorption [16, 17]. Strategies to improve the photocatalyst include surface modifications and defects, doping, cocatalyst with noble metals, morphologies, etc. [17]. Depending on the reaction parameters and catalyst modifications the reaction mechanism and selectivity of the product are not well established [17, 18].

Photocatalytic (PC) and photosynthetic (PS) processes are commonly discussed together as the CO_2 reduction reaction takes place similarly i.e., solar/light energy harvesting, separation, and transportation of the charges. However, the surface and recombination reactions take place differently. He and Janaky discussed that this is because the PC process reactions are thermodynamically favorable, and the reaction is accelerated by the catalyst. In a PS process, the reaction is thermodynamically unfavorable and requires photochemical energy to carry out the reaction. Thus, when the reduction takes place with the oxygen revolution reaction it should be defined as a PS process as it is thermodynamically unfavorable ($\Delta G > 0$) [12]. Osterloh highlighted that under optimized conditions, PC systems are only limited by the surface area whereas PS systems are limited by mass and charge transport or electrochemical selectivity [19]. This difference is important as both processes have different performances/

efficiencies. However, both PC and PS reactions have been carried out in similar reactors, thus differences will not be further highlighted.

Focusing on the reactors, PC systems have simple designs and are used where the driving force of the reduction reaction is only dependent on the solar energy/incident light. Thus, an ideal photoreactor is one that can uniformly distribute light across the entire system, ensuring all the photocatalysts present participate in the reaction [7]. The major drawback of this system is the oxidation and reduction reactions take place at the surface of the same photocatalyst as a result products produced are mixed and tend to further react. To overcome the drawback of this system, several reactors have been designed and analysed, that use various synthesis routes. They can be classified based on various parameters such as: operational modes (batch, flow, or recirculating), the type of catalyst bed (suspended or supported catalyst bed, or slurry, fixed-bed, and membrane photoreactors), the number of phases involves (gas-liquid, solid-gas-liquid systems), the light source (UV radiation and visible light source), etc. [20, 21]. The two key parameters that determine the type of reactor used is the phase of the system and the operating mode [7].

The performance of current reactors is still evaluated by quantum efficiency; however, the important step that needs to be considered for large-scale reactor design is the absorption of the light by the materials and the transport of the light to the active site [7]. To design feasible and efficient photoreactors the design process must consider several factors such as the construction materials, geometry and flowrates of the reactants, products and recycle streams. The general materials of construction for photoreactors are glass, quartz, Pyrex, and stainless steel [20]. Amongst the reactors present in the current market, the following reactors have been widely used: slurry photoreactor, optical photoreactor, monolith photoreactor, twin reactor (membrane photoreactor) and microreactor.

8.2.1.1 Slurry Photoreactor

The slurry photoreactor is the most common and conventional technology. The reactor is easier and cheaper to construct compared to other technologies [20]. It is a usually operated in a multiphase system (gas-liquid-solid), [22] favoring the generation of methane and methanol [23]. In this reactor, the catalysts are suspended in the liquid phase with the help of mechanical or gas-promoted agitation [22]. The reactor is then sealed, and carbon dioxide is injected. The reaction starts when the light is irradiated on the reactor, as displayed in Figure 8.3 [20]. In the system where catalyst powders are suspended in liquid medium (slurry) the rate of reaction is

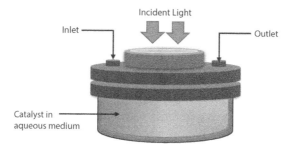

Figure 8.3 Schematic of a slurry photoreactor. Adapted from [7].

depended on the quantum efficiency of the catalyst, absorption properties of all the materials in the solution and surface light intensity [7, 23]. Quantum efficiency is an important parameter as it measures number of electrons reacted divided by the total absorbed photons by the photocatalyst [24, 25].

The photoreactor can be operated in either continuous or batch processes. Compared to continuous processes, the batch process obtains higher yields due to longer residence times. However, lower selectivity is observed due to accumulation of the products resulting in further side reactions with intermediates [26]. Moreover, a key disadvantage of this reactor is that post separation of the catalyst is required which can be time consuming and costly [23]. In certain applications where this technology is used at larger scale, for example in water treatment, the photocatalyst particles are separated in settling tanks or in continuous operations an additional crossflow filtration system is installed [27]. Previously, photocatalyst particle sizes between 30 nm to 2 mm were used due to the higher reaction efficiency obtained. However, due to difficulties in recovering the catalyst from the suspension larger particle sizes (ca. 10 to 100 mm) are used to allow sensible design of large-scale equipment [28]. Furthermore, due to the larger particle sizes the light radiation throughout the reactor is limited due to the absorptions, scattering, and shielding by the catalyst particles in the medium [23]. Due to the inefficacious light distribution in the reactor there is an issue of low specific surface area [20, 29].

Process intensification can be applied to photoreactors by improving the radiation intensity reaching all the catalyst surfaces. This can be achieved by intensification of the surface to volume ratio in the reactor while designing adequately illuminated surfaces to sure all exposed surfaces are illuminated [30]. In slurry reactors, the mechanical or gaseous mixing plays two process intensification roles. Firstly, the mixing increases the exposure of the catalyst to regions with higher light radiations which overcomes the effects of non-uniform light distribution. Secondly, in this system the catalyst particles tend to settle down thus mixing intensification reduces the mass and heat transfer resistances due to the movement of the catalyst particles [30]. However, depending on the type of catalyst used, excessive mixing can result in erosion of the catalyst [20]. Due to catalyst particle agglomeration in the reactor resulting into mass and proton transfer limitations, the processing capacity of this reactor is restricted. Studies show that a higher catalyst loading, higher radiation rates and lower flow rates could improve the mass transfer limitations thus, enabling further scale-up [31].

8.2.1.2 Fixed Bed Photoreactor

To overcome the limitations of the slurry photoreactor, the catalyst particles can be immobilized on a fixed bed where post catalyst recovery systems are not required, and continuous operation of the process is permitted [22]. A photocatalytic reactor with immobilized catalysts is hence referred to as fixed bed reactor where the catalyst is deposited on a support by chemical bonds or physical surfaces forces. In general, this reactor is a cylindrical vessel with a window made from quartz that allows light to radiate on a fixed catalyst support. Figure 8.4 shows the schematics of different illumination strategies for fixed bed reactor designs with optical fiber reactor design with side illumination and internally illuminated with top illumination [7]. As can be seen, illumination is a major issue during scale-up. Fixed bed reactors can operate in a two phase or three phase systems. The two-phase gaseous system is more commonly studied as a fixed bed photoreactor supports the formation of methane and carbon monoxide [26]. In reactors with liquid-solid or gas-liquid-solid, the mass transfer from the liquid medium to the catalyst surface plays a significant role in determining the rate of the reaction [23]. As a consequence, light distribution is a limiting factor in fixed bed reactors, as it depends on the geometry of the irradiation source and spatial distance between the light source and catalyst [7].

In conventional fixed bed reactors, the catalyst is coated on either the wall of the reactor or on the light source casing. However, a major drawback with this approach is the low surface-to-volume ratios, and absorption and scattering of the irradiation by the catalyst [23, 32]. To improve the performance of this reactor, Ray *et al.* designed a fixed bed reactor (for water purification application) that has hollow glass tubes coated with catalyst particles and conduct light. This configuration increased the surface-volume ratio and reduced the possibility of light scattering and absorption. To further improve the reactors several studies have been carried out on

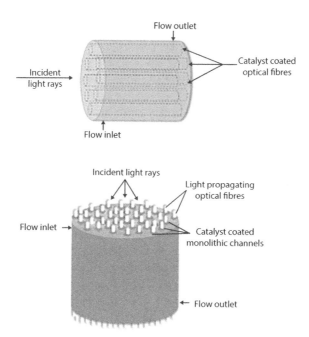

Figure 8.4 Schematic of fixed bed reactor design with optical fiber reactor design with side illumination (top) and internally illuminated reactor with top illumination (bottom) [7].

designing the supports of the photocatalyst. Amongst the various technologies, optical-fiber, and monolith photoreactors have displayed good performances [20].

In optical photoreactors, the photocatalyst is coated on optical fibers and mounted in the reactor. This setup maximizes the contact between the catalyst and reactants. The role of the fibers is to evenly distribute the light on the catalyst particles as a result these reactors tend to give high conversion rates [20]. Moreover, optical fibers can deliver the photoenergy to inaccessible reactive sites on the photocatalyst layer [33]. However, the drawbacks of this reactor are; there is a lower surface area available for coating support; nonefficient use of the entire reactor as the optical fibers only occupy approx. 20% to 30% of the reactor volume; the optical fibers are fragile and result in durability issues [20]. Bundling up more optical fibers in the reactor results in heat build-up leading to deactivation of the catalyst [33, 34]. Moreover, the durability of the fibers is dependent on the strength of the adhesion and thickness of the catalyst layer deposited. In systems with TiO_2 photocatalyst the adhesion is possible due to the electrostatic interaction. In large-scale continuous operation, this interaction is unable to withstand large flow conditions in both liquid and gaseous phases [33]. It is possible to increase the durability and adhesion of the fibers by roughening the surface of the fiber surface before depositing the photocatalyst; however, this results in uneven distribution of the catalyst and light [7, 33].

The use of 3D structures like honeycomb monoliths with channels that are straight and parallel has been used for various application in industry due to their ease of scale up through increasing the dimension and number of channels and ease of control of structural parameters [7]. Monolith reactors are commonly used in vehicle emission control and in power plants to reduce the NOx in the flue gas through catalytic reduction [35]. In CO_2 photoreduction systems, the unique feature of monolith photoreactors is that it has a network of channels impregnated with the photocatalyst on its walls. This feature allows light to penetrate well amongst the catalyst particles and increase the surface area about 10 to 100 times more compared to other catalyst support beds [20]. Moreover, in this reactor a low-pressure drop can be achieved at high flowrates. However, even though the monolith reactor overcomes the disadvantages of the optical reactor when the efficiencies are compared the optical reactors generate higher quantum efficiencies. This is due to the opacity of the channels there is a lower penetration of light in the reactor resulting in lower conversion rates compared to the optical fiber reactor [20, 34]. It should be noted that the operation of an internal illumined monolith reactor should be limited to the thin film flow regime as the presence of bubbles can result in refraction and reflection of the light in the reactor [36].

In comparison to the slurry reactor, the monolith reactor demonstrates higher conversion and quantum efficiencies when using the monolith as a catalyst carrier. This is due to the larger available surface area and elimination of uneven light distribution [37]. Furthermore, due to the promising results offered by both the optical and monolith reactors, several research groups have been carried out experiments in attempt to combine the higher surface area of the monolith and effective light distribution by the fiber optics. Results from Ali *et al.* and Lin *et al.* demonstrated higher quantum efficiencies were obtained [38, 39]. A study by Ola *et al.* displayed up to 23.5 times higher efficiencies than the slurry batch reactor [40].

8.2.1.3 *Twin Photoreactor (Membrane Photoreactor)*

In both the slurry and fixed bed reactors, the redox reactions occur on the surface of the catalyst. In this reaction environment, the hydrocarbons can further oxidize with the abundant oxygen produced during the water oxidation reaction. As a result, the product yields and selectivity are reduced. To overcome this limitation, the twin reactor was developed, as outlined in Figure 8.5 [41]. In a twin reactor, the water splitting and carbon dioxide reduction reactions take place in two compartments separated with a membrane (e.g., Nafion membrane, Neosepta membrane). The membrane allows exchange of hydrogen ions but isolates the oxygen to prevent further reactions with the products and reduces the chances of the backward reaction between the oxygen and hydrogen ions [42, 43]. The twin reactor illustrated in the figure below is also referred to as a modified water splitting Z-scheme system. In this system, two different types of catalysts are required (i.e., p-type and n-type) [41]. Twin reactors offer higher conversion efficiencies and selectivity compared to conventional reactors [20]. Yu *et al.* investigated the photocatalytic water splitting and hydrogenation of CO_2 in a twin photoreactor using IO_3^-/I^- as a redox mediator. Their results demonstrated that the quantum efficiency of the twin photoreactor improved by four-folds compared to using a single reactor, i.e., from 0.015% to 0.070% [44].

Twin reactors can operate in 2-phase or 3-phase systems. In the system, where the reduction takes place in aqueous phase, the target product is usually alcohols, such as methanol [26].

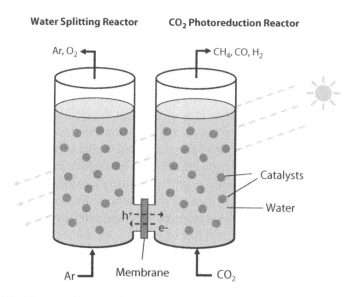

Figure 8.5 Schematic diagram of a twin reactor. Adapted from [41].

When CO_2 reduction takes place with water in liquid phase, the water splitting reaction competes with the reduction reactions lowering the yield of hydrocarbon products produced. Moreover, when CO_2 is bubbled in the liquid, the reactant transport is limited by the interfaces and liquid phase. The use of the membrane reactor resolves this problem as it is possible to separate the gas and liquid phase enhancing the carbon dioxide transport [29]. Xiong et al. investigated the performance of a twin photoreactor that carried out CO_2 reduction in gas phase and water splitting in liquid phase. The experimental results displayed an enhancement in the hydrogen evolution and carbon reduction reaction. This was because the twin reactor enabled better control of both reactions as the gas and liquid phase were present in different reactors separated using the Nafion membrane [45]. Another advantage of the twin membrane reactor is it allows continuous operation without loss of catalyst particles compared to the conventional slurry photoreactor [27]. Experimental studies have displayed that continuous operation significantly improves the reaction efficiencies. For example, Pomilla et al. studied the photocatalytic conversion of CO_2 to methanol and ethanol in a continuous membrane reactor, using C_3N_4 organic catalyst immobilized on a Nafion membrane. Their results showed that a 10 times higher conversion was achieved in the reactor compared to a batch reactor [46].

Moreover, photocatalytic membrane reactors improve mass transfer due to the forced transport of gas through the membrane's pores, where the photocatalyst is deposited [35]. This technology offers higher efficiencies and selectivity; however, large-scale applications have not been possible. This is due to the cost of the membranes and maintenance requirements. The membranes are prone to fouling and degradation due to operating conditions and ultraviolet light [35, 47]. Twin reactors can be configured as slurry or fixed depending on how the membrane is utilized [20]. In the slurry reactor the photocatalyst is suspended in the reaction environment and in the fixed reactor the photocatalyst is immobilized on the membrane. In systems where the catalyst is deposited on the membrane, fouling is avoided; however, the reaction efficiency is lower than in systems where the photocatalyst is in suspension [48]. The slurry configuration provides a larger active surface area as a result the reaction efficiencies are higher; however, the catalyst particles scatter the light and uniform distribution of the photocatalyst are the main challenges [48]. An advantage of the fixed system is the usage of commercially available membranes. The improvements in process efficiency, modularity, and ease of scale up obtained using photocatalytic membrane reactors has resulted in significant interest in this technology. However, for larger-scale applications the performance of the reactor still needs enhancement [13].

8.2.1.4 Microreactor

Microreactors are reactors with channels the order of micrometres since reducing the channel ensures that diffusion is the dominant mixing mechanism [49]. In heterogenous microreactors, the photocatalyst is usually deposited on the wall of the reacting channels for catalytic microreactor [50, 51]. Recapping the crucial factors that need to be considered when designing a photoreactor are the mass and proton transfer, homogenous distribution of light, and the available reaction surface area [29, 52]. Microreactors have favorable characteristics for improving mass and proton transfer compared to conventional reactors. Due to the small channel dimensions, microreactors achieve higher surface to volume ratios, homogenous distribution, and penetration of the light throughout the microreactor [53]. Reduction in the mass transport limitations in the reactor makes it easier to study the reaction kinetics in the reactor. This allows better understanding of the reactions that take place increasing the possibility to control the selectivity and yield of the desired products [35]. The shorter molecular distances and high photon densities lead to short resistance times compared to the larger photoreactors and offers good heat dissipation [30, 35]. Moreover, due to the small reaction volumes and short diffusion resistances the safety concerns, such as handling of explosive components (H_2 and O_2) and toxic components (CO), are minimized [30].

Incorporation of optofluidics into microreactor design has shown an improvement in the mass and photon transport, homogenous illumination, and better light penetration throughout the reactor. Figure 8.6 outlines the schematic of an optofluidic membrane microreactor. This technology has been widely used in processes such as photocatalytic

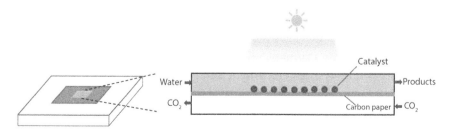

Figure 8.6 Schematic of an optofluidic membrane microreactor. Adapted from [51].

water splitting and photocatalytic fuel cells for wastewater treatments [54, 55]. A notable study in the intensification of microreactors for CO_2 photoreduction is the combination of optofluidics microreactors and membrane photoreactors for CO_2 photoreduction to methanol. Cheng *et al.* developed this reactor and displayed the performance of the reactor using the product yields obtained. Their system produced a max methanol product yield of 110 mmole/gcat h using TiO_2 catalyst loading of 4.5 mg/cm². These results demonstrated the feasibility and exemplary performance obtained using this reactor [29].

The advantages of this intensified reactor design compared to the previous reactor designs discussed are, compared to slurry reactors the difficult and costly catalyst separation process was eliminated as the catalyst was immobilized on the membrane. Additionally, the membrane separates the gas and liquid phases ensuring the CO_2 has direct contact with the catalyst, thus, overcoming the mass transfer issue of the reactants. Moreover, during the reaction, oxygen gas is produced which reoxidizes the products formed. Using the membrane ensures that the oxygen produced is removed from the gas phase side. Finally, the use of optofluidics increases the available reaction surface area to volume ratio, reduces the photon transport distance and gives a more homogenous light distribution [29]. The optofluidic microreactor can be used to perform kinetic studies quickly and economically for various photocatalysts and carry out experiments to determine the optimum reaction conditions [54].

The main drawback of microreactors is scaling up and processing larger quantities. Theoretically, this technology can be scaled up by numbering up the number of reactors without changing the geometry of the channels, but the problem arises when incorporating uniform illumination [29, 30, 50]. Additionally, microreactors have been used in applications where the catalyst is suspended in the reaction media, but this leaves potential for channel blockages [30]. To avoid blockages, Delacour *et al.* demonstrated that pulsed ultrasound can be employed [56]. Thus, further research on larger-scale applications is still necessary to assess the feasibility.

8.2.2 Electrochemical System

Electrochemical (EC) systems utilize catalytic conversion of CO_2 to other chemicals and fuels in an electrolytic cell powered by electrical energy, as outlined in Figure 8.1 [14]. The electrical energy is externally supplied, e.g., from solar photovoltaic (PV) cells, which can either be integrated into the system or coupled together with a DC-DC connection to the electrolytic cells [12]. Over the past decade, the power conversion efficiency (PCE) of PV technologies has continually improved while the cost continually decreases. PCE is the ratio between the incident solar energy to the electrical energy output. For multijunction cells, such as AlGaInP/AlGaAs/GaAs/GaInAs, the PCE is approximately 47.1% [12]. Between 2010 and 2020, the cost of electricity from utility scale solar PV has dropped by 85% [57]. Thus, utilizing PV to supply the electrical energy for the electrochemical conversion of CO_2 and producing liquid fuels is viewed as an appealing solution to store renewable solar energy [58].

An electrolytic cell generally has a negative electrode (cathode), positive electrode (anode), and an electrolyte. In the electrochemical reaction of CO_2 and H_2O, the water oxidation takes place at the anode (Equation 8.1) producing hydrogen ions, while the CO_2 reduction takes place at the cathode (Equation 8.2).

$$H_2O \rightarrow 4H^+ + O_2 + 4e^- \tag{8.1}$$

$$aCO_2 + bH^+ + be^- \rightarrow C_a H_{b-2} O_{2a-1} + H_2O \tag{8.2}$$

Electroreduction of carbon dioxide can be carried out in heterogenous and homogenous reactions. In heterogenous reactions, the main reaction takes place either on the solid-gas or solid-liquid interfaces. Focusing on the cathode side, electroreduction of CO_2 generally undergoes the following steps: first the CO_2 adsorbs on the electrode/catalyst, then the surface diffusion and the electron-proton transfer takes place, finally the products desorb of the electrode surface [59]. The structure of the catalyst surface plays a vital role in the performance of a heterogenous reduction reaction, such as high-index surfaces have a lower number of coordination atoms that enhance the catalysts capability [59]. On the other hand, a homogenous reaction system is often referred to as indirect electrolysis where the catalyst acts as the redox shuttle. During the reaction, the catalyst accepts electrons from the electrodes and enters a highly reduced state. The reduced catalyst then donates electrons to the CO_2 in the solution and returns to the original state [59, 60].

The technology readiness level of electrochemical reduction process is quite low (TRL 3-5). This is due to a number of challenges that need to be overcome for large-scale/commercial applications [14, 61]. Thermodynamically, the water oxidation reaction takes place at standard potential of −1.23 V while the CO_2 reduction reaction to common products such as CO, HCOOH takes place near 0 V as outlined in Table 8.2 [30, 62]. This makes the reduction half reactions difficult and can be overcome

Table 8.2 Standard potentials related to CO_2 reduction and water oxidation.

Reactions	E^0 vs. NHE at pH = 0 [62]
$2H^+ + 2e- \rightarrow H_2$	0V
$2H_2O \rightarrow O_2 + 4e^- + 4H+$	-1.23 V
$CO_2 + 2H^+ + 2e- \rightarrow HCOO^- + H+$	-0.31 V
$CO_2 + 2H^+ + 2e- \rightarrow CO + H_2O$	0.11 V
$2CO_2 + 12H^+ + 12e^- \rightarrow C_2H_5OH + 3H_2O$	0.08 V

using an overpotential. However, using an overpotential makes the reaction energy inefficient and the desired product selectivity decreases due to the multielectron transfer. Additionally, the hydrogen evolution reaction (which takes place at 0V) competes with the reduction reaction, which reduces the catalyst stability and Faradaic efficiency [59]. Thus, control over the reduction reaction activity and desired product selectivity while suppressing the hydrogen evolution reaction is important to improve the efficiency of the process [63].

The application of active and selective electrocatalysts is essential to improve the efficiency of the reduction reaction. The electrocatalyst utilized must possess a strong C-O bond energy but lower hydrogen bond energy [59]. Figure 8.7 shows the suggested reaction mechanisms of CO_2 during electrochemical reduction on metal electrode [14]. Depending on the desired product, various catalysts, such as noble metal, transition metal, organic polymers, and non-metallic catalysts, like carbon nanofibers, have been developed [59]. Ideal CO_2 reduction electrocatalysts need to consider the activity (sufficient exposed active sites, good electron conductivity),

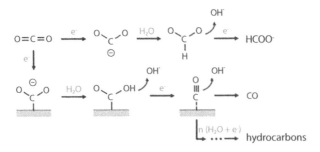

Figure 8.7 Suggested reaction mechanisms of CO_2 during electrochemical reduction on metal electrode [14].

stability (product desorption and ion attack resistance), and selectivity (appropriate intermediate adsorption and limited hydrogen adsorption) [59].

The cost of the catalyst is an important factor when considering large-scale applications. For example, carbon monoxide was synthetized using noble metals, such as Pd, Au, and Ag, at moderate potentials (0.6V), resulting with faradaic efficiencies up to 90%. However, due to high system and catalyst costs large-scale applications have not been successful [64]. When using metal electrocatalysts, HCOOH and CO are the most common products [14]. For other products such as methanol, methane and ethylene limited materials are available that can further reduce to C-O intermediate. According to various research articles, only Cu and Co catalysts have successfully produced other products. This is due to their high binding energy towards the C-O intermediate [65, 66]. However, these processes have either low faradaic efficiencies or need high cell voltages to achieve adequate current densities or conversion [67]. Nevertheless, further improvements of the electrocatalytic materials are continually witnessed. Recently, Azenha *et al.* carried out the electroreduction of CO_2 to methanol using CuO nanowires. The structure of the nanowires resulted in remarkable methanol selectivity and the hydrogen evolution reaction was suppressed. At an overpotential of 410 mV and atmospheric conditions, 1.27×10^{-4} molm^{-2} s^{-1} of methanol yield was obtained with a faradaic efficiency of 66% [63].

8.2.2.1 CO_2 Electrochemical Reactors

The most common lab-scale reactor for CO_2 electroreduction is the H-type cell as it is easy to operate, easy to setup, and has low assembly costs. Figure 8.8 shows a schematic illustration of H-cell [68]. Generally, a CO_2 electrolysis cell consist of two or three electrodes, an ion exchange membrane, the reactant CO_2 gas, and the electrolytes. The role of the ion exchange membrane is to separate the anode side form the cathode side to prevent mixing [61, 69]. Moreover, the catalyst material is either deposited onto the membrane or a gas diffusion layer substate [69]. Many electrochemical reactors use a two-electrode system (with a working and counter electrode). In such systems only, the current or voltage controllable. In a three-electrode system, the working electrode (cathode) and the reference electrode are present in the cathode side, while the counter electrode (anode) is present in the anode side [68]. Integrating the reference electrode allows the control of potential in the cathode side. Potential control is useful when using catalysts that experience chemical decomposition or surface property changes

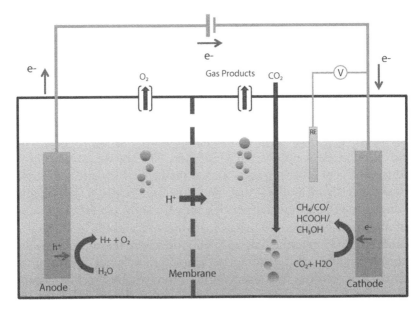

Figure 8.8 Schematic illustration of H-cell. Adapted from [78].

during continuous operation. It is possible to have reference electrodes in the cathode and anode side creating a four-electrode system. Such as system is used when the water oxidation reaction is difficult, thus control over both half-cells is necessary [70]. Comparing the performance of a two electrode and three electrode system, a study performed by Woldu et al. using a polycrystalline Cu cathode and hematite anode demonstrated that the two-electrode configuration displayed better performance at high bias – 0.8 V compared to the three-electrode system. They concluded that for practical applications where the electrical energy is powered using solar cells, the reactor with two electrodes was more desirable [71].

The H-type cell reactor is commercially available, offers a wide range of electrode selections and is normally used for screening catalysts in lab scale [68]. However, the experimental data obtained from this reactor is not relevant for a dynamic environment of a CO_2 electrolyser system. This is due to the low CO_2 solubility in the electrolyte limits the mass transport in the reactor. This results to low current densities of <100 mA cm^{-2} [58, 68]. Moreover, the limited available surface area of the electrodes and the large distance between the electrodes makes it difficult to apply this system on commercial applications. To improve the mass transfer limitations and the CO_2 electroreduction rate, continuous flow reactors have been developed (Figure 8.9). In this system, the reactants and products are continuously

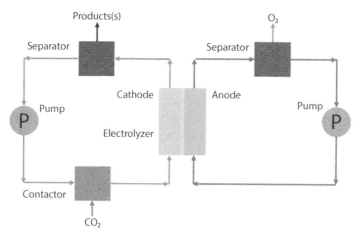

Figure 8.9 A typical CO_2 electrolyzer design with a gas/liquid contactor for CO_2 intake [61].

circulated to and from the electrodes. These reactors offer a higher control of the reagent delivery allowing gaseous CO_2 feed to the cathode side. Compared with the H-cell reactor, the thermodynamics and kinetics of the flow reactors are different as a higher concentration of CO_2 is present on the reacting surface thus generating higher current densities [58, 68]. In a typical system, CO_2 is introduced to the cathode through a gas/liquid contractor and saturates the aqueous catholyte. The CO_2 saturated catholyte is then pumped into the flow cell where it undergoes reduction and is followed by a product separation process using a gas-liquid separator. The same process occurs at the anode side; however, a gas/liquid contractor is not typically used [59].

Another advantage of flow reactors is due to the design of the continuous flow cells, they can be developed and studied at laboratory scale and then can be effectively scaled to the larger stacks like an electrolyzer unit. This approach has been successfully witnessed in commercial scale applications of hydrogen fuel cells and water electrolyzer. However, CO_2 flow reactors are not a mature industry as there are few studies present on the system and electrocatalysts applicable for CO_2 electroreduction [58, 68]. Depending on the operating temperatures and ion transport medium (e.g., membrane), several CO_2 flow reactors have been studied. For temperatures below 100°C the electrolyzer can be categorized into two primary architectures: membrane-based flow cells and microfluidic flow cells [58]

Compared with the H-cell reactor, both the above mentioned have demonstrated to achieve high current densities, i.e., > 200 mA cm^{-2}. The high current densities are achieved by supplying the CO_2 to the cathode side in gas phase rather than in the aqueous electrolyte.

Membrane-based flow cells, such as membrane electrode assemblies (MEA), have received significant attention in recent years. A typical membrane-based flow cell offers a sandwich like design with a polymer exchange membrane, cation/anion conducting polymer-based electrodes, anode electrolyte, cathode electrolyte, flow fields and gas diffusion electrodes [58, 67]. Gas diffusion electrodes contain immobilized electrocatalysts to overcome the mass transfer limitations between the three phases (gas-liquid-solids) and interphases in the cell [30, 58, 72, 73]. The membranes play an important role as they allow flexibility in using different electrolyte volumes and enhance the separation of products [73]. There are different membrane categories available such as the cation exchange membrane (CEM), anion exchange membrane (AEM) and bipolar membrane (BPM), each resulting different kinetics of ion transport pathways and electrolyte applicability [58]. Briefly, CEMs (also referred to as proton exchange membranes) allow the flow of positive ions (H^+) from the anode to the cathode. The common CEM is Nafion, which offers high conductivities, high chemical stabilities, and optical transparency. The disadvantages of using this membrane are the high costs and over a period the cell experiences a pH imbalance which increases the required voltage for the water splitting at the anode [74]. AEMs allow the flow of negative ions such as (OH^-, HCO_3^-) from the cathode to the anode. This mechanism limits the delivery of H^+ ions to the cathode compared to CEMs reducing the competing hydrogen evolution reaction. However, AEMs create basic conditions in the cell where OH^- react rapidly with CO_2 forming HCO_3^- and CO_3^{2-}. These ions are larger in size and inhibit ion transport in the membrane, thus reducing the CO_2 reduction reaction efficiencies [58, 75]. In BPMs, the anion and cation membranes are connected using an interfacial catalyst layer that allows the self-dissociation of water into H^+ and OH^- ions. These membranes transport the OH^- to the anode and the H^+ to the cathode. The main advantage of this membrane is it maintains a pH balance across the entire cell as a result inexpensive anode and cathode catalysts can be used [58].

A microfluidic cell uses gas-diffusion electrodes for the deposition of electrocatalysts. Studies on the properties of gas diffusion electrode materials for microfluidic cells are equally relevant to membrane-based flow reactors [58]. A microfluidic cell is designed to allow gaseous CO_2 flow on one side and liquid electrolyte flow on the other. By manipulating the three-phase boundary layers, the CO_2 gas can reach the catalyst surfaces without a hindrance of the electrolyte solubility limit resulting in high current densities (100 mA/cm² at 3V). Moreover, a basic electrolyte can be used as gaseous CO_2 does not come into contact with the catholyte [61]. Microfluidic reactors allow the ability to control and change reaction parameters such as the applied cell potential, residence time, pressure, temperature, and CO_2 concentration and flow rate [30, 69]. Moreover, an emerging type of CO_2 electrolysers is the zero-gap cells. This electrolyser is referred to as zero-gap as the anode and cathode are pressed together with an ion exchange membrane in between to create a zero gap with no flow channels formed between them. In this reactor, a humidified gaseous CO_2 stream is fed to the cathode where the reaction occurs at the boundary of the membrane and cathode. Compared to microfluidic cells, this device can handle higher local CO_2 concentrations by increasing the pressure resulting in higher current densities and reaction rates. Moreover, this device has lower ohmic losses and less risk of catalyst poisoning caused by the catholyte in comparison to membrane-based flow cells [76]. However, a drawback for this device is due to the proximity of the electrodes liquid products can flow back into the GDE pores as a result blocking the active catalyst sites. Thus, the liquid products can penetrate the gas flow channels and prevent further reactions [68]. Moreover, the exchange of positive ions only through the membrane can result in acidification of the catholyte increasing the production of H_2 over the CO_2 reduction reaction [70]. Other notable reactor design utilized for high-temperature applications is the solid oxide electrochemical cell (SOEC). At elevated temperatures (>400C), ceramic-based membranes are also used that conduct protons or O_2 species, depending on reactor configuration and utilize heat to lower overall reduction potentials [67]. This technology has shown potential to produce CO up to 200 Nm³ h⁻¹; however, the formation of higher value and longer chain hydrocarbons still faces technical challenges [67].

8.2.3 Photoelectrochemical (PEC) Systems

Photoelectrochemical (PEC) systems were first reported in 1978 by Halmann *et al.*, showing a novel route for photo-assisted electrocatalytic reduction of aqueous carbon dioxide producing formic acid, formaldehyde, and methanol [77]. The integration of an electrochemical reactor with a light source and a photocatalyst obtains a photoelectrochemical (PEC) reactor (Figure 8.1). Electrochemical CO_2 reduction has achieved high-energy efficiencies; however, the significant drawback for this system is the use of expensive metal electrodes combined with the cost of additional equipment, such as PV panels, converters, and wiring [78]. In principle, heterogenous PEC system reduces the requirement for additional electrical generation devices, resulting in lower system costs and electrical energy losses in energy transporting electricity to the electrolysis cell [79].

The heterogenous photoelectrochemical (PEC) cell generally consists of semiconducting photoelectrodes, and counter electrodes. In the configuration illustrated in Figure 8.1 (C), the radiation is harvested by the photocathode to generate charge carriers and reduce the CO_2. At the anode, water undergoes half-cell reactions to generate hydrogen ions and oxygen. Moreover, the anode and cathode are usually separated using a proton exchange membrane (e.g., Nafion) which ensures that the products do not get reoxidized.

In a PEC system, the CO_2 can be reduced to a number of products; CO, H_2, HCOOH, CH_3OH, CH_4. The extent of the reaction is dependent on the catalyst material, reaction media and the potential applied [79]. Thermodynamically the CO_2 reduction reaction is an uphill reaction and the spontaneity of the reaction to produce a certain product is determined by the reduction potential (E^o). Therefore, it is necessary to supply energy to overcome the negative potentials. In PC systems, the photocatalyst plays an important role in using the light as the input energy to carry out the redox reactions. Applying the same concept, the semiconductors in a PEC cell can convert the incident photons with energy that is equal or greater than the width of its band gap to generate electron-hole pairs. The bandgap represents the energy-states between the valence band, the highest energy-state filled with electrons, and the conduction band, the lowest unfilled energy-state. The incident photons with the sufficient energy promotes electrons from the valence band to the conduction band resulting in an increase of electrical mobility of the semiconductors. The electrons that were promoted in the valence band leave behind empty positions referred to as holes. This photo-excited electron and hole pairs is what carries out the oxidation and reduction reactions [78]. Like PC systems, PEC systems rely on the light harvesting; however, the advantage of this system is the configuration allows the application of external biases to support the photoreaction by lowering the energy need, thus resulting in higher reaction efficiency [6]. Theoretically, a variety of semiconductors can be used in PEC systems [78].

The reactor configurations of PEC are comparable to the EC systems where the gas diffusion electrodes (GDEs) and gas phase environmental conditions at the cathode side are preferred to avoid the limitations of solubility of CO_2 in aqueous environments. GDEs are generally made using a carbon fibre substrate, microporous layer, and catalyst layer [79].

Depending on the reaction conditions and cell configuration of PEC systems, the preferred ion transport membrane is the cation exchange membrane (CEM) where the protons/ H^+ ions are transferred from the anode to the cathode. Recently, the use of bipolar membranes (BPM), which allow anion and cation transfer, has displayed better performances. BPMs can separate the anode and cathode compartments while allowing selective anion and proton transfer. This maintains constant pH across the cell allowing the use of abundant metal anodes and highly active cathode, which is not possible with CEM and AEMs [79]. A notable study is by Zhou et al., where they carried out experiments to produce formate from CO_2 photo-electroreduction. They reported a 10% solar to fuels efficiency with a >94% faradaic efficiency using a bipolar membrane, GaAs/InGaP/TiO2/Ni photoanode and Pd/C nanoparticle-coated Ti mesh cathode. Comparing the results with a cell without a membrane it was evaluated that the BPM facilitated the redox reactions by separating both electrodes and producing lower overpotentials in the entire cell [80].

Compared to EC reactors, for PEC systems, an additional consideration comes into play, i.e., the illumination of the photocatalyst surface. The key considerations when designing PEC reactors are: the light source and homogenous distribution of the radiation on the photocatalyst surface; materials of construction as the light transmission is influenced by the construction materials; heat management due to the heat generated by the light source, crucial in gas-solid reactions; mixing and flow characteristics depending on the phases involved and mode of operation [79]. Continuous flow electrochemical reactors have displayed exemplary performance compared to batch reactors. Minimal studies have been performed on continuous PEC reactors to form liquid products as compared to batch reactors. The continuous flow PEC reactor developed by Homayoni et al. display a six-fold higher production rates compared to a batch PEC reactor. The CO_2 reduction was carried out using a CuO/Cu_2O nanorod array photocathode, Nafion membrane, and stainless steel-378 anode to produce alcohols. Additionally, the results displayed that longer chain alcohols (mainly ethanol and isopropanol) were produced in the continuous reactor [79, 81]. Moreover, the use of nanostructure photoelectrodes gives a larger specific surface area enhancing the current density [70].

There are three different PEC system configurations depending on which electrode acts as the photoelectrode (Figure 8.10) [78]. These systems include the photocathode – anode system, cathode–photoanode system and the photocathode- photoanode system. The photocathode and dark anode system, as afore described, is the commonly employed system where the photocathode is a p-type semiconductor with a high conduction

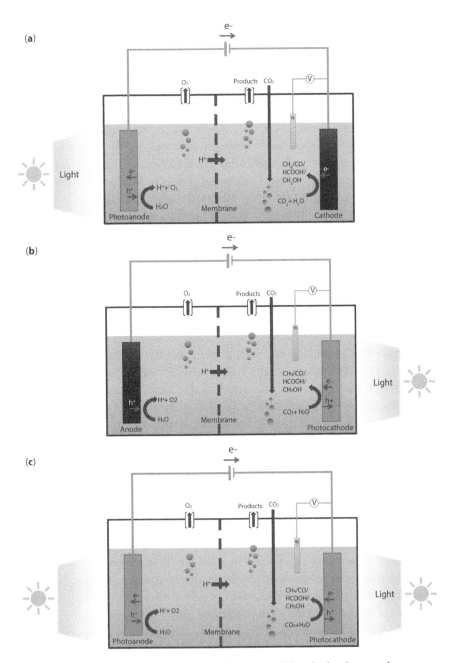

Figure 8.10 Three different PEC system configurations: (a) cathode-photoanode, (b) photocathode-anode, and (c) photocathode-photoanode system. Adapted from [78].

band and the anode is a metal [79]. The p-type semiconductor is doped with impurities that accept electrons from the valence band. Ideally, it should absorb a large part of the solar spectrum and its conduction band should lie at a negative potential more than that of the reduction reaction potential [82]. The commonly used photocathode is copper oxides due to their low cost, abundance, chemical stability, and toxicity [82]. Other photocathode materials used are expensive, toxic, and unstable; as a result, the overall efficiencies of this system remain low and improvement of the photocathode is challenging [79].

The photoanode with a dark cathode system uses n-type semiconductors such as TiO_2, ZnO $BiVO_4$ and WO_3 as photoanodes as they are abundant, stable, and affordable [79]. The n-type semiconductor is one which is doped with impurities that can donate excess electrons to the conduction band. The cathode consists of a metallic electrocatalyst for CO_2 reduction. The performance of a photoanode device depends on both the photoanode and cathode [78]. This is because, the photoanode plays an essential role in harvesting the solar light and lowering the external bias required for the reduction reaction resulting in a wide range of electrocatalysts that can be used on the cathode to improve the efficiency and control the selectivity [78, 79]. Comparing photocathode and photoanode driven devices, the photocathode driven devices can achieve higher efficiency and selectivities. This is because the p-type photocathode can be coupled with selectivity oxygen evolution reaction catalysts, such as platinum. While the n-type photoanode are coupled with a low selective electrode, such as Cu-based to carry out the CO_2 reduction, which experience lower efficiencies because of the competitive hydrogen evolution reaction [82].

Lastly, the photocathode-photoanode system contains a p-type semiconductor that carries out the CO_2 reduction reaction and n-type semiconductor that carries out the water oxidation reaction. In this system, both the photoelectrodes absorb the light generating electron hole pairs simultaneously. This system does not require an external potential, and the use of two different semiconductor system is known as a Z-Scheme as it mimics the electron excitation and transfer in natural photosynthesis [78]. The charge carries generated mostly recombine at the ohmic contact between the electrodes, leaving a minority of charge carriers to drive the redox reaction [78]. It should be noted that due to reaction overpotentials and parasitic losses some system report the need for additional external potential to carry out the reactions [79]. Currently, the PEC CO_2 reduction systems suffer from low solar-to-fuel (STF) efficiencies and the highest yet recorded is 1.2% [82]. Hence, a key factor for scale-up consideration is its improvement.

8.3 Design Considerations for Scale-Up

There are five main parameters that govern the kinetics of photocatalysis include the catalyst mass, reactant partial pressure/concentrations, wavelength of irradiation, radiation flux and system temperature [83]. When considering the catalyst there are several other factors that have been studied to have an influence on the reaction efficiencies. This includes the catalyst geometry, catalyst doping, catalyst morphology, catalyst support (e.g., immobile over mesh or monolithic support) [26]. Additionally, there are a wide range of operating parameters that affect the reaction efficiencies, for example, temperature, pressure, pH, reaction time, concentration of reactants and catalyst, voltage, and the reducing agent [26]. The effect of catalyst support, reactor geometry, mode of operation (batch or continuous flow) and phase of system was discussed in the previous sections. Figure 8.11 illustrates the key design considerations for scale-up of solar fuel production systems [26]. This section will focus on following general parameters that are applicable to PC, PEC, and PV+EC systems —irradiation, temperature, and pressure of the system.

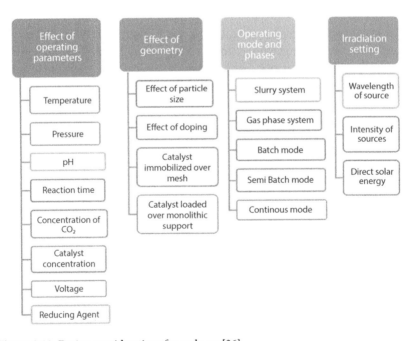

Figure 8.11 Design considerations for scale-up [26].

Irradiation is the driving force for solar fuel production. The wavelength of the light source and intensity is crucial for the reactions and will help determine the system and reaction efficiencies. For photocatalytic conversion of CO_2, the benchmark catalyst, such as TiO_2, is used under UV light conditions and are limited to the absorption of the entire light spectrum [26]. Moreover, optimally the minimal the distance the light travels from the lamp to the reactor and then to the active surface, the better the reaction yields as the intensity of the photons on the surface are higher [21]. Thus, the design of the reactors/systems needs to be based around the light source.

The impact of temperature and pressure on the reaction yields displays conflicting conclusions depending on the objectives of the studies. CO_2 photocatalytic reduction generally is conducted at ambient conditions, i.e., at room temperatures between 293K and 353K [7]. This is because the solubility of CO_2 in water increases at lower temperatures. High solubility impacts the targeted product selectivity for example, methanol formation is aided by high solubility of CO_2 [26]. Having said that, studies show that temperature has a positive influence in the increase of selectivity towards hydrogenated products such as methane and hydrogen gas [21]. Moreover, higher temperatures result in an increase of the reaction rate due to the increased collision frequency and diffusion rates. At low temperatures, the desorption rate of the products from the catalyst sites decreases resulting in reduced active sites. Therefore, the temperature variation has a direct impact of the efficiency of the reaction where an increase in temperature increases the reaction rate and catalyst availability, and a decrease in temperature increases the solubility and adsorption rate [26]. In terms of larger-scale production systems, increasing temperature does benefits the production rates; however, additional hurdles arise, such as the cost of setting up a sophisticated system that can maintain the high temperatures and sourcing the thermal energy [7].

For an electrochemical system, high temperatures offer better CO_2 transport, better kinetics, and higher current densities. However, the cell performance in terms of selectivity and solubility of CO_2 decreases. The normal range of operating temperature for CO_2 electroreduction processes is usually between 293 K and 313 K [84]. Increasing temperatures further than this range can result into drying and damage on the membrane electrode assembly resulting in increased cell resistance and performance losses [85].

According to Henry's Law, the solubility of CO_2 can also be increased by increasing the pressure, resulting in an increase of the product selectivity and reaction rate [7]. However, several studies report that initially as the pressure increases the product yield increases until a certain point from which the yield starts to decrease. In photocatalytic reactions, the decrease in yield is experienced at high pressures due to the CO_2 dominating the active sites over H_2O [21, 26]. Similarly, in electrochemical cells a high pressure enhances mass transfer due to the increase in the CO_2 solubility in the electrolytes. The pressure favors the reaction until a certain point after which the efficiencies start to decrease due to crossover of products in through the membrane and a decrease in pH because of the CO_2 dissolution that favor the hydrogen evolution reaction [86]. In the electrochemical cells a pressure balance is necessary between the anode and cathode sides. This is because pressure imbalances can result in mechanical stresses on the membrane resulting in crossover of reagents [85]. Moreover, higher operating pressures increases the cost and complexity of constructing the electrochemical reactors thus process optimization should consider the costs and product selectivity when designing CO_2 reduction reactors [84].

8.4 Future Systems and Large Reactors

Since 2000 there has been an exponential rise in the numbers of patents related to solar fuels (Figure 8.12). This is a key indicator that design issues related to solar fuel systems that affects the techno-economic considerations are being addressed. The obstacles when scaling up of photocatalytic and photoelectrochemical reactors is significantly complicated than scaling up conventional chemical reactors. The requirement for homogenous catalyst illumination is a crucial additional engineering element in the reactor design, besides the conventional reactor scale-up complications such as homogenous mixing and mass transfer, the reactant-catalyst interactions, reaction kinetics, and control of operating conditions, as outlined in the previous sections [23].

In addition, for scale-up feedstock supply and preparation needs to be considered. The availability of suitable CO_2 can be an issue if ultra-purity is needed to avoid operational issues, such as catalyst contamination. After the reactor, other issues including product separation and storage might be of issue depending on end use of fuel. Figure 8.13 shows the schematic of the commercial NASA solar-powered, thin-film devices for carbon dioxide conversion [87]. The water in the product stream is envisaged to be reused by the photocatalyst but any slippage can be condensed from the product stream where classic thermodynamic approaches can be used. Moreover, for all solar fuel production systems safety is of great concern after

Advances in Design and Scale-Up of Solar Fuel Systems 275

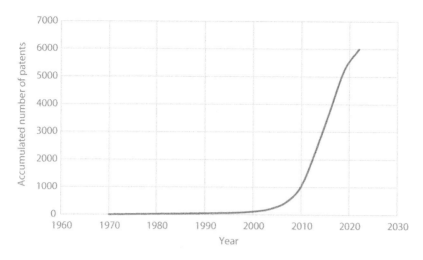

Figure 8.12 Number of patents related to solar fuels.

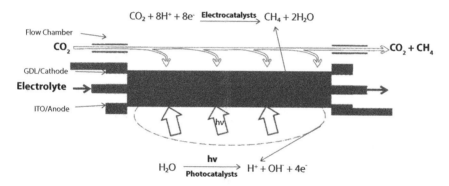

Figure 8.13 Commercial developed solar fuel system developed by NASA (Patent No: 9,528,192, [87]).

the reaction established regulations must be followed for handling the chemical/fuel of concern.

The design and scale-up of large reactors can be aided using modelling tools such as computational fluid dynamics (CFD) to understand the systems and optimize them while ensuring system parameters are maintained [88]. Although computational modelling saves both time and money compared to physical testing, such studies demand highly computational equipment depending on the available tools/models [32]. This has been demonstrated by several research studies, for example, Alvarado-Rolon *et al.* modelled the radiation field on a heterogenous photoreactor using the four fluxes method to determine the local volumetric rate energy absorption (LVREA). The numerical solutions obtained corresponded with experimental data and required less computational power in comparison to using models based on the discrete ordinated model (DOM) [89]. Furthermore, in photoreduction systems, it can be witnessed that certain photocatalysts can carry out the reduction reaction when a light source is not present, i.e., a "dark" reaction. Studies show that simulating special reactor designs, that can operate during periods of irradiation and thermal reactions can aid with improving the feasibility of photoreduction systems [32].

The design of conventional solar fuel production reactors, such as the slurry reactor, faces significant challenges related to the transmission and reflection of the light reducing the quantum and solar-to fuel efficiencies. The reactor geometry needs to account for better utilization of the light and the collection of the incident photons, as witnessed in optical fibre reactors. Additionally, the performance of the reactor can be enhanced by increasing the available surface area allowing the light to homogenously distribute across the photocatalyst. The operating parameters such as the temperature is an important parameter for CO_2 photoreduction, and the pH significantly affects the performance electrochemical CO_2 reduction systems. In reactors that involve stirring and mixing intensification reduces the mass and heat transfer resistances due to the movement of the catalyst particles [30]. However, rigorous mixing can result in erosion of the catalyst and other materials while resulting in insufficient mass transfer between the reactants and catalyst. Thus, for larger-scale reactors the flow characteristics need to be carefully incorporated. Furthermore, when considering the entire system commercial applications, additional studies need to be carried out to determine parameters such as: the system energy and efficiency losses, the sourcing of the photocatalyst (especially when utilizing noble metals), the stability, durability and costs of the catalyst and membranes [19].

8.5 Conclusions

Scale-up of solar fuel reactors is currently in its infancy, but great advancements in catalyst and supporting reactor design is leading to exponential growth in patents that indicate a bright future for commercial systems. Solar fuel production mimics the natural photosynthesis in plants and this can be carried out artificially using photocatalytic systems. The conventional photocatalytic reactor i.e., slurry photoreactor, allows efficient use of the catalyst however, a significant drawback is the costly and time-consuming process for post separation of the catalyst. To tackle this issue, fixed-bed reactors were developed where the catalyst is immobilized on supports either by chemical bonds or physical surfaces forces. A notable fixed-bed reactor design is the optical photoreactor. This reactor boasts higher conversion rates as the catalyst is coated on optical fibres which evenly distributes the light on the catalyst and maximises contact with reactants. However, in both the slurry and fixed bed photoreactors the redox reactions take place on the surface of the catalyst. In this environment, the oxygen produced from the water oxidation reaction can further reactor with the products produced during the reduction reaction. Twin reactors overcome this limitation carrying out the water oxidation and reduction reactions in separate compartments, yielding higher product selectivity and yield. Despite the advances in the design of photocatalytic reactors the system efficiency and product yields are significantly low. Therefore, photoelectrochemical and electrochemical systems have been studied to improve the CO_2 reduction reaction using electrical energy.

In comparison, electrochemical reduction of CO_2 using electrical energy from solar powered photovoltaics yields higher solar-to fuel efficiencies. However, the reactors developed in this system are complex, utilize expensive materials (such as noble metal electrocatalysts and membranes) and require significant energy. Thus, although several reactors have been developed/studied there is still room for improvement for large-scale and economic application of solar fuel production.

References

1. The Royal Chemistry, 13. How does climate change affect the strength and frequency of floods, droughts, hurricanes, and tornadoes? *R. Soc.*, Mar. 2020. https://royalsociety.org/topics-policy/projects/climate-change-evidence-causes/question-13/ (accessed Nov. 20, 2021).
2. Hernández, S., Farkhondehfal, M.A., Sastre, F., Makkee, M., Saracco, G., Russo, N., Syngas production from electrochemical reduction of CO_2: Current status and prospective implementation. *Green Chem.*, 19, 10, 2326–2346, May 2017.
3. Li, K., An, X., Park, K.H., Khraisheh, M., Tang, J., A critical review of CO_2 photoconversion: Catalysts and reactors. *Catal. Today*, 224, 3–12, Apr. 2014.
4. US Department of Energy, DOE explains..solar fuels, 2021. https://www.energy.gov/science/doe-explainssolar-fuels (accessed Aug. 14, 2021).
5. Jooss, C. and Tributsch, H., Solar fuels, in: *Fundamentals of Materials for Energy and Environmental Sustainability*, pp. 656–674, Jan. 2011.

6. Chang, X., Wang, T., Gong, J., CO_2 photo-reduction: Insights into CO_2 activation and reaction on surfaces of photocatalysts. *Energy Environ. Sci.*, 9, 7, 2161–2472, 2016.
7. Ola, O. and Maroto-Valer, M.M., Review of material design and reactor engineering on TiO_2 photocatalysis for CO_2 reduction. *J. Photochem. Photobiol. C Photochem. Rev.*, 24, 16–42, Jun. 2015.
8. Olivo, A. et al., CO_2 photoreduction with water: Catalyst and process investigation. *J. CO_2 Util.*, 12, 86–94, Dec. 2015.
9. Pachaiappan, R., Rajendran, S., Senthil Kumar, P., Vo, D.-V.N., Hoang, T.K.A., A review of recent progress on photocatalytic carbon dioxide reduction into sustainable energy products using carbon nitride. *Chem. Eng. Res. Des.*, Nov. 117, 304–320, 2021.
10. Ciamician, G., The photochemistry of the future. *Science*, 36, 926, 385–394, Aug. 1912.
11. Kajino, T. et al., Direct synthesis of organic compounds from CO_2, water and sunlight. *R&D Rev. Toyota CRDL*, 43, 2, 43–52, 2012. Accessed: Nov. 20, 2021. [Online]. Available: http://www.tytlabs.co.jp/review/.
12. He, J. and Janáky, C., Recent advances in solar-driven carbon dioxide conversion: Expectations versus reality. *ACS Energy Lett.*, 5, 6, 1996–2014, Jun. 12, 2020.
13. Molinari, R., Lavorato, C., Argurio, P., The evolution of photocatalytic membrane reactors over the last 20 years: A state of the art perspective. *Catalysts*, 11, 7, 775, Jun. 2021.
14. Kamkeng, A.D.N., Wang, M., Hu, J., Du, W., Qian, F., Transformation technologies for CO_2 utilisation: Current status, challenges and future prospects. *Chem. Eng. J. Elsevier B.V.*, 409, 128138, Apr. 01, 2021.
15. Liu, L. and Li, Y., Understanding the reaction mechanism of photocatalytic reduction of CO_2 with H_2O on TiO_2-based photocatalysts: A review. *Aerosol Air Qual. Res.*, 14, 2, 453–469, Mar. 2014.
16. Tan, J.Z.Y. and Maroto-Valer, M.M., A review of nanostructured non-titania photocatalysts and hole scavenging agents for CO_2 photoreduction processes. *J. Mater. Chem. A*, 7, 16, 9368–9385, Apr. 2019.
17. Shehzad, N., Tahir, M., Johari, K., Murugesan, T., Hussain, M., A critical review on TiO_2 based photocatalytic CO_2 reduction system: Strategies to improve efficiency. *J. CO_2 Util.* Elsevier Ltd, 26, 98–122, May 08, 2018.
18. Thompson, W.A., Sanchez Fernandez, E., Maroto-Valer, M.M., Probability Langmuir-Hinshelwood based CO_2 photoreduction kinetic models. *Chem. Eng. J.*, 384, 123356, 2020.
19. Osterloh, F.E., Photocatalysis versus photosynthesis: A sensitivity analysis of devices for solar energy conversion and chemical transformations. *ACS Energy Lett.*, 2, 2, 445–453, Feb. 2017.
20. Francis, A., Shanmuga Priya, S., Harish Kumar, S., Sudhakar, K., Tahir, M., A review on recent developments in solar photoreactors for carbon dioxide conversion to fuels. *J. CO_2 Util.*, 47, 101515, 2021.

21. Thompson, W.A., Fernandez, E.S., Maroto-Valer, M.M., Review and analysis of CO_2 photoreduction kinetics. *ACS Sustain. Chem. Eng.*, 8, 12, 4677–4692, Mar. 2020.
22. Mazierski, P., Bajorowicz, B., Grabowska, E., Zaleska-Medynska, A., Photoreactor design aspects and modeling of light, in: *Heterogeneous Photocatalysis*. pp. 211–248, Springer, Berlin, Heidelberg, 2016.
23. Ray, A.K. and Beenackers, A.A.C.M., Development of a new photocatalytic reactor for water purification. *Catal. Today*, 40, 73–83, 1998.
24. Braslavsky, S.E., Glossary of terms used in photochemistry, 3rd edition (IUPAC recommendations 2006). *Pure Appl. Chem.*, 79, 3, 293–465, 2007.
25. Serpone, N., Relative photonic efficiencies and quantum yields in heterogeneous photocatalysis. *J. Photochem. Photobiol. A Chem.*, 104, 1–3, 1–12, Apr. 1997.
26. Variar, A.G., Ramyashree, M.S., Ail, V.U., Shanmuga, P.S., Sudhakar, K., Tahir, M., Influence of various operational parameters in enhancing photocatalytic reduction efficiency of carbon dioxide in a photoreactor: A review. *J. Ind. Eng. Chem.*, 99, 19–47, Jul. 25, 2021.
27. Hossain, M.F., Water, in: *Sustainable Design and Build*, pp. 301–418, Jan. 2019.
28. Bickley, R., II, Slater, M.J., Wang, W.J., Engineering development of a photocatalytic reactor for waste water treatment. *Process Saf. Environ. Prot.*, 83, 3, 205–216, May 2005.
29. Cheng, X. *et al.*, Optofluidic membrane microreactor for photocatalytic reduction of CO_2. *Int. J. Hydrog. Energy*, 41, 4, 2457–2465, Jan. 2016.
30. Adamu, A., Russo-Abegão, F., Boodhoo, K., Process intensification technologies for CO_2 capture and conversion–a review. *BMC Chem. Eng.*, 2, 1, 1–18, Jan. 2020.
31. de los Milagros Ballari, M., Alfano, O.M., Cassano, A.E., Mass transfer limitations in slurry photocatalytic reactors: Experimental validation. *Chem. Eng. Sci.*, 65, 17, 4931–4942, Sep. 2010.
32. Bouchy, M. and Zahraa, O., Photocatalytic reactors. *Int. J. Photoenergy*, 5, 191–197, 2003.
33. Du, P., Carneiro, J.T., Moulijn, J.A., Mul, G., A novel photocatalytic monolith reactor for multiphase heterogeneous photocatalysis. *Appl. Catal. A Gen.*, 334, 1–2, 119–128, Jan. 2008.
34. Marinangeli, R.E. and Ollis, D.F., Photoassisted heterogeneous catalysis with optical fibers: I. Isolated single fiber. *AIChE J.*, 23, 4, 415–426, Jul. 1977.
35. Oliveira De Brito Lira, J., Riella, H.G., Padoin, N., Soares, C., An overview of photoreactors and computational modeling for the intensification of photocatalytic processes in the gas-phase: State-of-art. *J. Environ. Chem. Eng.*, 9, 2, 105068, Apr. 2021.
36. Carneiro, J.T., Berger, R., Moulijn, J.A., Mul, G., An internally illuminated monolith reactor: Pros and cons relative to a slurry reactor. *Catal. Today*, 147(SUPPL.), S324–S329, Sep. 2009.

37. Ola, O. and Maroto-Valer, M.M., Copper based TiO_2 honeycomb monoliths for CO_2 photoreduction. *Catal. Sci. Technol.*, 4, 6, 1631–1637, Feb. 2014.
38. Ali, S. *et al.*, Gas phase photocatalytic CO2 reduction, 'a brief overview for benchmarking. *Catalysts*, 9, 9, 727, Aug. 2019.
39. Lin, H. and Valsaraj, K.T., An optical fiber monolith reactor for photocatalytic wastewater treatment. *AIChE J.*, 52, 6, 2271–2280, Jun. 2006.
40. Ola, O., Maroto-Valer, M., Liu, D., MacKintosh, S., Lee, C.W., Wu, J.C.S., Performance comparison of CO_2 conversion in slurry and monolith photoreactors using Pd and Rh-TiO_2 catalyst under ultraviolet irradiation. *Appl. Catal. B Environ.*, 126, 172–179, Sep. 2012.
41. Lu, X., Luo, X., Tan, J.Z.Y., Maroto-Valer, M.M., Simulation of CO_2 photoreduction in a twin reactor by multiphysics models. *Chem. Eng. Res. Des.*, 171, 125–138, Jul. 2021.
42. Lee, W.H., Liao, C.H., Tsai, M.F., Huang, C.W., Wu, J.C.S., A novel twin reactor for CO_2 photoreduction to mimic artificial photosynthesis. *Appl. Catal. B Environ.*, 132–133, 445–451, Mar. 2013.
43. Cheng, Y.H., Nguyen, V.H., Chan, H.Y., Wu, J.C.S., Wang, W.H., Photoenhanced hydrogenation of CO_2 to mimic photosynthesis by CO co-feed in a novel twin reactor. *Appl. Energy*, 147, 318–324, Jun. 2015.
44. Yu, S.H., Chiu, C.W., Wu, Y.T., Liao, C.H., Nguyen, V.H., Wu, J.C.S., Photocatalytic water splitting and hydrogenation of CO_2 in a novel twin photoreactor with IO3−/I−shuttle redox mediator. *Appl. Catal. A Gen.*, 518, 158–166, May 2016.
45. Xiong, Z. *et al.*, Enhanced CO_2 photocatalytic reduction through simultaneously accelerated H2 evolution and CO_2 hydrogenation in a twin photoreactor. *J. CO_2 Util.*, 24, 500–508, Mar. 2018.
46. Pomilla, F.R. *et al.*, CO_2 to liquid fuels: Photocatalytic conversion in a continuous membrane reactor. *ACS Sustain. Chem. Eng.*, 6, 7, 8743–8753, Jul. 2018.
47. Sundar, K.P. and Kanmani, S., Progression of photocatalytic reactors and it's comparison: A review. *Chem. Eng. Res. Des.*, 154, 135–150, Feb. 2020.
48. Molinari, R., Lavorato, C., Argurio, P., Szymański, K., Darowna, D., Mozia, S., Overview of photocatalytic membrane reactors in organic synthesis, energy storage and environmental applications. *Catalysts*, 9, 3, 239, Mar. 2019.
49. Lokhat, D., Domah, A.K., Padayachee, K., Baboolal, A., Ramjugernath, D., Gas–liquid mass transfer in a falling film microreactor: Effect of reactor orientation on liquid-side mass transfer coefficient. *Chem. Eng. Sci.*, 155, 38–44, Nov. 2016.
50. Odiba, S., Olea, M., Hodgson, S., Adgar, A., *Computational Fluid Dynamics for Microreactors Used in Catalytic Oxidation of Propane*, 2013.
51. Heggo, D. and Ookawara, S., Multiphase photocatalytic microreactors. *Chem. Eng. Sci.*, 169, 67–77, Sep. 2017.
52. Matsushita, Y., Ohba, N., Kumada, S., Sakeda, K., Suzuki, T., Ichimura, T., Photocatalytic reactions in microreactors. *Chem. Eng. J.*, 135, SUPPL. 1, S303–S308, Jan. 2008.

53. MacDowell, N. et al., An overview of CO_2 capture technologies. *Energy Environ. Sci.*, 3, 11, 1645–1669, Nov. 2010.
54. Saad Ahsan, S., Gumus, A., Erickson, D., Redox mediated photocatalytic water-splitting in optofluidic microreactors. *Lab. Chip*, 13, 409, 49–414, 2012.
55. Li, L. et al., Optofluidics based micro-photocatalytic fuel cell for efficient wastewater treatment and electricity generation. *Lab. Chip*, 14, 17, 3368–3375, Sep. 2014.
56. Delacour, C., Lutz, C., Kuhn, S., Pulsed ultrasound for temperature control and clogging prevention in micro-reactors. *Ultrason. Sonochem.*, 55, 67–74, Jul. 2019.
57. IRENA, *Renewable power generation costs in 2020*. International Renewable Energy Agency, Abu Dhabi, 2021, Accessed: Feb. 16, 2022. [Online]. Available: /publications/2021/Jun/Renewable-Power-Costsin-2020.
58. Weekes, D.M., Salvatore, D.A., Reyes, A., Huang, A., Berlinguette, C.P., Electrolytic CO_2 reduction in a flow cell. *Acc. Chem. Res.*, 51, 4, 910–918, Apr. 2018.
59. Liang, F., Zhang, K., Zhang, L., Zhang, Y., Lei, Y., Sun, X., Recent development of electrocatalytic CO_2 reduction application to energy conversion. *Small*, 17, 44, 2100323, Jun. 2021.
60. Zhang, S., Fan, Q., Xia, R., Meyer, T.J., CO_2 reduction: From homogeneous to heterogeneous electrocatalysis. *Acc. Chem. Res.*, 53, 1, 225–264, Jan. 2020.
61. Lu, Q. and Jiao, F., Electrochemical CO_2 reduction: Electrocatalyst, reaction mechanism, and process engineering. *Nano Energy*, 29, 439–456, Nov. 2016.
62. Kalamaras, E., Maroto-Valer, M.M., Andresen, J.M., Wang, H., Xuan, J., Thermodynamic analysis of the efficiency of photoelectrochemical CO_2 reduction to ethanol. *Energy Procedia*, 158, 767–772, Feb. 2019.
63. Azenha, C., Mateos-Pedrero, C., Alvarez-Guerra, M., Irabien, A., Mendes, A., Enhancement of the electrochemical reduction of CO_2 to methanol and suppression of H2 evolution over CuO nanowires. *Electrochim. Acta*, 363, 137207, Dec. 2020.
64. Krause, R. et al., Industrial application aspects of the electrochemical reduction of CO_2 to CO in aqueous electrolyte. *Chem. Ing. Tech.*, 92, 1–2, 53–61, Jan. 2020.
65. Shen, J. et al., Electrocatalytic reduction of carbon dioxide to carbon monoxide and methane at an immobilized cobalt protoporphyrin. *Nat. Commun.*, 6, 1, 1–8, Sep. 2015.
66. Schouten, K.J.P., Kwon, Y., van der Ham, C.J.M., Qin, Z., Koper, M.T.M., A new mechanism for the selectivity to C1 and C2 species in the electrochemical reduction of carbon dioxide on copper electrodes. *Chem. Sci.*, 2, 10, 1902–1909, Sep. 2011.
67. Grim, R.G., Huang, Z., Guarnieri, M.T., Ferrell, J.R., Tao, L., Schaidle, J.A., Transforming the carbon economy: Challenges and opportunities in the

convergence of low-cost electricity and reductive CO_2 utilization. *Energy Environ. Sci.*, 13, 472, 2020.
68. Ma, D., Jin, T., Xie, K., Huang, H., An overview of flow cell architecture design and optimization for electrochemical CO_2 reduction. *J. Mater. Chem. A*, 9, 20897–20918, Aug. 2021.
69. Yang, W., Dastafkan, K., Jia, C., Zhao, C., Design of electrocatalysts and electrochemical cells for carbon dioxide reduction reactions. *Adv. Mater. Technol.*, 3, 9, 1700377, Sep. 2018.
70. Endrődi, B., Bencsik, G., Darvas, F., Jones, R., Rajeshwar, K., Janáky, C., Continuous-flow electroreduction of carbon dioxide. *Prog. Energy Combust. Sci.*, 62, 133–154, Sep. 2017.
71. Woldu, A.R., Shah, A.H., Hu, H., Cahen, D., Zhang, X., He, T., Electrochemical reduction of CO_2: Two-or three-electrode configuration. *Int. J. Energy Res.*, 44, 1, 548–559, Nov. 2019.
72. Lu, Q. and Jiao, F., Electrochemical CO_2 reduction: Electrocatalyst, reaction mechanism, and process engineering. *Nano Energy*, 29, 439–456, Nov. 2016.
73. Merino-Garcia, I., Alvarez-Guerra, E., Albo, J., Irabien, A., Electrochemical membrane reactors for the utilisation of carbon dioxide. *Chem. Eng. J.*, 305, 104–120, Dec. 2016.
74. Smith, W.A., Photoelectrochemical cell design, efficiency, definitions, standards, and protocols, in: *Photoelectrochemical Solar Fuel Production: From Basic Principles to Advanced Devices*, S. Giménez and J. Bisquert (Eds.), pp. 163–197, Springer, Cham, 2016.
75. Hori, Y., Ito, H., Okano, K., Nagasu, K., Sato, S., Silver-coated ion exchange membrane electrode applied to electrochemical reduction of carbon dioxide. *Electrochim. Acta*, 48, 18, 2651–2657, Aug. 2003.
76. Liu, Y., Li, F., Zhang, X., Ji, X., Recent progress on electrochemical reduction of CO_2 to methanol. *Curr. Opin. Green Sustain. Chem.*, 23, 10–17, Jun. 2020.
77. Halmann, M., Photoelectrochemical reduction of aqueous carbon dioxide on p-type gallium phosphide in liquid junction solar cells. *Nature*, 275, 5676, 115–116, Sep. 1978.
78. Kalamaras, E., Maroto-Valer, M.M., Shao, M., Xuan, J., Wang, H., Solar carbon fuel via photoelectrochemistry. *Catal. Today*, 317, 56–75, Feb. 2018.
79. Castro, S., Albo, J., Irabien, A., Photoelectrochemical reactors for CO_2 utilization. *ACS Sustain. Chem. Eng.*, 6, 12, 15877–15894, Oct. 2018.
80. Zhou, X. *et al.*, Solar-driven reduction of 1 atm of CO_2 to formate at 10% energy-conversion efficiency by use of a TiO_2-protected III-V tandem photoanode in conjunction with a bipolar membrane and a Pd/C cathode. *ACS Energy Lett.*, 1, 4, 764–770, Oct. 2016.
81. Homayoni, H., Chanmanee, W., de Tacconi, N.R., Dennis, B.H., Rajeshwar, K., Continuous flow photoelectrochemical reactor for solar conversion of carbon dioxide to alcohols. *J. Electrochem. Soc.*, 162, 8, E115–E122, May 2015.

82. Kalamaras, E., Belekoukia, M., Tan, J.Z.Y., Xuan, J., Maroto-Valer, M.M., Andresen, J.M., A microfluidic photoelectrochemical cell for solar-driven CO_2 conversion into liquid fuels with CuO-based photocathodes. *Faraday Discuss.*, 215, 329–344, Jul. 2019.
83. Herrmann, J.M., Heterogeneous photocatalysis: State of the art and present applications in honor of Pr. R.L. Burwell Jr. (1912–2003), former head of ipatieff laboratories, Northwestern University, Evanston (Ill). *Top. Catal.*, 34, 1, 49–65, May 2005.
84. Garg, S. *et al.*, Advances and challenges in electrochemical CO_2 reduction processes: An engineering and design perspective looking beyond new catalyst materials. *J. Mater. Chem. A*, 8, 4, 1511–1544, Jan. 2020.
85. Tufa, R.A. *et al.*, Towards highly efficient electrochemical CO_2 reduction: Cell designs, membranes and electrocatalysts. *Appl. Energy*, 277, 115557, Nov. 2020.
86. Ramdin, M. *et al.*, High pressure electrochemical reduction of CO_2 to formic acid/formate: A comparison between bipolar membranes and cation exchange membranes. *Ind. Eng. Chem. Res.*, 58, 5, 1834–1847, Feb. 2019.
87. NASA, Solar powered carbon dioxide (CO_2) conversion, 2016. Accessed: Apr. 06, 2022. [Online]. Available: http://technology.nasa.gov/.
88. Ješić, D., Lašič Jurković, D., Pohar, A., Suhadolnik, L., Likozar, B., Engineering photocatalytic and photoelectrocatalytic CO_2 reduction reactions: Mechanisms, intrinsic kinetics, mass transfer resistances, reactors and multi-scale modelling simulations. *Chem. Eng. J.*, 407, 126799, Mar. 2021.
89. Alvarado-Rolon, O., Natividad, R., Romero, R., Hurtado, L., Ramírez-Serrano, A., Modelling and simulation of the radiant field in an annular heterogeneous photoreactor using a four-flux model. *Int. J. Photoenergy*, 2018, 1–16, 2018.

Part IV
SOLAR-DRIVEN WATER SPLITTING

9

Photocatalyst Perovskite Ferroelectric Nanostructures

Debashish Pal[1], Dipanjan Maity[2], Ayan Sarkar[3] and Gobinda Gopal Khan[1]*

[1]*Department of Material Science and Engineering, Tripura University (A Central University), Suryamaninagar, Agartala, Tripura, India*
[2]*Department of Condensed Matter Physics and Material Sciences, S. N. Bose National Centre for Basic Sciences, Salt Lake, Kolkata, India*
[3]*Department of Chemistry, National Taiwan University, Taipei, Taiwan*

Abstract

Photocatalysis and photoelectrocatalysis have attracted intense research attention considering the direct conversion of renewable solar energy into chemical energy through sustainable and pollution-free routes. Recently, perovskite-type ferro/piezo-electric oxide semiconductor nanostructures have revealed significant potential as photocatalysts and photoelectrocatalysts, considering their attractive semiconducting property coupled with ferro/piezo-electric and optoelectronic properties. The integration of nanoscale dimension-related properties like shape, size effect, surface area enlargement, and quantum confinement effects with the tuneable ferro/peizo-electric properties of ferroelectric semiconductors provide ample opportunity to design novel photo(electro)catalysts. This chapter summarizes the current state of perovskite nanostructured ferroelectric photo(electro) catalysts for solar fuel production and pollutant degradation. The fundamental art of piezo/ferroelectric photo(electro)catalysis is explained with a detailed review of various synthesis and design strategies employed for the nanostructured ferroelectric semiconductor photo(electro)catalysts. The mechanism of the photocatalytic and photoelectrochemical activities of the ferroelectric photo(electro)catalysts are illustrated, explaining the light-harvesting, photocarrier generation, photocarrier separation/recombination, and transportation activities and the effect of ferro/piezo-potential. Finally, the outlook for future research and development related to ferroelectric photo(electro)catalysis is highlighted.

---

*Corresponding author: gobinda.gk@gmail.com; gobindakhan@tripurauniv.ac.in

Keywords: Perovskite, ferro/piezoelectric, semiconductor nanostructures, photocatalysis, photoelectrochemical, solar fuel, pollutant degradation

9.1 Introduction

Proper utilization of abundant renewable solar energy has been a key target in the present time, considering the environmental impact and depletion of fossil fuels. Photocatalysis or photoelectrocatalysis has drawn considerable attention for harvesting solar energy into chemical energy through green and sustainable pathways. Photo(electro)catalysis directly involves the solar light for several chemical processes like solar fuel production through water splitting, carbon dioxide fixation, and degradation of pollutants. Hence, photo(electro)catalysis is an attractive, sustainable approach considering present global energy demands, and environmental crises [1–3]. The direct conversion of solar to chemical energy, mimicking the natural photosynthesis process to produce fuels using water and carbon dioxide under solar radiation, demands a rather simple system, where the semiconductor photocatalysts can be utilized [3–5]. Semiconductors are attractive materials considering solar light absorption for photocarrier generation, photocarrier separation, and the necessary redox reactions, which are the key requirements for photocatalysis [6]. However, the performance of the semiconductor photo(electro)catalysts is far below the target for practical applications. Therefore, several innovative strategies are adopted to improve the performance of the semiconductor photo(electro) catalysts.

Recently, ferroelectric semiconductors, where the properties of the semiconductors are coupled with the ferro/piezoelectric property of tunable polarization charges or electric potential, have attracted significant attention as photo(electro)catalysts [7]. Perovskite-type ferroelectric materials are mostly oxide semiconductors, where the polarization charges and piezopotential are reported to be an attractive strategy to improve the bulk photogenerated carrier separation and transportation for the surface photoelectrochemical (PEC) reactions [8–10]. Moreover, the tuning of electric polarization enables controlling the band alignment at the ferroelectric semiconductor/electrolyte interface, governing the separation and transportation of the photocarriers for enhanced PEC reaction kinetics [6, 7, 11]. Recently, the study of the ferrotronics or piezotronics properties based on photo(electro)catalysis of the nanoscale dimension ferroelectric semiconductors has drawn intense research interest because of the advantageous features of nanomaterials. Ferroelectric nanostructures exhibit interesting photocatalytic activities due to the quantum size, shape, and confinement effects coupled with a large surface area of the nanomaterials and the intrinsic ferro/piezoelectric properties [6, 12, 13]. Ferroelectric semiconductor nanostructures have consequently emerged as promising candidates for photo(electro) catalysis.

Against this backdrop, this chapter will elaborate on several aspects of the nanostructured ferroelectric semiconductor photo(electro)catalysts. The chapter starts with the fundamentals of the ferroelectric properties and the related perovskite-type (ABO_3) ferroelectric semiconductors, followed by the mechanisms of photocatalysis and photoelectrocatalysis. Subsequently, details on how the piezo/ferroelectric properties can tune the photocarrier migration and band bending at the semiconductor/electrolyte interface to govern the overall photo(electro)catalysis process have been discussed. Following this section, we have described the significance of nanostructured ferroelectric semiconductors for photo(electro) catalysis. The details of the various synthesis routes employed to fabricate the ferroelectric nanostructures photocatalysts are elaborated in the next section. The last section of this chapter elaborates the discussion on photo(electro)catalytic water splitting and photocatalytic degradation of dye/pollutant activity of various perovskite-type ferroelectric semiconductor nanostructures.

9.2 Ferroelectric Properties and Materials

The ferroelectric properties were first reported in 1921 for the single crystal Rochelle salt [14]. Later, the ferroelectric properties of various ceramic-based bulk polycrystalline materials were explored gradually, considering their numerous commercial applications [15, 16]. Ferroelectricity [14, 16] is defined as the spontaneous electric polarization induced as the inherent characteristics of certain materials with asymmetric crystal structures. The asymmetric crystal structures of certain materials result in the nonoverlapping of inherent charges leading to spontaneous electric polarization because of the in-built electric dipoles. Moreover, the spontaneous electric polarization in certain crystals can be tuned by applying an external electric field. The concept of ferroelectricity cannot be clarified without explaining phenomena like piezoelectricity and pyroelectricity. Piezoelectricity is defined as the phenomenon where equal and opposite

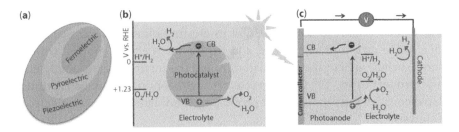

Figure 9.1 (a) Venn diagram showing the ferroelectric materials belong to the subgroup of piezoelectric and pyroelectric materials. The diagrams for (b) photocatalytic and (c) photoelectrocatalytic water splitting (on a photoanode) processes.

electric charges are generated at the surface of certain dielectric materials subjected to an applied force along its asymmetry direction parallel to the induced electrical field [16]. Piezoelectricity arises due to the lack of central symmetry in dielectric crystals, where the polarity of charge can be tuned by changing the direction of applied mechanical force. In pyroelectric materials, the temperature variation in certain crystals generates the surface electric polarization along the specific crystallographic directions. Therefore, all the ferroelectric materials belong to the subgroup of pyroelectric and piezoelectric materials (Figure 9.1a). The ferroelectric materials are mostly in the form of inorganic oxides, inorganic nonoxides, and also organic materials, including polymers. However, with regard to photo(electro)catalysis, the inorganic oxide-based semiconducting ferroelectric materials with ABO_3 perovskite-type crystal structures have significant potential. As ferroelectric perovskite-type oxide semiconductors, bismuth ferrite ($BiFeO_3$), barium titanate ($BaTiO_3$), strontium titanate ($SrTiO_3$), potassium, sodium and lithium niobates ($KNbO_3$, $NaNbO_3$, and $LiNbO_3$), lanthanum ferrite ($LaFeO_3$), zinc stannate ($ZnSnO_3$), bismuth manganite ($BiMnO_3$), yttrium ferrite ($YFeO_3$), lead titanate ($PbTiO_3$), and zirconate titanate (PZT, $PbZr_xTi_{(1-x)}O_3$) have been investigated as potential ferroelectrics for photocatalysis.

9.3 Fundamental of Photocatalysis and Photoelectrocatalysis

9.3.1 Photocatalytic Production of Hydrogen Fuel

The photocatalysis starts with the absorption of the photons by the photocatalysts. The semiconductor photocatalysts absorb light whenever the energy of the incident photons is higher than the bandgap energy of the semiconductor photocatalysts. Absorbing suitable photon energy, the electrons at the valence band (VB) of the semiconductor photocatalysts are excited to the conduction band (CB), leaving photogenerated holes at the VB. The process is known as photoexcitation or electron-hole pair generation. The photogenerated electron-hole pairs at the depletion region of the photocatalysts are readily separated because of the built-in potential at the semiconductor/electrolyte interface and migrate toward the surface of the photocatalysts from the bulk. In due course, the photogenerated electrons and holes participate in surface redox reactions at the photocatalyst/electrolyte interface. The photogenerated electrons and holes reduce and oxidize the water molecules to form H_2 and O_2, respectively, at the surface of the photocatalysts whenever water is used as an electrolyte (Figure 9.1b) [17]. This process is known as light-induced or photocatalytic water-splitting through which hydrogen fuel is produced. However, the ability of the semiconductor photocatalyst for water-splitting depends on the CB and VB position of the semiconductor; as for the optimum water splitting, the semiconductor bandgap should straddle the reduction and oxidation potentials of the water. Generally, the bottom of CB has to be more negative than the redox potential of H^+/H_2 (0 V vs. reversible hydrogen electrode (RHE)), and the top of the VB has to be more positive than the redox potential of O_2/H_2O (1.23 V vs. RHE) to facilitate optimum water splitting [6]. As the redox potential difference of water is 1.23 V, hence, theoretically, any semiconductor with a bandgap ≥ 1.23 eV is suitable for water splitting [18]. Photocatalysis takes place in the photocatalyst bath, where the semiconductor photocatalysts are suspended in the water.

9.3.2 Photoelectrocatalytic Hydrogen Production

In photoelectrocatalysis, the photocatalytic process, as described above, takes place in the presence of an externally applied bias, which drives the photogenerated charge carriers in a photoelectrochemical cell [6, 18, 19]. The photoelectrochemical cell consists of the semiconductor photocatalyst as the working electrode (photoanode or photocathode) and a counter electrode separated by an electrolyte (or immersed in two different electrolytes) connected by an external electric wire [6]. The *n*-type semiconductor photocatalyst acts as a photoanode (Figure 9.1c), where the photogenerated electrons and holes are driven by an external bias to reach the counter electrode (cathode) and the surface of the photoanode, respectively, to initiate the water-splitting redox reactions. Similarly, the *p*-type semiconductor photocatalyst acts as a photocathode and produces hydrogen at the surface through water reduction.

9.3.3 Photocatalytic Dye/Pollutant Degradation

The photocatalyst suspended in the electrolyte can also degrade dyes and pollutants present in the electrolyte. Here, the photogenerated holes in the semiconductor can be captured by the donors in the electrolyte and oxidize the organic molecules/dyes or form the hydroxyl (·OH) radicals (OH$^-$/·OH = 2.8 V vs. NHE) [20], which work as the oxidizing agent to oxidize dyes and pollutants. Because of the strong nonselective oxidizing ability, the hydroxyl radicals play a crucial role in the photocatalytic elimination of different organic substances [20]. Similarly, the photogenerated electrons in photocatalyst reduce the dye or pollutants or can be captured by the acceptors in the electrolyte to produce $O_2^{·-}$ superoxide anions (O_2/$O_2^{·-}$ = −0.33 V vs. NHE), which reduce pollutants [21]. As the direct formation of the hydroxyl radicals requires high energy, the hydroxyl radicals are often generated through different pathways involving the superoxide anions [20].

9.4 Principle of Piezo/Ferroelectric Photo(electro)catalysis

The effective photo(electro)catalysis mainly depends upon three fundamental steps: i) efficient light absorption and generation of photocarriers (electron-hole pairs), ii) effective separation and transportation of photocarriers to the surface, and iii) surface redox reaction kinetics. In this background, the electric polarization in ferroelectrics and the piezoelectric effect are found to be effective to generate a polar charge-dependent electric field to control the band bending at the semiconductor liquid junction (SCLJ), photogenerated bulk carrier separation, photocarrier transfer and migration, and the surface redox reaction kinetics [6, 22].

Whenever the *n*-type semiconductor ferroelectric photocatalyst is suspended in an electrolyte, the movement of photogenerated free carriers into the electrolyte induces an upward band bending at the semiconductor surface (shown by dotted lines in Figure 9.2a). This upward band bending further acts as a potential barrier for the photoinduced electrons to migrate toward the surface of the semiconductor to initiate a photocatalytic reduction reaction. However, this band bending favors the rapid migration of

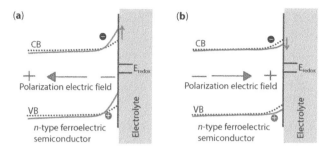

Figure 9.2 The band structure of *n*-type ferroelectric semiconductor photocatalyst/electrolyte interface at (a) negative and (b) positive polarizations. The original band position of the ferroelectric semiconductor is shown by dotted lines.

photogenerated holes toward the surface for the oxidation reaction. Hence, the ferroelectrics or piezoelectric photocatalysts with the internal spontaneous electric field because of polarization can further tune the band bending at the photocatalyst/electrolyte interface. For example, the accumulation of negative polarization charges at the surface of the ferroelectric photocatalyst will further enhance the upward band bending, leading to the suppression of electron transfer toward electrolyte and boosting the hole transfer kinetics for the oxidation reaction (Figure 9.2a) [6, 23]. Similarly, the accumulation of the positive polarization charges at the surface of the ferroelectric photocatalyst reduces the band bending (downward bending), facilitating the electron migration toward the electrolyte, although this will reduce the reductive potential a bit (Figure 9.2b) [6, 24]. This downward bending also enhances the oxidation kinetics as the top of the valance band becomes more positive. Most importantly, the band bending at the ferroelectric photocatalyst/electrolyte interface can be easily tuned with the direction and strength of the spontaneous electric field.

For ferroelectric photoelectrocatalysis, the above band bending related to polarization charges is also applicable. For ferroelectric/piezoelectric semiconductor photoanode, the accumulation of negative polarization charges (or negative piezopotential) at the photoanode/electrolyte interface will further induce upward band bending (Figure 9.3a). This will help in the easy migration of photogenerated electrons at the conduction band toward the current collector and transfer electrons to the counter electrode through the external circuit. Simultaneously, the holes at the valence band will drift toward the ferroelectric photoanode surface for an enhanced surface oxidation reaction [6, 7]. Similarly, the induction of positive piezopotential (accumulation of positive polarization charges) at the surface of the ferroelectric photoanode (Figure 9.3b) will induce band edge flattening

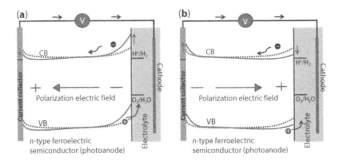

Figure 9.3 The band structure of current collector/n-type ferroelectric semiconductor photoanode/electrolyte interface at negative (a) and positive (b) polarization conditions for water splitting. The original band structure of the ferroelectric semiconductor photoanode is shown by dotted lines.

(downward movement of band edge), leading to a reduced possibility for rapid photocarrier separation for photoelectrocatalytic reactions [6].

9.5 Ferroelectric Nanostructures for Photo(electro) catalysis

Because of the advancement of nanoscience, low-dimensional ferroelectric nanostructures have attracted significant attention considering the factors related to the nanoscale dimension like large surface area, quantum confinement, quantum shape and size effects, the shape-dependent charge transportation, etc [6]. These factors have played a dominant role in determining the photo(electro)catalytic response of nanoscale dimensional ferroelectric materials. The nanoscale dimensional ferroelectric materials include the zero-dimensional (0-D) nanoparticles (NPs), one dimensional (1D) nanofibers (NFs), nanotubes (NTs), and nanobelts (NBs), nanowires (NWs), and two dimensional (2D) nanofilms (NFs), nanoflakes. The ferroelectric nanoparticles anchored with semiconductor photoanode boost light absorption, charge separation, and surface reaction kinetics for the photoelectrochemical (PEC) water splitting [12]. The 1D ferroelectric nanostructures are reported to exhibit enhanced PEC activity resulting from large surface area and enhanced charge separation arising because of rapid unidirectional carrier transportation along the axis of the nanostructure [13]. The 1D ferroelectric nanostructures also help in the rapid transfer of photogenerated charge carriers to the surface for PEC reactions [13]. The enhanced piezoelectric polarization in 1D ferroelectric nanostructures is found to induce a significant driving force for photocarrier separation, enhancing the redox process for piezophototronic photocatalytic hydrogen production [25]. The 2D ferroelectric semiconductor nanosheets exhibit significantly improved piezo-/ferroelectric response because of the quantum shape and size effect, enhancing photoelectrocatalytic activity over other nanostructures of the same material [26]. The 2D piezoelectric nanostructures also develop enhanced piezo-phototronic actives because of the distortion of inversion symmetry due to the 2D quantum confinement effect [27, 28]. Overall, the dimension, morphology, and shape of the ferroelectric nanostructures define the built-in electrical polarization in the piezo/ferroelectrics resulting in various catalytic properties. The large surface area associated with nanostructures is preferable to enhance surface-related catalytic reactions.

9.6 Synthesis and Design of Nanostructured Ferroelectric Photo(electro)catalysts

Considering the significance of nanoscale piezo/ferroelectric materials for catalysis, several strategies have been employed to synthesize various ferroelectric photo(electro)catalysts having different structures, morphology, and composition. This section will focus on a brief overview of the various synthesis routes for the nanostructured ferroelectric catalysts.

9.6.1 Hydrothermal/Solvothermal Methods

The hydrothermal/solvothermal method is the most popular route for fabricating different types of nanomaterials as this method is comparatively easy, scalable, less skill-dependent, and low-cost. Moreover, various types of inorganic oxide semiconductors, including the perovskite oxides nanostructures, can be synthesized by this technique. This method involves the heterogeneous chemical reaction of the precursor solutions confined in an autoclave at high pressure and temperature, higher than the room temperature. The temperature, pressure, and the addition of necessary surfactants are the key parameters to control the reaction kinetics for the growth of nanostructures. This bottom-up approach starts with the homogeneous nucleation and growth of the materials. Hence, depending upon the time, the size and shape of the nanomaterials can be tuned. The hydrothermal method has been extensively employed to synthesize various types of nanostructured ferroelectric catalysts.

Recently, Li *et al.* [29] synthesized single-crystalline BiFeO$_3$ NWs (Figure 9.4a and b) by hydrothermal route with photocatalytic activity under visible light. Firstly, the precursor was filtered out from the mixed salts solution of Bi(NO$_3$)$_3$·5H$_2$O and FeCl$_3$·6H$_2$O in acetone with ammonia solution under stirring. Afterward, the water-suspended co-precipitate was added with NaOH and put into an autoclave. The BiFeO$_3$ NWs were obtained by conducting the hydrothermal treatment at 140°C for 48 hours. BiFeO$_3$ nanocubes photocatalyst were prepared by solvothermal route [30]. In this method, the precursor solution was first prepared by adding Bi(NO$_3$)$_3$·5H$_2$O, Fe(NO$_3$)$_3$·9H$_2$O, Na$_2$CO$_3$, and KOH in distilled water. This solution was transferred into a Teflon reactor and placed into a microwave oven to obtain the BiFeO$_3$ nanocubes after the heat treatment at 180°C for 30 min.

NaNbO$_3$ NWs were also fabricated via the hydrothermal route [31]. In this technique, Nb(OC$_2$H$_5$)$_5$ was added in the P123 solution with the necessary amount of NaOH to prepare the precursor solution, which was autoclaved at 200°C for 24 hours to prepare the Na$_2$Nb$_2$O$_6$·H$_2$O NWs precursor. The orthorhombic phase of NaNbO$_3$ NWs was finally obtained through

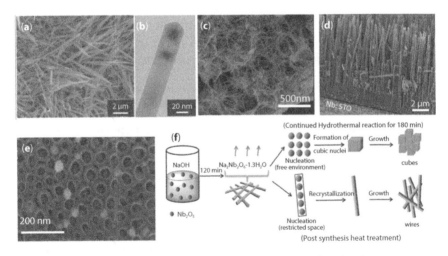

Figure 9.4 (a) SEM and (b) TEM images of BiFeO$_3$ NWs. Reproduced with permission [28]. Copyright 2013, The Royal Society of Chemistry. (c) FESEM image of NaNbO$_3$ NWs. Reprinted with permission from [32]. Copyright (2015), American Chemical Society. (d) SEM image of KNbO$_3$ NWs. Reproduced with permission [34]. Copyright 2008, Elsevier. (e) SEM image of Al/BaTiO$_3$ Nano-heterostructure. Reproduced with permission [10]. Copyright 2019, Elsevier. prepared via hydrothermal methods. (f) Schematic of the synthesis of NaNbO$_3$ NWs and Nanocubes via hydrothermal methods. Reprinted with permission from [32]. Copyright (2017), American Chemical Society.

the calcination of the NWs precursor powder at 550°C for 4 hours. Gu *et al.* have reported a slightly modified hydrothermal technique to prepare the cotton-like shape $NaNbO_3$ NWs (Figure 9.4c) assembly for photocatalytic H_2 production [32]. Here, the mixture of NaOH in ethylene glycol added with Nb_2O_5 was used as the precursor for the hydrothermal process. The autoclave was kept at 200°C for 4 hours, and the temperature of the autoclave was suddenly reduced in between 150 and 170°C for 5 min for a regular interval of time to prepare the cubic structure of $NaNbO_3$ NWs.

The NRs structure of $NaNbO_3$ with orthorhombic phase was fabricated by Singh *et al.* [33]. The precursor solution containing Nb_2O_5 dispersed in deionized water with NaOH added to it, and it was then transferred to an autoclave for the hydrothermal treatment at 150°C for 48 hours to prepare the piezo-photocatalyst $NaNbO_3$ NRs. Finally, the $NaNbO_3$ NRs were dispersed on flexible indium-doped tin oxide (ITO) coated polyethylene terephthalate (PET) substrates to prepare the Piezo-Photoanode. Liu *et al.* [34] have reported the synthesis of single-crystalline $NaNbO_3$ nanocubes and NWs from the precursor of $Na_2Nb_2O_6 \cdot 1.3H_2O$ NWs, prepared by hydrothermal route using Nb_2O_5 and NaOH in the autoclave at 180°C with different hydrothermal reaction times and post-synthesis heat ttreatment (Figure 9.4f). The $NaNbO_3$ nanoparticle coated on conducting fluorine-doped tin oxide (FTO) coated substrate was also used as photoelectrode with enhanced performance [35]. Here, the $NaNbO_3$ NPs were fabricated by solvothermal reaction at 150°C for 48 hours in the autoclave using the mixture of Nb_2O_5 and NaOH in water. The $NaNbO_3$ nanoparticle film was prepared by coating the nanoparticles on the FTO substrate.

Single-crystalline $KNbO_3$ NWs arrays (Figure 9.4d) were directly fabricated on the Nb-doped $SrTiO_3$ substrate by hydrothermal route [36]. The niobium metal powder dissolved in KOH solution was used as a precursor for the autoclave. The solution was heat-treated at 150°C for 12 hours with the substrate dipped into the solution to grow the vertically aligned arrays of $KNbO_3$ NWs photoelectrode. Ding *et al.* [37] have reported the synthesis of $KNbO_3$ NWs using a hydrothermal method where the precursor solution was prepared by mixing Nb_2O_5 and KOH in water. The solution was loaded in an autoclave and heated at 150°C for one week to obtain the $KNbO_3$ NWs for photocatalytic water splitting. A similar hydrothermal method was also reported to synthesize the $KNbO_3$ NPs using the same precursor solution [38]. However, the hydrothermal reaction was conducted at 150°C only for 4 hours to obtain the $KNbO_3$ NPs. Synthesis of $KNbO_3$ Nanosheets and Nanocubes with ferro-photoelectrochemical activity was also reported by Yu *et al.* [26]. The $KNbO_3$ nanosheets were fabricated by hydrothermal reaction of the mixture of KOH and Nb_2O_5 dissolved in the ethylene glycol in an autoclave heated at 200°C for 12 hours. For the growth of $KNbO_3$ Nanocubes, a mixture of KOH and Nb_2O_5 was dissolved in water and heated in a sealed autoclave at 180°C for 48 hours.

$BiFeO_3$ nanoparticles and luminescent lanthanide (Ln^{3+})-doped upconverting nanoparticles/$BiFeO_3$ core/shell nano-heterostructures have been fabricated via hydrothermal route [39]. First, the EDTA-Bi/Fe(III) complex was fabricated using $Bi(NO_3)_3 \cdot 5H_2O$ and $Fe(NO_3)_3 \cdot 9H_2O$ in HCl solution with EDTA dissolved in it. The EDTA-Bi/Fe(III) complex dissolved in water was mixed with urea and H_2O_2 and put into an autoclave to heat the solution at 200°C for 18 hours to prepare the $BiFeO_3$ nanoparticles. The hydrothermal process was conducted in the presence of a specific amount of the as-prepared luminescent lanthanide (Ln^{3+}) doped upconverting nanoparticles to prepare the core/shell nano-heterostructures.

A modified hydrothermal synthesis route for the growth of LN-type R3c $ZnSnO_3$ NWs was reported recently [25] using the ZnO thin films as the seed layer grown on the FTO substrate. $SnCl_4 \cdot 5H_2O$ solution was prepared in water, and it was mixed with cetyltrimethylammonium bromide and a suitable amount of NaOH to adjust the pH to 12.5. The ZnO seed layer coated FTO was placed in the solution in an autoclave at 220°C for 24 hours to prepare the ferroelectric $ZnSnO_3$ NWs for piezocatalysis study. Similarly, $SnCl_4 \cdot 5H_2O$ and ZnO solution mixed with the required amount of NaOH was also used as the precursor solution to grow $ZnSnO_3$ NWs through the hydrothermal route [40]. The precursor solution containing the ZnO seed layer coated FTO was heated in a sealed autoclave at 200°C for 20 hours to prepare the crystalline R3c $ZnSnO_3$ NWs. Hydrothermal synthesis of piezophotocatalyst $ZnSnO_3$ NWs was also reported by Lo et al. [41]. In this method, the aqueous precursor solution was prepared by mixing $SnCl_4 \cdot 5H_2O$ and ZnO powder with polyethylene glycol (PEG) and NaOH. The $ZnSnO_3$ NWs were grown on FTO at 160°C for 1 hour of heat treatment of the solution in an autoclave. Recently, S-doped $ZnSnO_3$ NPs were also fabricated by hydrothermal route [42]. In this method, the as-prepared $ZnSnO_3$ powder was mixed in the CH_3CSNH_2 solution prepared in ethanol and heated at 180°C for 12 hours in a sealed autoclave to obtain the S-doped $ZnSnO_3$ NPs. Similarly, $ZnSnO_3$ NPs with photocatalytic properties were fabricated via hydrothermal route [43]. In this typical method, the homogeneous precursor solution was prepared by mixing zinc acetate and tin tetrachloride in an aqueous NaOH solution. The solution was heated at 160°C for 11 hours in a sealed autoclave to obtain the $ZnSnO_3$ NPs.

Single-crystalline $PbTiO_3$ nanoplates with improved ferroelectric photocatalytic activity were fabricated by hydrothermal route [44]. In this process, the precursor solution was prepared by mixing anatase TiO_2 and $Pb(NO_3)_2$ powder in the aqueous solution of KOH. The solution was heated at 200° C for 24 hours in a sealed Teflon-lined autoclave. Finally, the precipitate was washed and dried to obtain $PbTiO_3$ nanoplates.

Aligned arrays of Al NPs coupled $BaTiO_3$ NTs (Figure 9.4e) were grown on Ti foam substrate by the similarly modified hydrothermal synthesis route [10]. Here, first, the arrays of TiO_2 NTs were fabricated via anodization of Ti foam. Afterward, the TiO_2 NTs/Ti foam was transferred to an autoclave containing the precursor solution of barium hydroxide and KOH prepared in water. The autoclave was transferred to an oven heated at 180°C for 3 hours to convert the TiO_2 NTs into $BaTiO_3$ NTs for photocatalytic H_2 production. The hydrothermal synthesis of $BaTiO_3$ nanoparticles with tetragonal crystal structure was reported by Li et al. [45]. In this typical method, $Ba(NO_3)_2$ and $Ti(SO_4)_2$ mixed together with KOH was used as the precursor solution in the autoclave. The cube-shaped nanoparticles of $BaTiO_3$ were obtained through hydrothermal treatment at 180°C for 12 hours.

The $BaTiO_3/TiO_2$ shell/core NRs arrays were fabricated by *in situ* hydrothermal methods by converting the surface layer of TiO_2 to $BaTiO_3$ [11, 46]. In this process, the TiO_2 NRs were first grown on the FTO substrate by the solvothermal route. Afterward, the as-prepared TiO_2 NRs was placed in an autoclave filled with the mixture of $Ba(OH)_2 \cdot 8H_2O$, diethylene glycol, ethanol, isopropanol, tetrabutylammonium hydroxide solution (40 wt%), and deionized water and heated at 170°C for 4 hours to obtain the nanoheterostructured $BaTiO_3/TiO_2$ NRs with enhanced piezophototronic activity. Similarly, a hydrothermal synthesis route was employed to fabricate the $BaTiO_3/TiO_2$ shell/core NRs arrays, where the thickness of the $BaTiO_3$ layer was tuned by controlling the reaction temperature in between 150 and 210°C for 2 hours [47].

Similarly, the TiO_2–$SrTiO_3$ core–shell NWs arrays with enhanced photoelectrochemical property were fabricated by using the rutile TiO_2 NWs arrays grown on FTO substrate as the source of Ti to fabricate $SrTiO_3$ [9]. The as-prepared TiO_2 NWs were immersed in the solution containing $Sr(OH)_2 \cdot 8H_2O$, diethylene glycol, ethanol, 2-propanol, tetrabutylammonium hydroxide solution, and deionized water in an autoclave was heated at 160°C for the different time duration (1–4 hours) to control the growth of $SrTiO_3$ shell layer on TiO_2 NRs. The $TiO_2/SrTiO_3$ heterostructure NTs arrays were also fabricated by using the as-papered TiO_2 NTs as the source and template material in the hydrothermal reaction bath [48]. The arrays of TiO_2 NTs were placed in the autoclave filled with strontium hydroxide octahydrate solution in deionized water. The sealed autoclave was heated in the oven at 180°C for different reaction times to control the morphology of the $TiO_2/SrTiO_3$ nano-heterostructure. Cheng et al. [49] also have fabricated $TiO_2/SrTiO_3$ nano-heterostructure using the as-prepared TiO_2 NTs as the starting materials. The TiO_2 NTs were dipped into $Sr(OH)_2 \cdot 8H_2O$ and NaOH solution prepared in water and sealed into the autoclave for hydrothermal reaction at 150°C for 45 min to prepare the $TiO_2/SrTiO_3$ nano-heterostructure photoanode.

9.6.2 Sol-Gel Methods

The sol-gel route has been very popular for synthesizing various complex perovskite oxides nanostructures as this is an easy, scalable, and low-cost technique. In this method, the precursor sol is prepared by dissolving the required salts in water or other solvents. After forming a homogeneous sol, heat treatment of the sol is conducted to form a dried gel structure through evaporation of the solvent. The dried gel structure is calcined and annealed at high temperatures as necessary to form the crystalline nanomaterials. Sometimes, different kinds of organic nanostructured template materials are added to the sol to use as the template for synthesizing focused nanomaterials. The organic nanostructured templates are finally removed through high-temperature calcination to obtain the pure phase of the perovskite oxides nanostructures.

Zhang et al. have reported the synthesis of $BiFeO_3$ nanoparticles (NPs) by sol-gel route [50]. The $Bi(NO_3)_3 \cdot 5H_2O$ and $Fe(NO_3)_3 \cdot 9H_2O$ with a 1:1 molar ratio were dissolved in DMF solvent to prepare a homogeneous precursor solution. The precursor gel is formed through heat treatment of the solution at 100°C. The as-obtained gel was heated at 350°C for 60 min to evaporate the organic components. The crystalline sample is prepared through calcination at 550°C for 120 min. Finally, the $BiFeO_3$ NPs were obtained through grounding the crystalline samples for 2 hours. The $BiFeO_3$ NPs thin-film photoelectrode was prepared by drop-casting the NPs on substrates.

The PZT ($Pb(Zr_{0.20}Ti_{0.80})O_3$) nanostructured thin films were grown on the ordered arrays of Au nanostructures grown on ITO substrate through the sol-gel route [51]. The precursor solution was prepared by dissolving the required amount of lead acetate in acetic acid, in which titanium isopropoxide and zirconium isopropoxide were added slowly. Afterward, 2-methoxyethanol was added to the solution to prepare a clear yellow sol. The sol was coated on the patterned Au nanostructured/ITO substrate and dried at 150°C for 5 min in air. Finally, annealing was conducted to obtain the polycrystalline PZT nanostructured thin film for photoelectrochemical applications. The PZT nanocrystal powder was also fabricated by sol-gel methods [52]. First, the PZT precursor solution was prepared by adding titanium isopropoxide solution prepared in isopropanol with zirconium acetylacetonate dissolved in acetic acid and lead acetate trihydrate dissolved in acetic acid. The solution was heat-treated at 120°C to prepare the dried gels. Finally, PZT nanocrystal powder was obtained through grinding the gel followed by annealing at 600°C for 3 hours.

The orthorhombic nano-crystalline $LaFeO_3$ perovskite and transition metal (Mn, Co, Cu) doped $LaFeO_3$ with photocatalytic hydrogen production ability was synthesized via the sol-gel route [53, 54]. In this typical $LaFeO_3$ NPs synthesis route, the aqueous solution was prepared by mixing $La(NO_3)_3 \cdot 6H_2O$, and $Fe(NO_3)_3 \cdot 6H_2O$ with citric acid. The solution was transformed into the gel using ultrasonication energy. The gel was dried in the oven and heated at 500°C for 2 hours to remove the organic components. Finally, the nano-crystalline $LaFeO_3$ was prepared through grinding and heat treatment of the dried sample. Highly pure phase cubic perovskite $LaFeO_3$ NPs were prepared through the ionic-liquid-based sol-gel route, where the starting solution was prepared by mixing the aqueous solution of $La(NO_3)_3 \cdot 6H_2O$, and $Fe(NO_3)_3 \cdot 6H_2O$ with 1-ethyl-3-methylimiamidizolium acetate [55]. Finally, the $LaFeO_3$ NPs were obtained through dehydration of the sol followed by high-temperature calcination of the gel at 900°C. The orthorhombic $LaFeO_3$ NPs were also prepared by Shen *et al.* by sol-gel route [56]. Firstly, a solution was prepared by dissolving $La(NO_3)_3 \cdot 6H_2O$ in glacial acetic acid. Simultaneously another solution was prepared by mixing $Fe(NO_3)_3 \cdot 9H_2O$ in a water and glacial acetic acid. After evaporating the water of the $Fe(NO_3)_3 \cdot 9H_2O$ solution, $La(NO_3)_3 \cdot 6H_2O$ solution was mixed with it and heated at 110°C. After cooling down the solution, water, lactic acid, glycerol, and ethylene glycol were added, respectively, into the above mixture. The resultant sol was dried to form the gel and calcinated at high temperature to prepare the $LaFeO_3$ NPs for photocatalytic application.

The $NaNbO_3$ NWs (Figure 9.5a) with photocatalytic properties were prepared by sol-gel route by Saito *et al.* [57]. An equimolar mixture solution of $(NH_4)_3[NbO-(ox)_3]\cdot 3H_2O$ and NaOH was heated at 573 K for 2.5 hours in trioctylamine to obtain a residue. The precipitate was filtered and calcined in air at 773 K for 5 hours to prepare the $NaNbO_3$ NWs.

The hexagonal $YFeO_3$ NPs assembled films (Figure 9.5b) were fabricated by electrophoretic deposition of the $YFeO_3$ NPs prepared by the sol-gel route [58]. A solution was prepared by mixing $Fe(NO_3)_3 \cdot 9H_2O$, $Y(NO_3)_3 \cdot 6H_2O$, and citrate in water. The necessary heat treatments were employed to transform the solution into a porous gel. The gel was calcined at a higher temperature to obtain the final product of $YFeO_3$ NPs. Finally, the $YFeO_3$ NPs

Figure 9.5 (a) SEM image of NaNbO$_3$ NWs grown by sol-gel route. Reprinted with permission from [57]. Copyright (2010), American Chemical Society. (b) SEM image of hexagonal YFeO$_3$ NPs assembled film fabricated by sol-gel route. Reproduced with permission [55]. Copyright 2017, The Royal Society of Chemistry. And (c) Top view and cross-sectional (inset) SEM images of the LaFeO$_3$/ZnO NRs grown on FTO. Reproduced with permission [58]. Copyright 2021, The Royal Society of Chemistry. Fabricated by sol-gel routes. (d) FESEM image of the nanostructured LaFeO$_3$ thin film prepared by solution process. Reprinted with permission from [60]. Copyright (2020), American Chemical Society. (e) FESEM image of the BiFeO$_3$/TiO$_2$ nano-heterostructure arrays fabricated by wet chemical route. Reprinted with permission from [12]. Copyright (2015), American Chemical Society. (f) SEM image of nanocrystalline YFeO$_3$ fabricated via solution method. Reproduced with permission [61], Copyright 2017, The Royal Society of Chemistry.

were deposited on FTO to prepare the photoanode. Synthesis of YFeO$_3$ NPs was also reported through dehydration followed by high temperature (1000°C for 2 hours) calcination of the aqueous solution of Y(NO$_3$)$_3$·6H$_2$O, Fe(NO$_3$)$_3$·9H$_2$O, and 1-ethyl-3- methylimiamidizolium acetate [59].

Porous LaFeO$_3$ NPs were fabricated by the sol-gel route for photocatalytic application [60]. In this typical synthesis, the starting solution was prepared by adding a stoichiometric amount of La(NO$_3$)$_3$·6H$_2$O and Fe(NO$_3$)$_3$·9H$_2$O into the mixture of ethylene glycol, ethanol, and deionized water. Afterward, amino-functionalized carbon nanospheres were added to the solution. Finally, the solution was dried at 80°C to form a gel in an oven. The carbon nanosphere temple was removed through heat treatment at 400°C and the porous LaFeO$_3$ NPs were obtained through post-annealing treatment at 600°C.

Perovskite LaFeO$_3$ NPs and the LaFeO$_3$ NPs decorated ZnO NRs (Figure 9.5c) nano-heterostructure were also prepared to study their water oxidation properties [61, 62]. The highly crystalline LaFeO$_3$ NPs were fabricated by the sol-gel auto-combustion technique. The aqueous precursor solution prepared by mixing La(NO$_3$)$_3$·6H$_2$O, Fe(NO$_3$)$_3$·9H$_2$O, and citric acid was heated to prepare the yellowish sol, which was further heated at 130°C overnight to form the brownish-black solid dry gel. Finally, the LaFeO$_3$ NPs were prepared by grinding and high-temperature annealing of the dry gel. The LaFeO$_3$ NPs anchored ZnO NRs photoanodes were prepared by spin coating the as prepared LaFeO$_3$ NPs dispersed solution on the ZnO NRs array substrate followed by calcination at 300°C.

9.6.3 Wet Chemical and Solution Methods

Various other wet chemical and solution methods are employed to fabricate perovskite-type multiferroic oxide nanostructures as this is an easy, economical, and less skill-dependent route. The solution methods are mostly famous for growing the nanostructured thin films or nanoscale coating on the substrates. It is also possible to grow nanostructures of multiferroic oxides on the surface of a nanoscale template structure through the wet chemical and solution methods.

An innovative wet solution-based technique was reported by Liu *et al.* to synthesize LaFeO$_3$ NRs photoanode on FTO substrate [63]. In this method, the β-FeOOH NRs arrays were first fabricated on FTO by the chemical bath deposition route. After that, an aqueous solution of La(NO$_3$)$_3$ was drop-casted on the as-prepared β-FeOOH NRs and dried. Finally, the LaFeO$_3$ NRs were obtained through annealing at 650°C followed by etching the surface La$_2$O$_3$ layer. The nanostructured LaFeO$_3$ thin film (Figure 9.5d) photoelectrode was prepared by the spray pyrolysis method on the FTO substrate [64, 65]. In this typical method, the iron oxide/hydroxide precursor precipitate was first prepared by adding aqueous ammonia in the solution containing Fe(NO$_3$)$_3$·9H$_2$O dissolved in methanol. Afterward, the precipitate was mixed with Lanthanum(III) isopropoxide, methanol, and trifluoroacetic acid to prepare the precursor solution for the spray pyrolysis. The spray pyrolysis was conducted, taking the solution in a 50 mL syringe on the surface of FTO glass heated at 150°C to prepare the nanostructured LaFeO$_3$ thin film.

Sarkar *et al.* have reported the wet chemical synthesis of three-dimensional nanoarchitecture of BiFeO$_3$ NPs anchored TiO$_2$ NTs arrays (Figure 9.5e) photoanode [12]. The as-prepared TiO$_2$ NTs template was immersed into the solution of Bi(NO$_3$)$_3$·5H$_2$O and Fe(NO$_3$)$_3$·9H$_2$O prepared in 2-methoxyethanol. Subsequently, the sol wetted TiO$_2$ NTs template was dried and annealed at 480°C to obtain the BiFeO$_3$ NPs/TiO$_2$ NTs heterostructure photoanode.

Tang et al. have reported the microwave-assisted solution synthesis of nanocrystalline monophasic perovskite YFeO$_3$ (Figure 9.5f) with visible-light photocatalytic properties [66]. In this synthesis method, a homogeneous solution was first prepared by dissolving Fe(NO$_3$)$_3$·9H$_2$O, Y(NO$_3$)$_3$·6H$_2$O, and polyvinyl alcohol into water added with urea. Eventually, the nanocrystalline YFeO$_3$ powder was obtained through microwave-assisted combustion of the precursor solution.

The LiNbO$_3$ NWs were also prepared via the solution synthesis route [67]. First, the precursor solution was prepared by mixing (NH$_4$)$_3$[NbO(Ox)$_3$]·H$_2$O and LiOH in trioctylamine. The solution was heated at 613 K for 2.5 hours to obtain the LiNbO$_3$ NWs precipitate.

Dong et al. have reported the synthesis of ZnSnO$_3$ hollow nanospheres through the wet solution route by using the as-prepared carbon spheres as templates [68]. In this method, the as-prepared carbon spheres were dispersed into the solution containing SnCl$_2$·2H$_2$O and Zn(CH$_3$COO)$_2$·2H$_2$O dissolved in HCl. The dark precursor was filtered and dried overnight in an oven at 100°C. Finally, the ZnSnO$_3$ hollow nanospheres photocatalysts were obtained through heat treatment at 600 ºC for 2 hours.

Mesoporous BiFeO$_3$ nanostructures were prepared by wet solution process for water oxidation application [69]. The precursor solution was prepared by mixing Bi(NO$_3$)$_3$, Fe(NO$_3$)$_3$, tartaric acid, ethanol in a concentrated HNO$_3$ solution. This solution was added dropwise into the nanoporous carbon dispersed in anhydrous hexane, and finally, this solution was stirred until the full evaporation of the solvents. The mesoporous BiFeO$_3$ nanostructures were obtained after annealing of the resulting powder in N$_2$ at 400°C followed by calcination at 350°C in air. Solution-based synthesis of BiFeO$_3$ thin film with enhanced ferroelectric photoelectrochemical properties was also reported by Liu et al. [70]. In this typical method, the BiFeO$_3$ thin film was prepared through spin-coating of the solution containing Bi(NO$_3$)$_3$·5H$_2$O and Fe(NO$_3$)$_3$·9H$_2$O dissolved in 2-methoxyethanol on the Si substrate followed by annealing at 400°C.

The porous anodic aluminum oxide (AAO) template-assisted solution-based synthesis of PbTiO$_3$ NTs was reported by Ahn et al. [71]. In this method, the PbTiO$_3$ precursor solution was prepared by mixing Pb(C$_2$H$_3$O$_2$)$_2$·3H$_2$O and Ti[O(CH$_2$)$_3$CH$_3$]$_4$ in 2-methoxy ethanol, where acetylacetone was added as a chelating agent. The sol was spin-coated on the AAO template on the FTO substrate followed by drying at 150°C. The dry gel was calcined at 350°C and annealed at 650°C to prepare the PbTiO$_3$ NTs arrays within AAO. The PbTiO$_3$ NTs were separated by dissolving

AAO in phosphoric acid solution. Kooshki et al. have reported the synthesis of $PbTiO_3$ NPs through the solution synthesis route [72]. In this route, the solution was prepared by adding various capping agents in the aqueous $Pb(NO_3)_2 \cdot 6H_2O$, where tetraethyl orthotitanate was added as the source of Ti. The as-prepared solution was heated to obtain the precipitate, which was calcined and annealed at high temperature to obtain the $PbTiO_3$ NPs. Vertically aligned core–shell $PbTiO_3$-TiO_2 NTs arrays have also been fabricated through spin-coating of the $Pb(C_2H_3O_2)_2 \cdot 3H_2O$ solution onto the surface of the as-prepared TiO_2 NTs followed by high-temperature annealing at 550°C for 1 hour [73].

9.6.4 Vapor Phase Deposition Methods

The vapor phase deposition is also widely used to synthesize various semiconductor thin films and nanostructures on different substrates. In this technique, the material to be deposited is vaporized and then condensed through chemical reactions to deposit thin films on a particular substrate. Vapor phase deposition is categorized as physical vapor deposition (PVD) and chemical vapor deposition (CVD) methods. Thin films of different ferroelectric semiconductors with photoelectrocatalytic activity are also fabricated through vapor phase deposition methods.

Ji et al. have reported the synthesis of ferroelectric $BiFeO_3$ thin films through the radio frequency (RF) magnetron sputtering route on the $SrTiO_3$ (STO) (001) substrate [74]. The $BiFeO_3$ thin films were deposited at a power of 120 W of the RF magnetron sputtering using the mixed gas atmosphere, where the ratio of 7:1 between Ar and O_2 was maintained. The STO (001) substrate was kept at a temperature of 680°C for the deposition of $BiFeO_3$ thin films. The pulsed laser deposition (PLD) synthesis of ferroelectric $BiFeO_3$ films on the as-prepared arrays of TiO_2 NTs was reported by Lee et al. [75]. In this typical process, the KrF excimer laser with a wavelength of 248 nm was used, where the energy fluence of 150 mJ/cm^2 was employed for PLD. The thickness of the $BiFeO_3$ thin films on TiO_2 NTs was controlled by tuning the film deposition time. May et al. also have reported the photoelectrochemical water splitting performance of $LaFeO_3$ thin films grown by the PLD technique [76]. In this work, the $LaFeO_3$ thin films were grown on Nb-doped $SrTiO_3$ substrate by operating a KrF excimer laser at 10 Hz and 1.5 J/cm^2. The substrate was placed at 645°C under 200 mTorr of oxygen pressure to produce thin nanoscale films by controlling the number of pulses used.

Recently, $LaFeO_3$ thin film was grown on FTO as well as on g-C_3N_4 via direct current (DC) magnetron co-sputtering using both iron and lanthanum targets [77]. The sputtering was conducted at an atmospheric pressure of 0.58 Pa containing 10 vol.% H_2 and 90 vol.% Ar. The power to Fe and La targets were 80 and 65 W, respectively, to grow the stoichiometric $LaFeO_3$ thin films. The film thickness was controlled via deposition time, and the substrate was rotated at 28 rpm for the uniform growth of the amorphous film, which was later crystallized through post-annealing treatment at 500°C. Yu et al. also reported the synthesis of $LaFeO_3$ thin film on ITO substrate by PLD technique using the sintered $LaFeO_3$ powder as the target material [78]. PLD of $LaFeO_3$ thin film was conducted at 650°C on ITO substrate.

Similarly, $BiMnO_3$ nanostructures were grown on the Nb-doped $SrTiO_3$ substrate by the PLD technique to study photoelectrochemical properties [79]. The $BiMnO_3$ nanostructures were fabricated using the $Bi_{1.4}MnO_3$ as target materials under the pressure of 1×10^{-6} mbar. The KrF excimer laser with a wavelength of 248 nm and pulse duration of 15.4 ns was used. A laser fluence of 3 J.cm^{-2} with a pulse frequency of 10 Hz was used, where the substrate temperature was 620°C under a partial oxygen pressure of ~10 mTorr.

9.6.5 Electrospinning Methods

Recently, the electrospinning technique has attracted significant attention to fabricate 1D fibers of various ceramics and oxide semiconductors. In this method, a high voltage is applied between the substrate and the specially designed metallic needle through which the precursor solution is ejected. The external applied electric field deforms the shape of the droplet coming out from the tip of the needle, and the repulsion between the surface charges helps to generate a stable jet that undergoes deformation to produce a 1D fibers-like structure on the substrate. It is possible to grow inorganic perovskite oxide micro and nanoscale fibers and other structures on the substrate by tuning the parameters of the electrospinning methods.

Recently, Das et al. have reported the synthesis of ferroelectric $BiFeO_3$ nanostructured photocathodes by electrospinning technique [13]. In this method, a homogeneous precursor solution was first prepared by dissolving the required amount of bismuth nitrate and iron nitrate in N, N–dimethylformamide added with polyvinylpyrrolidone to control the viscosity. The solution was injected at a constant flow rate of 0.18 ml/h under the applied voltage of 12 kV between the needle and the FTO substrate, on which the $BiFeO_3$ nanostructures were grown. The FTO substrate was 10 cm away from the needle and rotated with a speed of 1000 rpm. Different nanostructures and thin films of $BiFeO_3$ were fabricated by changing the relative humidity. The as-prepared nanostructures were heated at 250°C for 15 min to evaporate the solvent, and finally, the crystalline structure was obtained through annealing at 520°C for 2 hours.

9.7 Photo(electro)catalytic Activities of Ferroelectric Nanostructures

9.7.1 Photo(electro)catalytic Activities of $BiFeO_3$ Nanostructures and Thin Films

$BiFeO_3$ is a widely explored multiferroic perovskite oxide material, which shows robust ferroelectric properties [80]. Extensive research has been conducted exploring the photo(electro)catalytic activity of $BiFeO_3$ thin films and nanostructures due to its interesting ferroelectric semiconducting properties and narrow bandgap energy (2.2–2.7 eV), suitable for efficient solar light harvesting [80–84]. The valance and conduction band of $BiFeO_3$ straddles the water redox potentials, which make it suitable for unassisted solar water splitting under visible light [74]. The ferroelectric polarization of $BiFeO_3$ affects the band bending, which has a crucial role in photogenerated carrier separation and transportation and hence in the photo(electro)catalytic activity. Based on the fabrication routes, $BiFeO_3$ can be both n-type and p-type, and thereby it can serve as a photoanode and a photocathode for PEC water splitting, respectively [81, 83–85]. Despite having remarkable potential, two major bottlenecks limit its application as a photoelectrode material. $BiFeO_3$ suffers severe and rapid recombination of photogenerated charge carriers, which is undesirable for achieving a large current density and enhanced PEC activity. Secondly, the fast degradation of $BiFeO_3$ under photocatalytic conditions also limits its application [80, 85].

In earlier work, Chen *et al.* [84] prepared amorphous and polycrystalline, n-type $BiFeO_3$ thin film on $Pt/TiO_2/SiO_2/Si$ substrate by pulsed laser deposition, where the polycrystalline $BiFeO_3$ exhibited two-fold photocurrent density over the amorphous $BiFeO_3$. However, the photocurrent density was only limited to 0.159 mA.cm^{-2} at 1.5V vs. SCE under the illumination of a 400 W Xe lamp with a 420 nm cut-off filter in a 0.2 M Na_2SO_4 aqueous solution (pH 7.5). Similar behavior of epitaxially grown $BiFeO_3$ thin film was observed by Ji *et al.* [74]. This work showed how the film thickness and ferroelectric polarization altered the PEC property of $BiFeO_3$ thin films. The water oxidation onset potential was observed at 0.21 $V_{Ag/AgCl}$ for a film of a thickness of 223 nm. The onset potential was shifted cathodically to

−0.036 V vs. Ag/AgCl when the film thickness reduced to 112 nm. The reduced film thickness minimized the Ohmic loss, which resulted in a cathodic shift in onset potential. For a polarized $BiFeO_3$ film with negative charges on its surface, 0.016 V cathodic shift of the onset potential was noticed. The positive and negative polarization charges tuned the downward and upward band bending, facilitating the hydrogen and oxygen evolution reactions, respectively. However, the effect of polarizations on the photocurrent enhancement of $BiFeO_3$ thin films was found insignificant by Ji et al. [74], which may be due to the poor control of electric polarization in the aqueous solution. A more significant change of PEC activity by electric polarization was reported by Liu et al. [70], in a p-type nanostructured polycrystalline (average grain size 70-150 nm) $BiFeO_3$ film grown on $Pt/Ti/SiO_2/Si(100)$ substrate by sol-gel method. Furthermore, they decorated the film surface with Ag NPs to enhance the optical absorbance and minimize the electron-hole pair recombination. The optimized $BiFeO_3$ film (thickness 200 nm) exhibited a photocurrent density of 15 $\mu A.cm^{-2}$, which was higher than the epitaxial $BiFeO_3$ films (10 $\mu A.cm^{-2}$ at 0.64 V vs. Ag/AgCl, 300 W Xe lamp) [74]. The polarized $BiFeO_3$ film with negative and positive charges on the surface exhibited −0.12 V and 0.08 V onset potential, respectively. Furthermore, the coupling of Ag NPs with $BiFeO_3$ films induced a strong localized surface plasmon resonance (LSPR) effect, which enhanced the overall optical absorbance of the photoelectrode and generated a large number of electron-hole pairs. Further, the Fermi level mismatch at Ag NPs/$BiFeO_3$ interface led to a built-in potential, facilitating photocarrier separation. The Ag NPs/$BiFeO_3$ film exhibited an enhanced photocurrent density (35 $\mu A.cm^{-2}$ at 0.64 V vs. Ag/AgCl) and photostability [74].

Gu et al. [85] have argued that the limited photocurrent density (<100 $\mu A.cm^{-2}$) of $BiFeO_3$ may arise because of the poor surface catalytic activity of $BiFeO_3$ for water oxidation or reduction reactions. Therefore, they loaded the surface of $BiFeO_3$ with Pt catalyst, which is the best catalyst for H_2 evolution reaction (HER). However, the $BiFeO_3$/Pt junction is not suitable for the transfer of electrons toward the electrolyte as the work function of $BiFeO_3$ (4.8 eV) is much below the work function of Pt 5.4 eV. Therefore, the upward barrier height at the $BiFeO_3$/Pt junction drives the electron away from the electrode/electrolyte interface (Figure 9.6a). They came up with a solution by introducing a conductive layer of porous carbon between $BiFeO_3$ and Pt. Since the work function of carbon is 4.2 eV, smaller than that of $BiFeO_3$, it will prevent the formation of the Schottky junction between $BiFeO_3$ and Pt. The downward band bending at $BiFeO_3$/porous carbon interface drove the electron toward the electrolyte, whereas

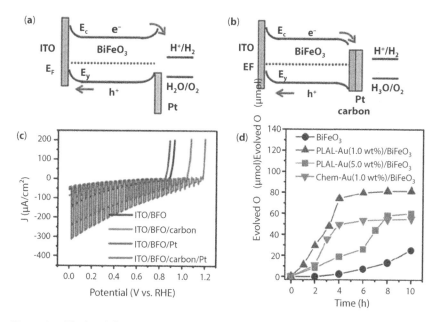

Figure 9.6 The band diagrams of the (a) ITO/BiFeO$_3$/Pt and (b) ITO/ BiFeO$_3$/porous Carbon/Pt electrodes. (c) Photocurrent-voltage plots of the ITO/BiFeO$_3$(BFO)/porous Carbon/Pt electrodes with positively polled the BiFeO$_3$. Reproduced with permission [81]. Copyright 2017, The Royal Society of Chemistry. (d) Oxygen evolution activity of various BiFeO$_3$ NWs based photoanodes. Reproduced with permission [28]. Copyright 2013, The Royal Society of Chemistry.

the highly conducting porous carbon layer helped in electron tunneling through the heterojunction (Figure 9.6b). The performance of the BiFeO$_3$/porous carbon/Pt electrode was further enhanced by introducing positive poling in BiFeO$_3$. The BiFeO$_3$/porous carbon/Pt photocathode exhibits a photocurrent density of −235.4 μA.cm^{-2} at 0 V$_{RHE}$ (Figure 9.6c) and an open-circuit voltage of 1.19 V$_{RHE}$ with the stability of up to 2 hours, under 100 mW.cm^{-2} illumination from a Xe lamp. Shen et al. [86] further investigated this material system by replacing porous carbon with a thin layer of amorphous TiO$_2$ layer having a work function of 4.4 eV, which is lower than that of BiFeO$_3$. Like porous carbon, the TiO$_2$ layer prevents the Schottky barrier formation between BiFeO$_3$ and Pt, and the favorable band alignment between BiFeO$_3$/TiO$_2$/Pt promotes the photoelectron transfer toward the electrolyte. A photocurrent density of −460 μA.cm^{-2} at 0 V vs. RHE was achieved with BiFeO$_3$/TiO$_2$/Pt photocathode. An acidic environment is suitable for HER. However, BiFeO$_3$ is very unstable under an acidic environment. Within 15 min of operation, BiFeO$_3$ started degrading, whereas the BiFeO$_3$/TiO$_2$/Pt electrode was found to be stable up to 10 hours under an acidic environment. Therefore, the TiO$_2$ layer not only promoted the electron transfer process but also acted as a surface passivating over layer on BiFeO$_3$ to provide long-term stability.

Recently, Das et al. [13] reported the PEC activity of nanostructured BiFeO$_3$ photocathodes, where the as-prepared BiFeO$_3$ thin film, nanowebs, and nanofibers electrodes exhibited a significant photocurrent density of −56.5 μA.cm^{-2}, −64.9 μA.cm^{-2}, and −86.2 μA.cm^{-2} at −0.4V vs. Ag/AgCl, respectively. The highly exposed electrochemical surface area of the nanofibres and the improved charge separation due to the rapid diffusion of charge carriers along the axis resulted in remarkably enhanced photocurrent density in BiFeO$_3$ nanofibers compared to other BiFeO$_3$ photocathodes. The hydrogen-treated BiFeO$_3$ NPs photoanode exhibited more than three-fold enhanced photocurrent density compared to the pristine BiFeO$_3$ NPs. Such enhanced PEC performance was attributed to the formation of mid band gap oxygen vacancy defects favoring water splitting activity [50]. The Au NPs decorated single-crystalline BiFeO$_3$ NWs were found to exhibit significantly higher photocatalytic water oxidation activity [29]. The Au NPs, synthesized by pulsed laser ablation in the liquid phase (PLAL), anchored with BiFeO$_3$ NWs, exhibited enhanced visible-light-driven oxygen evolution through water splitting, which produced ~81 μmol of oxygen gas under 10 hours of operation (Figure 9.6d). The PLAL-Au/ BiFeO$_3$ NWs with 1 wt% Au NPs anchoring provided more than 30 times higher photocatalytic activity over the pristine BiFeO$_3$ NWs under the first 4 hours of operation because of efficient photogenerated carrier separation.

The mesoporous BiFeO$_3$ nanostructures were found to exhibit enhanced photocatalytic oxygen evolution activity under UV-visible radiation ($\lambda >$ 380 nm) with an average oxygen evolution rate of 66 mmol. h^{-1}g^{-1}. The OER performance of the mesoporous BiFeO$_3$ further increased because of the anchoring of Au NPs, which scavenged the photogenerated electron from BiFeO$_3$, resulting in a long-lived photogenerated hole to take part in the water oxidation reaction. The optimized Au/BiFeO$_3$ nanostructure exhibited an enhanced oxygen evolution rate of 586 mmol. h^{-1}g^{-1} and long-time durability for water splitting [69]. The TiO$_2$/BiFeO$_3$ nano-heterostructure photoanode achieved over 480% enhancement in the photocurrent density at 0.4 V vs. Ag/AgCl because of the coupling of low bandgap BiFeO$_3$, which formed a favorable band alignment with TiO$_2$ NTs boosting the photocarrier separation at the heterojunction [12].

Tandem cell design is one of the most attractive approaches toward PEC water splitting. Cheng et al. constructed Si/ITO/Au/BiFeO$_3$ hybrid photocathode and catalyzed its surface by MoS$_2$/Pt, which provided excellent photocurrent density and stability [87]. In this tandem cell arrangement, Si serves the role of a narrow bandgap material. A p/n junction was constructed by Al^{3+} doping in an n-type Si substrate. This p/n junction ensured the movement of electrons toward the front surface and hole toward the back Al electrode, and thereby carrier separation was achieved. Further, an n^+ layer was constructed by phosphorus doping on the n-type front surface, which bent the surface band of Si downward and promoted electron transportation. An ultrathin (2 nm) layer of Al_2O_3 was deposited, which served the role of the surface protecting layer and the surface passivating layer. On the top of Al_2O_3, ITO and a thin layer of Au NPs were deposited. Finally, $BiFeO_3$ film was deposited over the Au-spattered ITO surface. The ITO served the role of an electrode during the poling of ferroelectric $BiFeO_3$ film. The Au nanoparticles injected the hot electrons into $BiFeO_3$ because of LSPR. In this architecture, the $BiFeO_3$ served the role of a wide bandgap absorber and photoelectron extractor from the Si surface by depolarization of the electric field. Further, the surface of $BiFeO_3$ was anchored with MoS_2/Pt to function as the HER catalyst. This hybrid electrode exhibited a photocurrent density of -9.1 $mA.cm^{-2}$ at 0 V vs. RHE under 100 $mW.cm^{-2}$ Xe lamp illuminations. Nearly 125 μmol of H_2 was generated at 0 V vs. Ag/AgCl with an enhanced photostability under two hours of operation.

The manipulation of the ferroelectric polarization direction was found to significantly control the PEC activity of $BiFeO_3$ ferroelectric photoelectrodes grown on ITO substrate. The photocurrent density and the open-circuit potential were tuned remarkably from 0 to 10 mA. cm^{-2} and from 33 to 440 mV, when the poling direction was switched between −8 to +8 V [88]. The polarization switching was found to tune the band bending at electrode/electrolyte interface controlling photoinduced carrier separation and transportation. The positive poling was found to reduce the band bending at $BiFeO_3$/electrolyte and boost the photoelectron transfer toward the electrolyte resulting in enhanced external quantum yield. The positive poling in $BiFeO_3$/ITO helped in the rapid migration of photogenerated holes from Rhodamine B dye, enhancing the photocatalytic dye degradation process.

9.7.2 Photo(electro)catalytic Activities of $LaFeO_3$ Nanostructures

$LaFeO_3$ is another perovskite ferroelectric oxide semiconductor that has attracted sufficient attention from researchers in the field of solar light-driven water splitting and photocatalysis due to its robust stability under solar illumination in an aqueous medium, abundant quantity of the constituent rare earth element, and its multifunctional physical properties [80]. The ferroelectric $LaFeO_3$ is a narrow bandgap semiconductor in the range 2–2.6 eV, ideal for harvesting visible solar illumination [80]. The valence band of $LaFeO_3$ originated from the hybridized O 2p and Fe 3d orbitals, and the conduction band originates from the La 4f orbital [89]. This band arrangement enables the lattice Fe to participate in the hole transport mechanism similar to other ferrites [90]. The dual character of $LaFeO_3$ as light-absorbing material and electrocatalyst heavily influenced the researchers to fabricate single material-based photoelectrodes. Both the n-type and p-type conductivity of $LaFeO_3$ has been reported in the literature. In spite of having such great potential to be a photoelectrode material, the sluggish OER kinetics and poor carrier transportation limit its application possibility. The unavailability of sophisticated fabrication techniques is also responsible for the poor photocatalytic activity of $LaFeO_3$ [65].

The high quality and pure phase of $LaFeO_3$ nanoparticles were synthesized by Celorrio *et al.* through a novel ionic-liquid-based method, and the NPs were further assembled on the FTO through screen printing. A cathodic photocurrent response was observed with an onset potential at 1.1 V vs. RHE. However, the reported photocurrent response was significantly low [55]. Spontaneous solar H_2 production without applied bias using nanostructured $LaFeO_3$ photocathode has been reported [65]. The nanostructured $LaFeO_3$ photocathode exhibited a photocurrent density of 0.16 mA.cm^{-2} at 0.26 V_{RHE} with stability up to 21 hours. Production of 0.18 µmol.cm^{-2} H_2 was reported for nanostructured $LaFeO_3$ during the first 6 hours of operation. However, the current-voltage curves under chopped illumination showed large photocurrent spikes, which indicated that a sufficient number of electrons were accumulating at the electrode/electrolyte junction and recombined with the holes before injecting into the electrolyte [64, 65].

This problem of carrier recombination at the $LaFeO_3$/electrolyte interface due to lack of exposed surface area of the electrode was resolved by fabricating nanoporous p-type $LaFeO_3$ through electrodeposition [91]. The nanoporous $LaFeO_3$ photocathode achieves a photocurrent density of −100µA.cm^{-2} with an onset potential of 1.41 V_{RHE}, which was nearly close to its flat band potential. More interestingly, no photocurrent spikes were observed in LSV measurements under chopped light, which indicated the surface recombination of charge carriers was reduced significantly. However, the photocurrent density is still far below what is suggested by its bandgap energy. The bulk recombination of photocarriers due to poor separation of electron-hole pair is the major performance-limiting parameter for nanoporous $LaFeO_3$ photocathode.

On the other hand, transition metal (Mn, Co, Cu) doping seemed to have a great influence in enhancing the carrier density and charge transport and thereby the PEC activity of $LaFeO_3$ NPs photoanode [54]. The 10% Mn, Co and Cu doped $LaFeO_3$ showed a photocurrent density of 0.06, 0.21 and 0.21 mA.cm^{-2}, respectively, over the poor photocurrent density of pristine $LaFeO_3$ (0.01mA.cm^{-2}) at 0.5 $V_{Ag/AgCl}$. The onset potential was also shifted from 0.48 V to 0.34 V for Mn doping and 0.27 V for Co and Cu doping. However, the metal doping did not have any impact on the bandgap of $LaFeO_3$ but greatly enhanced the carrier density, which resulted in the improved photoconversion efficiency.

PEC activity of nanostructured $LaFeO_3$ photoanodes was further improved by Liu et al. by incorporating the oxygen vacancy defects in $LaFeO_3$ and introducing the $NiFeO_x$ surface catalytic overlayer [63]. The oxygen defects activated the surface of $LaFeO_3$ NRs toward OER. Additionally, a surface passivating layer of $NiFeO_x$ was deposited to suppress the surface carrier recombination. The nanostructured $LaFeO_3$ photoanode delivered a benchmark photocurrent density of 0.4 mA cm^{-2} at 1.23 V_{RHE}. This study also suggested that the PEC performance of defect engineered $LaFeO_3$ NRs photoanodes were limited due to the bulk recombination of charge carriers and ultrashort hole diffusion length (<5 nm). The visible-light-driven photocatalytic hydrogen production from nano-crystalline $LaFeO_3$ was investigated, where the rate of hydrogen production is 3315 µmol. g^{-1} h^{-1} under the illumination [53]. The coupling of $LaFeO_3$ nanostructures with aligned arrays of ZnO NRs has been found to be effective for solar water splitting. The Fe-doped $LaFeO_3$ nanostructure anchored ZnO NRs photoanode delivered a photocurrent density as high as 2.21 mA.cm^{-2} at 1.23 V_{RHE} [61]. Similarly, the $LaFeO_3$/ZnO NRs coupled with Co-Al layered double hydroxide photoanode produced a photocurrent density of 2.46 mA.cm^{-2} at 1.23 V_{RHE} and an applied bias photoconversion efficiency over 0.76% [62]. The favorable heterojunction formation between $LaFeO_3$ and ZnO NRs enabled the rapid separation of the photogenerated hole into the $LaFeO_3$ leading to enhanced charge separation efficiency for PEC water splitting.

Apart from that, the enhanced visible-light-driven degradation of hazardous 2,4-DCP was reported for Bi-doped porous $LaFeO_3$ coupled ZnO nanocomposite [60]. The effective utilization and separation of the photogenerated charge carriers in this nano-heterostructure made it highly photoactive. The highly active ·OH and $O_2^{·-}$ radicals were found to be responsible for 2,4-DCP degradation over Bi-doped $LaFeO_3$ and Bi-doped $LaFeO_3$/ZnO hybrid, respectively. The photocatalytic degradation of MB and MO dyes in the presence of $LaFeO_3$ NPs has also been reported.

Complete degradation of the dyes was achieved through photocatalysis under visible light radiation after 240 min [56].

9.7.3 Photo(electro)catalytic Activities of $BaTiO_3$ Nanostructures

$BaTiO_3$ is a wide bandgap (3.2 eV), n-type piezoelectric, and ferroelectric semiconductor, in which ferroelectricity was first discovered by Wul and Goldman in 1945 [92]. In spite of the wide band gap energy suitable for absorbing only the UV solar radiation, the high electrochemical stability, suitable band positions against water redox potential, and defect controlled electronic properties make it a potential photoelectrode material for solar water splitting [47]. $BaTiO_3$, as a classic piezoelectric semiconductor, also exhibits modulated photocarrier dynamics (generation, separation, and transport) under the variation of induced piezoelectric polarization [10].

The PEC water oxidation, using a polycrystalline $BaTiO_3$ electrode, was first reported by Kennedy et al. [93], including the effect of temperature and pH on the photocurrent response. In order to promote the light absorption of $BaTiO_3$ in the visible region and enhance the PEC activity, the synthesis of doped $BaTiO_3$ nanostructures was suggested by Upadhyay et al. [94]. The Fe-doped $BaTiO_3$ NPs observed a 0.39 eV redshift of the absorption edge for 2% Fe doping. The pristine $BaTiO_3$ NPs exhibited 0.07 $mA.cm^{-2}$ current density at 0.5V vs. RHE which increased to 2.55 $mA.cm^{-2}$ after 2% Fe-doping. The first-principles density functional theory (DFT) calculations by Huang et al. suggested that 5 atomic % Te doping in $BaTiO_3$ can reduce the bandgap and consequently enhance the PEC activity of $BaTiO_3$ [95]. The Ce-doped $BaTiO_3$ nano-assemblies have been reported to exhibit enhanced photocatalytic and PEC activities because of doping and tuning of the built-in ferro/piezoelectric field [96]. Ce-doping in $BaTiO_3$ nano-assemblies enhanced the carrier density by forming oxygen vacancy and reduced the bandgap energy, resulting in an increase in the photocurrent density. The externally induced ferro/piezoelectric polarization in Ce-doped $BaTiO_3$ nano-assemblies was found to tune the favorable band bending and built-in potential boosting the photocarrier separation and transport. The optimized Ce-doped $BaTiO_3$ nano-assemblies photoanode exhibited a photocurrent density of 1.45 $mA.cm^{-2}$, onset potential of -0.504 V, and a hydrogen gas evolution rate of 22.50 $\mu mol. h^{-1} cm^{-2}$. The Ce-doped $BaTiO_3$ nano-assemblies photocatalyst also exhibited enhanced dye degradation activity with the rate constant of 0.0139 m^{-1} and 0.0147 m^{-1} for methylene blue and methyl violet, respectively.

The formation of nano-heterojunction is also very effective in enhancing the efficiency of the $BaTiO_3$ nanostructure-based photocatalysts. For example, core/shell type crystalline/amorphous black $BaTiO_3$ nanocube fabrication has proved to be an effective strategy to enhance the photocatalytic and PEC activity of $BaTiO_3$ under visible light [45]. The oxygen vacancy in the amorphous $BaTiO_3$ layer greatly enhanced the visible light absorption; however, the bandgap remained the same with the pristine $BaTiO_3$. It also shows the improved separation of charge carries for which the black $BaTiO_3$ nanocubes exhibited 13 times higher photocurrent density than pristine $BaTiO_3$ at 1.23 V vs. RHE. The photocatalytic dye degradation efficiency was also improved due to black $BaTiO_3$ fabrication. Under visible light illumination, black $BaTiO_3$ degraded 62.4% MB dye within 5 hours compared to only 7.7% dye degradation by pristine $BaTiO_3$. Li et al. [97] fabricated a type-II p/n heterojunction photoanode through electrochemical deposition of Cu_2O NPs over $BaTiO_3$ NRs. The low bandgap Cu_2O NPs enhanced the overall absorbance, the interfacial electric field at the p/n nano-junction promoted the charge separation, reducing carrier recombination and the type-II band alignment boosted the charge transport in the heterojunction photoanode. The Cu_2O NPs incorporation over $BaTiO_3$ NRs synergistically enhanced the applied bias photon to current conversion efficiency from 0.05% to 0.11% at 0.72 V vs. RHE under 100 $mW.cm^{-2}$ illuminations.

Along with the heterojunction formation, the introduction of ferroelectric polarization in $BaTiO_3$ was found to further boost the charge separation efficiency in $BaTiO_3/TiO_2$ NRs heterojunction photoanode [47]. The $BaTiO_3/TiO_2$ NRs heterojunction photoanode exhibited a 67% increase in photocurrent density over pristine TiO_2 NRs, because of heterojunction formation. Further, the control of the polling direction in $BaTiO_3$ tuned the band bending at $BaTiO_3/TiO_2$ NRs interface, which drove the holes rapidly toward the electrode/electrolyte junction leading to enhanced photocurrent for the water oxidation reaction. The photogenerated carrier separation was improved in the n-TiO_2 and p-Ag_2O nano-heterostructure by introducing a polarized nanolayer of $BaTiO_3$ in between [11]. The piezoelectric internal electric field in the $BaTiO_3$ layer was tuned to function as the driving force for photocarriers to reduce the electron-hole pair recombination leading to improved PEC activity. The optimized $TiO_2/BaTiO_3/Ag_2O$ nano-heterostructure after poling exhibited a photocurrent density of 1.8 $mA.cm^{-2}$ at 0.8 V versus Ag/Cl, which was about 2.6 times higher than TiO_2 NRs.

The coupling of the local surface plasmon resonance (LSPR) and piezo-polarization effects was found to enhance the water splitting and pollutant degradation activity of Al/BaTiO$_3$ nanoheterostructures [10]. The LSPR enabled the Al/BaTiO$_3$ nanoheterostructures photoelectrode to exhibit an enhanced H$_2$ evolution rate of 327 µmol.h^{-1}cm^{-2}. Moreover, the strong piezo-electric polarization induced in the Al/BaTiO$_3$ nanoheterostructures through the external magnetic field boosted the photocarrier separation and transfer process at the plasmonic Al/BaTiO$_3$ heterojunction. As a result, the piezo-polarization coupled with LSPR helped to achieve a 50% enhancement in H$_2$ production rate and 4-nitrophenol degradation efficiency for the polarized Al/BaTiO$_3$ nanoheterostructures. Under visible light illumination, the LSPR effect generated hot electrons in Al NPs that migrated into the BaTiO$_3$ nanostructures because the band bending at Al/BaTiO$_3$ interface left holes in Al NPs (Figure 9.7). Moreover, the built-in piezopotential in the same direction with the electric field with the space-charge region further promoted the separation of the photocarrier for enhanced PEC activity in Al/BaTiO$_3$ nanoheterostructures (Figure 9.7). The TiO$_2$/BaTiO$_3$/Au multilayered coaxial NRs arrays have been reported to exhibit enhanced photocatalytic bacteria-killing activity because of the production of reactive oxygen species like superoxide (O$_2$·$^-$) and hydroxyl (·OH) radicals under UV-visible light [46]. The application of suitable piezoelectric polarization in the BaTiO$_3$ sandwich layer further tunes the interfacial band bending at the heterojunction and facilitates the photocarrier separation for photocatalytic activity. The hot electrons generated in Au NPs can easily migrate into the TiO$_2$ side through BaTiO$_3$ because of band alignment and in-built piezo-potential, resulting in enhanced

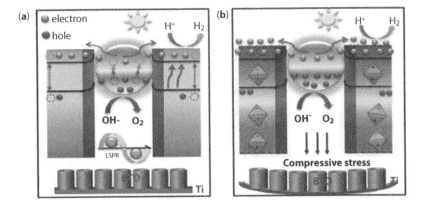

Figure 9.7 Schematic of the photocarrier generation/separation mechanism of the plasmonic photocatalysis (a) and piezophotocatalysis (b) process in Al/BaTiO$_3$ (BTO) nano-heterostructures. Reproduced with permission [10]. Copyright 2019, Elsevier.

photocarrier separation leading to 90% killing efficiency to the E. coli and gram-positive bacterium S. Aureus.

9.7.4 Photo(electro)catalytic Activities of $SrTiO_3$ Nanostructures

$SrTiO_3$ is another n-type wide bandgap, ferroelectric perovskite oxide semiconductor. Over the years, $SrTiO_3$ has been widely investigated as a promising photoanode material due to its favorable band position and superior stability in the electrolyte under illumination. $SrTiO_3$ has a relatively more negative conduction band than TiO_2, making it a potential material for total water splitting [98]. However, the wide bandgap (3.2 eV) and significant carrier recombination limit its usefulness in PEC application. Since the first report of water splitting of $SrTiO_3$ under solar light by Wrighton *et al.* [99], enormous effort has been made to tune its bandgap and carrier recombination to make $SrTiO_3$ more useful for photocatalysis.

The synthesis of self-assembled Ir-doped $SrTiO_3$ nanopillar (Ir:$SrTiO_3$ nanocomposite) was reported with excellent visible light harvesting property (up to λ~700 nm) for enhanced PEC reaction [100]. The Ir-doping and formation of Ir/ $SrTiO_3$ junction reduced the bandgap energy and improved photocarrier transportation in the Schottky space charge regions, whereas the nanopillars of Ir:SrTiO3 provided the charge transfer pathways. The Ir:$SrTiO_3$ nanocomposite exhibited 80% photocarrier harvesting efficiency under visible light (400–600 nm). Surface oxygen vacancy creation is an effective technique to enhance the absorbance of a semiconductor. Surface oxygen defect-rich $SrTiO_3$ nanocrystals have exhibited better absorbance in the visible sunlight region due to the presence of mid-bandgap defect states originating from the oxygen vacancy [101]. The pristine $SrTiO_3$ nanocrystals showed an H_2 evolution rate of 130.4 µmol.h^{-1}, which increased to 202.8 µmol.h^{-1} after oxygen vacancy formation. Similarly, the first principle calculation by Wang *et al.* [102] has reported bandgap reduction of $SrTiO_3$ by mono or co-doping of Mo and N. The Mo and N codoped $SrTiO_3$ have been reported having the most negative conduction band minima with a bandgap of 2.02 eV, suitable for water splitting under solar light.

The coupling of $SrTiO_3$ with suitable semiconductors is an effective strategy to boost the PEC activity of $SrTiO_3$ nanostructures. $SrTiO_3$ is an ideal candidate for the formation of heterojunction with TiO_2 as its conduction band is 200 mV more negative than the conduction band of TiO_2 [48]. $SrTiO_3$ nanocrystallites deposited TiO_2 nanotubes exhibited an enhanced PEC activity with 6.7% IPCE compared to 3.6% IPCE of the pristine TiO_2 NTs arrays [48]. Similarly, a 10 nm thick $SrTiO_3$ shell layer over TiO_2 NWs yielded a 1.43 mA.cm^2 current density at 1.23 V vs. RHE with the 87.7% charge separation efficiency, which was found to be 83% and 79% higher than pristine TiO_2 NWs, respectively [103]. This study further reported the effect of ferroelectric polarization on the PEC activity of the $TiO_2/SrTiO_3$ core/shell NWs photoanode. It was found that the favorable polarization direction amplifies band bending at $TiO_2/SrTiO_3$ interface, leading to enhanced photocarrier separation boosting overall photocurrent density. However, an unfavorable polarized electric field was found to reduce the depletion width at $TiO_2/SrTiO_3$ interface and hence suppressed the photocarrier generation [103]. The charge separation and transportation ability of 1D $TiO_2/SrTiO_3$ nano-heterojunction photoanode were further improved by doping [49]. The Cr^{3+}/Ti^{3+} dual doping in $TiO_2/SrTiO_3$ ($Cr:TiO_{2-x}/Cr:SrTiO_{3-x}$) enhanced the charge injection efficiency to 88% at 0.6 V vs. SCE over the TiO_2 NTs with charge injection efficiency of 54% only [49]. The $Cr:TiO_{2-x}/Cr:SrTiO_{3-x}$ NTs photoanode exhibited a photocurrent density of 4.05 mA.cm^{-2} at 0.6 V vs. SCE, which was 100 times higher than that of pristine TiO_2. The nanohybrid photoanode also generated a significantly larger amount of H_2 (139.26 µmol of at 0.6 V vs. SCE in 2 hours) under visible light than the pristine TiO_2 NTs (1.69 µmol). Similarly, the formation of novel 0D/2D $SrTiO_3$ nanoparticle/$SnNb_2O_6$ nanosheet heterojunction photoelectrode was also reported [104], where the 20 wt % $SrTiO_3$ NPs loaded hydride generated the highest amount of H_2 evolution (17.16 µmol), which was 298 times higher than that of bare $SrTiO_3$. The formation of type-II heterojunction between $SrTiO_3/SnNb_2O_6$ enabled rapid photogenerated electron migration into the $SrTiO_3$ NPs, resulting in enhanced H_2 evolution reaction. It was suggested that an interfacial interaction between $SrTiO_3$ and $SnNb_2O_6$ could result in inefficient charge separation and enhanced H_2 generation activity, which was confirmed by photoelectrochemical analysis.

Recently, the synthesis of different morphology and exposed facets $SrTiO_3$ nanostructures was reported with enhanced PEC activity [105]. The $SrTiO_3$ nanostructures with large surface area exposed [001] facets and reduced photocarrier recombination were addressed to explain the enhanced Rhodamine B dye degradation and H_2 production efficiency of the nanostructures. Similarly, degradation of methylene blue dye was reported using the $Pt/SrTiO_3$ NPs photocatalysts [106]. The enhanced photocatalytic activity of the $Pt/SrTiO_3$ NPs was because of the charge separation between the heterojunction and the presence of highly active Pt on the surface.

9.7.5 Photo(electro)catalytic Activities of $YFeO_3$ Nanostructures

$YFeO_3$ is another ferroelectric perovskite oxide semiconductor with a narrow bandgap of 2.3–2.4 eV. Few reports are available on the photocatalytic activity of $YFeO_3$ nanostructures [66, 107]. The narrow bandgap energy and magnetically separatable (recyclable) behavior make $YFeO_3$ a potential ferroelectric photocatalyst [66, 107]. However, only a few works have been conducted on the $YFeO_3$ nanostructure-based photoelectrodes. The photoanodic behavior of $YFeO_3$ NPs assembled film with a wide spectrum response was first demonstrated by Guo et al. [58]. The n-type $YFeO_3$ NPs photoanode exhibited a photocurrent density of 0.025 mA.cm^{-2} at 1.23 V_{RHE} with an onset potential of 0.95 V under AM 1.5G solar light. This study also suggested that the carrier recombination due to trapping and discharging effects for the charge carriers resulted in the limited photocurrent of nanostructured $YFeO_3$ photoanode. García et al. synthesized sintered nanostructured (NPs) and compact $YFeO_3$ film photocathodes and showed the effect of grain boundaries and material disorder on the charge carrier dynamics during HER [59]. The $YFeO_3$ photocathode exhibited an onset potential around 1.05 V vs. RHE. However, the incident photon to current conversion efficiency remained low due to the limited carrier collection efficiency.

The magnetically recoverable nanocrystalline $YFeO_3$ was reported to exhibit enhanced visible-light-driven photocatalytic activity, which could efficiently degrade Rhodamine B dye under irradiation. Nearly 75% of the $YFeO_3$ photocatalyst was recovered by applying an external magnetic field due to its magnetic behavior, which did not require any conventional filtration step [66]. The hexagonal $YFeO_3$ synthesized by a complex-assisted sol-gel technique exhibited a lower optical band gap (1.81 eV) than the orthorhombic structure (2.1 eV) [107]. The hexagonal $YFeO_3$ caused a 57.5% degradation of MO dye under visible light compared to 33.1% of orthorhombic $YFeO_3$.

9.7.6 Photo(electro)catalytic Activities of $KNbO_3$ Nanostructures

$KNbO_3$ perovskite exhibits interesting physicochemical ferroelectric and piezocatalytic properties. Due to its chemical stability, abundance, low toxicity, inertness, and suitable band position for the water redox potential, $KNbO_3$ has gained considerable attention toward water splitting and photocatalysis. However, the wide bandgap (3.06–3.24 eV) of $KNbO_3$ limits

its photocatalytic activity within the UV region. Hence, the engineering of $KNbO_3$ nanostructures to enhance its photoactivity under visible light is a major researcher target.

Engineering of morphology and crystal structure of the material is a conventional and effective technique to enhance photocatalytic efficiency. Crystallinity and exposed facets play a major role in the charge carrier dynamics during the photocatalytic process. In an earlier study, Ding et al. [37] demonstrated that $KNbO_3$ NWs exhibited an enhanced photocatalytic water splitting rate (1.03 mmol. h^{-1}) compared to $KNbO_3$ nanocubes (0.42 mmol. h^{-1}) and powder (0.79 mmol h^{-1}) under the radiation of 400 W mercury lamp.

The tuning of ferroelectric polarization and the shape of $KNbO_3$ nanostructures also helped to control the photocarrier dynamics to boost overall PEC activity. Li et al. [36] reported the enhanced photocurrent density of epitaxially grown vertically aligned $KNbO_3$ NWs photoanodes from 0.7 to 11.5 $\mu A.cm^{-2}$ at 0 V vs. Ag/AgCl under AM 1.5G illumination through the manipulation of ferroelectric polarization. The polarization effect also shifted the onset potential from −0.32 to −0.46 V vs. Ag/AgCl in $KNbO_3$ NWs. Yu et al. [26] further studied the morphology-dependent piezo-photocatalytic activity of the $KNbO_3$ nanostructures. The 2D $KNbO_3$ nanosheets exhibited better piezo-photocatalytic organic dye degradation efficiency compared to the $KNbO_3$ nanocubes. Ferroelectric negative poling in 2D $KNbO_3$ nanosheets resulted in a 55% enhancement in photocurrent density compared to 25% enhancement of the $KNbO_3$ nanocubes structure (Figure 9.8a and b).

Doping is an effective technique to tune the bandgap of a semiconductor by pinning the dopant level near the bottom/top of the conduction/valance band. The first-principles calculations by Maarouf et al. [108] have shown that substantial doping of Ag and Mn at the Nb site effectively reduces the bandgap of $KNbO_3$ and consequently enhances its optical and photocatalytic water splitting and CO_2 gas reduction activity. A similar study by Modak et al. [109] demonstrated that N and W co-doping in $KNbO_3$ enhanced its spectral response by introducing impurity states near the top of the valence band and the bottom of the conduction band. However, both the dopants may have promoted electron-hole recombination. Therefore, individual doping was not effective in enhancing the photocatalytic activity of $KNbO_3$. However, the N and W co-doping made $KNbO_3$ suitable for total water splitting.

Zhang et al. [110] demonstrated that Ag NPs loading on $KNbO_3$ NWs promotes light absorbance through the LSPR effect. Due to the electron transfer from $KNbO_3$ NWs to Ag NPs, carrier recombination reduced

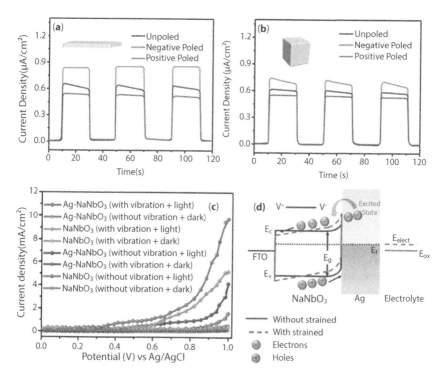

Figure 9.8 Transient photocurrent responses (at 0 V vs. Ag/AgCl) of the different polled $KNbO_3$ nanosheet (a) and nanotubes (b). Reproduced with permission [25]. Copyright 2019, Elsevier. (c) Current density –potential plots for the different $KNbO_3$/Ag nano-heterostructure photoanodes. (d) Schematic of the photocarrier dynamics of $KNbO_3$/Ag nano-heterostructure photoanode under poling conditions. Reproduced with permission [105]. Copyright 2021, Elsevier.

significantly, which resulted in excellent photodecomposition of organic pollutants under UV-visible light. The photocatalytic reaction rate of the optimal Ag NPs/$KNbO_3$ NWs for rhodamine B degradation under UV light was 13 times higher compared to $KNbO_3$ NWs. The anchoring of Ag NPs was also found to tune the bandgap energy of the monohybrid for light absorption. The formation of $KNbO_3$/α-Fe_2O_3 nanoheterostructures was also found to boost the photocatalytic dye degradation efficiency of $KNbO_3$ [38]. The $KNbO_3$/α-Fe_2O_3 nanoheterostructure exhibited 89% removal activity of methylene blue under visible light compared to the poor photocatalytic efficiency of $KNbO_3$ (11%). The improved photocatalytic activity of the $KNbO_3$/α-Fe_2O_3 nanoheterostructure was because of the enhanced light absorption and efficient photocarrier separation because of the formation of heterojunction.

9.7.7 Photo(electro)catalytic Activities of NaNbO$_3$ Nanostructures

NaNbO$_3$ is a ferroelectric perovskite, piezo-phototronic material [35]. NaNbO$_3$ is a promising alternative to replace the conventional water splitting photoanode materials like ZnO, TiO$_2$, etc., due to its low cost, nontoxicity to the environment, favorable band edge position, long term durability, and good photostability. The large bandgap (~3.4 eV) of NaNbO$_3$ limits its visible light absorption efficiency. However, the ferroelectric and piezoelectric control over the charge separation and transportation makes it an interesting photocatalytic material.

Ferroelectric polarization tuning the photoelectrochemical activity of NaNbO$_3$ NPs coated film type photoanode was demonstrated by Singh *et al.* [35]. A negative poling enhanced the photocurrent density from 0.31 to 0.51 mA.cm^{-2} at 1 V vs. Ag/AgCl due to the modification of band bending at the semiconductor/electrolyte interface. This led to efficient charge transfer efficiency for PEC reactions. Kumar *et al.* [111] have reported the plasmonic and piezo-phototronic controlled photocatalytic and photoelectrochemical water splitting activity of Ag-NaNbO$_3$ nanocomposite photoanode with enhanced photocurrent density (Figure 9.8c). In the Ag NPs coupled NaNbO$_3$ NRs photoanode, the LSPR effect of Ag NPs enhanced the photocarrier generation, whereas the built-in electric field in NaNbO$_3$ NRs because of the piezo-photoelectric effect boosted the carrier separation and transportation. The piezoelectric polarization also helped to develop a favorable band bending at the electrode/electrolyte interface, facilitating the rapid hole transfer for water oxidation (Figure 9.8d). The coupling of LSPR and piezo-photoelectric effects resulted in ten times efficient MB dye degradation efficiency and nine times enhanced photocurrent density in Ag-NaNbO$_3$ nanocomposite compared to pristine NaNbO$_3$ NRs. The efficient coupling between mechanical and electrical properties under light makes NaNbO$_3$ nanostructures an efficient photocatalyst [33]. The piezo-phototronic effect augmented the photocatalytic and photoelectrochemical properties of NaNbO$_3$ nanostructures. Over 115% enhancement in the photocatalytic degradation of methylene blue was recorded for NaNbO$_3$ nanostructures. The NaNbO$_3$ nanostructure photoanode also exhibited an 8% enhancement in photoconversion efficiency.

The coupling of 1D NaNbO$_3$ nanostructures with narrow bandgap materials like reduced graphene oxide (RGO) and g-C$_3$N$_4$ was reported to be very effective for enhancing the PEC water splitting efficiency of NaNbO$_3$ under visible light. Due to enhanced visible light absorption, type-II band formation, carrier separation, and transportation, g-C$_3$N$_4$/

NaNbO$_3$ nanofibers composite photoanode exhibited three-fold higher photocurrent density than pristine NaNbO$_3$ nanofibers. The optimal g-C$_3$N$_4$/NaNbO$_3$ nanocomposite photoanode generated a photocurrent density of 12.55 mA.cm^{-2} at 1 V vs. Ag/AgCl [112]. Similarly, the RGO/NaNbO$_3$ NRs nanocomposite exhibited over six times enhanced photocatalytic activity toward the degradation of MB compared to that of NaNbO$_3$ NRs [113]. The RGO/NaNbO$_3$ NRs photoanode also exhibited four times improved photocurrent density over the pristine NaNbO$_3$ NRs. Formation of ternary NaNbO$_3$/CdS/NiS$_2$ nanoheterostructure photocatalyst was also reported to boost the hydrogen production rate to 4.698 mmol. g^{-1} h^{-1} with enhanced long-time photostability [114]. The control on the microstructure of the catalyst and the formation of multiheterojunctions helped to achieve effective photocarrier separation for improved PEC water splitting.

9.7.8 Photo(electro)catalytic Activities of LiNbO$_3$ Nanostructures

LiNbO$_3$ is one of the most studied perovskite ferroelectric materials, widely used in linear and nonlinear optical applications like modulator and holographic memory [115]. Due to its stability and preferable band position, LiNbO$_3$ is also a suitable material for photocatalytic applications. Moreover, the ferroelectric polarization also provides additional opportunities to control the photocarrier recombination characteristics of LiNbO$_3$. However, only a few reports are available in the literature on the photocatalytic and PEC activity of the wide bandgap semiconductor (3.78 eV) LiNbO$_3$ nanostructures.

Saito et al. reported the morphology-controlled photocatalytic H$_2$ evolution activity of LiNbO$_3$ NWs [67]. The LiNbO$_3$ NWs having a highly exposed surface area of 28 m^2.g^{-1} exhibited a 47 μmol. h^{-1} H$_2$ evolution rate compared to its bulk counterpart, which had a hydrogen production rate of only 18 μmol.h^{-1} due to its considerably small surface area. The ferroelectric polarization was also used to tune the photocatalytic water splitting activity of LiNbO$_3$ single crystal. Fu et al. [116] reported that c+ and c− LiNbO$_3$ single crystal exhibited 47.3 and 153.6 μA.cm^{-2} photocurrent density at 1.23 V vs. RHE, respectively.

9.7.9 Photo(electro)catalytic Activities of PbTiO$_3$ Nanostructures

PbTiO$_3$ is an *n*-type perovskite semiconductor, widely studied because of its high ferroelectricity. The visible light active narrow bandgap (2.7 eV)

$PbTiO_3$ with advantageous conduction and valance band positions and ferroelectric polarization controlled carrier dynamics are advantageous for the photocatalytic and photoelectrochemical splitting of water. However, only a few studies have been conducted applying $PbTiO_3$ nanostructures as photocatalyst and photoelectrode for photo(electro)chemical water splitting.

The coupling of Pt catalyst on the {001} facets of $PbTiO_3$ nanoplates with positive or negative polarization surface changes was found to be effective for photocatalytic hydrogen production [44]. The selectively deposited Pt catalysts/$PbTiO_3$ nanoplates exhibited 9.4 times improved photocatalytic hydrogen production rate over the randomly deposited Pt catalysts/$PbTiO_3$ nanoplates. The study showed that the photogenerated electrons and holes accumulated on the positively charged and negatively charged $PbTiO_3$ {001} facet participate in water reduction and oxidation reactions, respectively. A recent study by Ahn et al. showed that the Pt-nanodot deposited $PbTiO_3$ NTs photoanode exhibited a photocurrent density of 64 $\mu A.cm^{-2}$ at 1.05 V vs. RHE, which was eight times higher than that of pristine $PbTiO_3$ NTs [71]. The Pt-dots/$PbTiO_3$ NTs also exhibit enhanced H_2/O_2 evolution rates, over seven times higher than that of $PbTiO_3$. Such enhanced water splitting performance of the Pt-dots/$PbTiO_3$ NTs was possible because of the effective separation of hole and electron reaction sites on the inner and outer surface of the $PbTiO_3$ NTs because of the Pt nanodots deposition.

Several studies indicate that nanocomposite formation with TiO_2 is also an efficient technique to enhance the photocatalytic activity of $PbTiO_3$ [117]. Especially, 1D nanostructured materials are advantageous for PEC activity. Jang et al. fabricated a vertically aligned one-dimensional, core/shell, $PbTiO_3/TiO_2$ NTs type heterojunction photoanode for PEC application [73]. The 1D structure provides a short hole diffusion length along the axial direction of the NTs. A type-II-like band alignment was formed between $PbTiO_3$ and TiO_2, promoting charge separation and transportation for PEC activity. As a result, $PbTiO_3/TiO_2$ NTs photoanode exhibited dramatically enhanced photocurrent density than pristine TiO_2 NTs. Yin et al. reported the fabrication of octahedral-shaped $PbTiO_3/TiO_2$ nanocomposites exhibiting a photocatalytic hydrogen evolution rate of 630.51 $\mu mol.h^{-1}$ under UV radiation, which is much higher compared to that of bare $PbTiO_3$ (17.49 $\mu mol.h^{-1}$) [118].

Transition metal doping has proved to be an effective approach to enhance the photocatalytic activity of $PbTiO_3$ thin films. Doping of transition metals changes the band position of $PbTiO_3$ thin films, which facilities the photoinduced carrier transportation and thereby boosts the photocatalytic and PEC performance. The 1 wt. % Cu-doped $PbTiO_3$ thin films exhibited a 2.5 times higher photocatalytic H_2 evolution activity under visible light ($\lambda > 400$ nm) compared to pristine $PbTiO_3$ thin films [119]. Similarly, Hu et al. [120] have shown that the Fe-doped $PbTiO_3$ thin films photoelectrode exhibited an enhanced photocurrent density of 220 $\mu A.cm^{-2}$ compared to pristine $PbTiO_3$ thin films (38 $\mu A.cm^{-2}$).

The nanocomposite formation through the coupling of $PbTiO_3$ with other low bandgap nanomaterials is also an effective strategy to boost the photocatalytic activity of $PbTiO_3$. For example, the rGO/$PbTiO_3$ demonstrated excellent photocatalytic H_2 evolution and rhodamine blue (RhB) degradation [121]. The $PbTiO_3$/carbon quantum dots nanohybrid was reported with excellent (99%) photocatalytic degradation efficiency for RhB dye under visible light illumination [72]. The enhanced photocatalytic activity of the nanohybrid is because of the improved visible light absorption and effective photocarrier separation because of heterojunction formation.

9.7.10 Photo(electro)catalytic Activities of $ZnSnO_3$ Nanostructures

$ZnSnO_3$ is another ferroelectric semiconductor of face-centered perovskite oxide class in which each ZnO_6 octahedron shares a face with SnO_6 octahedron. $ZnSnO_3$ is a lead-free piezocatalysts, belonging to the R3c space group, having several advantages like low cost and earth-abundant constituent elements and high chemical stability with a spontaneous polarization ~ 50.2 $\mu C.cm^{-2}$ along z-axis [122]. The noncentrosymmetric structure of $ZnSnO_3$ leads to strong piezoelectric polarization. High electron mobility, suitable band positions, biocompatibility, eco-friendly polymorphous, and tunable piezoelectric polarization make $ZnSnO_3$ a promising material for photocatalytic and photoelectrochemical water splitting applications. However, the wide bandgap (3.4 eV) and high recombination of photogenerated charge carriers are two major drawbacks of $ZnSnO_3$ for photocatalytic applications.

Hence, oxygen vacancy formation and tuning of ferro/piezo-polarization are found to be effective strategies to enhance the optical absorbance and promote carrier mobility in $ZnSnO_3$ nanostructure photocatalysts. Chang et al. [40] fabricated oxygen vacancy-rich $ZnSnO_3$ NWs photoanodes having a lower bandgap (~2.9 eV), as well as higher absorbance in the visible light region compared to pristine $ZnSnO_3$. As a result, the oxygen vacancy rich $ZnSnO_3$ NWs exhibited a high photocurrent density of 9.6 $mA.cm^{-2}$ at 0.3 V vs. Ag/AgCl, with a 4.8% applied bias to photoconversion efficiency (ABPE) compared to 3.2 mA.cm^{-2} photocurrent density and 1.5% ABPE of pristine ZnSnO$_3$ NWs. In another report, Wang et al. [25] demonstrated the piezocatalysis of oxygen vacancy enriched ZnSnO$_3$ NWs. The oxygen vacancy formation by H$_2$ annealing extended carrier lifetime approximately up to 8.3 ns. The piezoelectricity-induced photocatalysis in H$_2$ treated ZnSnO$_3$ NWs showed an H$_2$ production rate of 3453.1 μmol. g^{-1}h^{-1} without light radiation and a maximum RhB dye degradation efficiency of 92%. The mechanical force-induced polarization helped in easy photocarrier separation and drove the photogenerated electron and hole in opposite directions for PEC reactions. The oxygen vacancies acted as the exciton trapping centers, which helped electrons to reach the conduction band under illumination.

The piezophotocatalytic dye degradation activity of rhombohedral ZnSnO$_3$ NWs was studied by Lo et al. [41]. The ZnSnO$_3$ NWs with piezoelectric voltages, generated and tuned with applied stress (bending), exhibited substantially enhanced photodegradation efficiencies for methylene blue dyes, which was over 27% compared to the other samples (Figure 9.9a). The enhanced photocatalytic activity was attributed to the tuning of piezopotentials at the two ends of the ZnSnO$_3$ NWs, which led to favorable band bending for photocarrier separation and transfer, resulting in enhanced redox reaction kinetics (Figure 9.9b).

Fabrication of heterojunction with other semiconductors is another effective technique to improve the photocatalytic activity of ZnSnO$_3$ nanostructures. The ZnSnO$_3$/g-C$_3$N$_4$ nano-heterostructures fabricated assembling ZnSnO$_3$ NPs on the g-C$_3$N$_4$ nanosheets were reported to exhibit enhanced photocatalytic degradation of tetracycline [43]. The 10 wt.% ZnSnO$_3$ NPs/g-C$_3$N$_4$ photocatalyst showed almost 85% photocatalytic

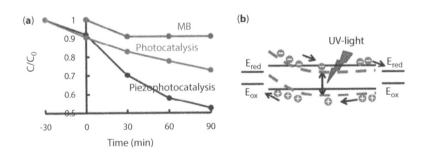

Figure 9.9 (a) The piezophotocatalytic methylene blue degradation activity of the ZnSnO$_3$ NWs. (b) The schematic of the piezopolarization induced band bending in ZnSnO$_3$ NWs, favoring the charge carrier separation. Reprinted with permission from [39]. Copyright (2015), American Chemical Society.

degradation efficiency compared to pristine $ZnSnO_3$ during 2 hours of visible light illumination. The favorable band alignment at the $ZnSnO_3$ NPs/g-C_3N_4 interface enables the rapid separation of photoinduced electrons from g-C_3N_4 to $ZnSnO_3$ NPs, leaving the long-lived holes in g-C_3N_4. Hence, the enhanced photocarrier lifetime boosted the photocatalytic activity of the nanoheterstructure. The synthesis of $ZnSnO_3$ hollow nanospheres/reduced graphene oxide (RGO) nano-heterostructure was reported by Dong et al. [68] with above 30% improvement of the photocatalytic activity of $ZnSnO_3$ hollow nanospheres due to RGO coupling. The $ZnSnO_3$ nanospheres/RGO nanocomposite exhibited enhanced photodegradation of metronidazole under visible light irradiation because of the enhanced light absorption and introduction of more active reaction sites duo to RGO anchoring. Guo et al. [123] fabricated core/shell type hollow $ZnSnO_3$/$ZnIn_2S_4$ nanoheterostructure between $ZnSnO_3$ and $ZnIn_2S_4$ (2.46 eV), which exhibited high photocatalytic water splitting efficiency with the H_2 evolution rate of 16340.18 µmol. $h^{-1}g^{-1}$ under visible light. The amount of H_2 gas was 2.3 times higher than that of pristine $ZnIn_2S_4$ (6939.28 µmol. $h^{-1}g^{-1}$). Meanwhile, pristine $ZnSnO_3$ cubes hardly showed any water-splitting activity under visible light.

The doping of suitable elements in $ZnSnO_3$ nanostructures was found to be a potential strategy to improve visible-light-driven photocatalysis because of enhanced light absorption due to the reduction of the bandgap energy. For example, the Sulfur doped $ZnSnO_3$ NPs exhibited nearly 90% photocatalytic degradation efficiency for rhodamine B (RhB) under visible light illumination [42]. The reason behind this was the reduced bandgap energy of $ZnSnO_3$ NPs from 3.7 eV to 2.4 eV because of S-doping, leading to enhanced visible light-harvesting efficiency.

9.8 Conclusion and Perspective

The direct conversion of renewable solar energy through several chemical methods in photo(electro)catalysis has recently earned growing interest. In the last decade, several novel materials and strategies have been innovated to improve the performance of photo(electro)catalyst materials. The development of various ferro/piezo-electric semiconductor photo(electro)catalysts provides an ideal opportunity for coupling the ferroelectric properties with the semiconductor properties for solar light harvesting. To date, the design of novel nanostructured ferro/piezo-electric semiconductor photo(electro)catalysts has been imposing, and significant advances have also been reported. The fundamental development of ferro/

piezo-electric semiconductor photo(electro)catalysts is centered around tuning the ferroelectric polarization or piezo potential in a favorable way to achieve suitable band bending at semiconductor/electrolyte interface and enhanced bulk charge separation and transfer kinetics for the PEC reactions. Therefore, the recent development of the synthesis of various nanostructured ferroelectric semiconductor photo(electro)catalysts with enhanced performance has been illustrated here. The nanoscale ferroelectric semiconductors are attractive because of several interesting features related to the nanoscale dimension like high surface-to-volume ratio, size, shape, and quantum confinement effects. Among the various methods applied to synthesize nanostructured ferroelectric semiconductors, the hydrothermal, sol–gel, and wet chemical methods have attracted intense focus as such methods are easy, cost effective and scalable.

Moreover, such methods are also applicable for the mass production of highly crystalline stoichiometric nanostructured photocatalysts. Indeed, the ferroelectric nanostructures have been impressive for photocatalytic and photoelectrocatalytic water splitting and the photocatalytic degradation of pollutants. There are also tremendous opportunities to tune the design and material of the photoelectrodes based on nanostructured ferro/piezo-electric semiconductors for harvesting solar energy. However, various challenges and issues are also there related to the development of nanostructured ferroelectric photocatalysts. The large surface area of nanostructures is advantageous in catalysis. However, the distribution and strength of the ferro/piezo-electric field might vary depending upon the surface area resulting in different catalytic activities. Hence, the nanoscale ferroelectric photocatalysts should be designed carefully.

Furthermore, in many of the reports, the photo(electro)catalysis activity of ferroelectric nanostructures is reported based on the advantages of nanoscale semiconductors, where the effect of ferroelectric polarization was not included or studied. Moreover, the design and understanding of 3D ferroelectric nanomaterials photocatalysts are not much focused till date. Similarly, other novel ferroelectric materials should be explored for photocatalysis. Recent works on the 2D nanostructured ferroelectrics have revealed enhanced ferroelectric effects; hence, such structures should be designed and studied extensively. The pyroelectric property-based nanostructured photocatalysis might be another area to investigate in detail in the future. Along with photocatalysts, the novel ferroelectric polarization-based electrocatalysts can also be designed and studied. Novel ferroelectric nanostructures can also be examined for other environmental and energy fields like the fixation of CO_2 and N_2 and the production of H_2O_2. Above all inherent mechanism of photocarrier generation, separation, and migration governed by the electric field in the ferroelectrics should be understood both experimentally and theoretically to design the efficient photo(electro)catalysts. It is evident that the coupling of ferroelectric nanostructures with other semiconductors or other ferroelectric materials provides an enormous opportunity to design high-performing photocatalysts. Therefore, the barrier potential at the interface, the space charge region, distribution of depletion layer width, interfacial band alignment modified by ferroelectric polarization, etc. should be investigated in detail to design the effective ferroelectric photo(electro)catalysts. Till date, different areas related to ferroelectric photocatalysts remain unexplored, where true attention should be paid in future research.

References

1. Wang, X., Song, J., Liu, J., Wang, Z.L., Direct-current nanogenerator driven by ultrasonic waves. *Science*, 316, 102, 2007.
2. Chen, S., Takata, T., Domen, K., Particulate photocatalysts for overall water splitting. *Nat. Rev. Mater.*, 2, 17050, 2017. https://doi.org/10.1038/natrevmats.2017.50.
3. Wang, Q. and Domen, K., Particulate photocatalysts for light-driven water splitting: Mechanisms, challenges, and design strategies. *Chem. Rev.*, 120, 919, 2020. https://doi.org/10.1021/acs.chemrev.9b00201.
4. Fujishima, A. and Honda, K., Electrochemical photolysis of water at a semiconductor electrode. *Nature*, 238, 37, 1972. https://doi.org/10.1038/238037a0.
5. Pan, L., Ai, M., Huang, C., Yin, L., Liu, X., Zhang, R., Wang, S., Jiang, Z., Zhang, X., Zou, J.-J., Mi, W., Manipulating spin polarization of titanium dioxide for efficient photocatalysis. *Nat. Commun.*, 11, 418, 2020. https://doi.org/10.1038/s41467-020-14333-w.
6. Pan, L., Sun, S., Chen, Y., Wang, P., Wang, J., Zhang, X., Zou, J., Wang, Z.L., Advances in piezo-phototronic effect enhanced photocatalysis and photoelectrocatalysis. *Adv. Energy Mater.*, 10, 2000214, 2020. https://doi.org/10.1002/aenm.202000214.
7. Wang, Z.L., Progress in piezotronics and piezo-phototronics. *Adv. Mater.*, 24, 4632, 2012. https://doi.org/10.1002/adma.201104365.
8. Morris, M.R., Pendlebury, S.R., Hong, J., Dunn, S., Durrant, J.R., Effect of internal electric fields on charge carrier dynamics in a ferroelectric material for solar energy conversion. *Adv. Mater.*, 28, 7123, 2016. https://doi.org/10.1002/adma.201601238.
9. Ma, W., Huang, K., Wu, X., Wang, M., Feng, S., Surface polarization enables high charge separation in TiO_2 nanorod photoanode. *Nano Res.*, 14, 4056, 2021. https://doi.org/10.1007/s12274-021-3340-0.

10. Guo, L., Zhong, C., Cao, J., Hao, Y., Lei, M., Bi, K., Sun, Q., Wang, Z.L., Enhanced photocatalytic H_2 evolution by plasmonic and piezotronic effects based on periodic Al/BaTiO$_3$ heterostructures. *Nano Energy*, 62, 513, 2019. https://doi.org/10.1016/j.nanoen.2019.05.067.
11. Liu, Z., Wang, L., Yu, X., Zhang, J., Yang, R., Zhang, X., Ji, Y., Wu, M., Deng, L., Li, L., Wang, Z.L., Piezoelectric-effect-enhanced full-spectrum photoelectrocatalysis in p–n heterojunction. *Adv. Funct. Mater.*, 29, 1807279, 2019. https://doi.org/10.1002/adfm.201807279.
12. Sarkar, A., Singh, A.K., Sarkar, D., Khan, G.G., Mandal, K., Three-dimensional nanoarchitecture of BiFeO$_3$ anchored TiO$_2$ nanotube arrays for electrochemical energy storage and solar energy conversion. *ACS Sustain. Chem. Eng.*, 3, 2254, 2015. https://doi.org/10.1021/acssuschemeng.5b00519.
13. Das, S., Fourmont, P., Benetti, D., Cloutier, S.G., Nechache, R., Wang, Z.M., Rosei, F., High performance BiFeO$_3$ ferroelectric nanostructured photocathodes. *J. Chem. Phys.*, 153, 084705, 2020. https://doi.org/10.1063/5.0013192.
14. Liang, L., Kang, X., Sang, Y., Liu, H., One-dimensional ferroelectric nanostructures: Synthesis, properties, and applications. *Adv. Sci.*, 3, 1500358, 2016. https://doi.org/10.1002/advs.201500358.
15. Haertling, G.H., Ferroelectric ceramics: History and technology. *J. Am. Ceram. Soc.*, 82, 797, 1999. https://doi.org/10.1111/j.1151-2916.1999.tb01840.x.
16. Martin, L.W. and Rappe, A.M., Thin-film ferroelectric materials and their applications. *Nat. Rev. Mater.*, 2, 16087, 2017. https://doi.org/10.1038/natrevmats.2016.87.
17. Maeda, K., Photocatalytic water splitting using semiconductor particles: History and recent developments. *J. Photochem. Photobiol. C Photochem. Rev.*, 12, 237, 2011. https://doi.org/10.1016/j.jphotochemrev.2011.07.001.
18. Hisatomi, T., Kubota, J., Domen, K., Recent advances in semiconductors for photocatalytic and photoelectrochemical water splitting. *Chem. Soc. Rev.*, 43, 7520, 2014. https://doi.org/10.1039/C3CS60378D.
19. Khaselev, O. and Turner, J.A., A monolithic photovoltaic-photoelectrochemical device for hydrogen production via water splitting. *Science*, 280, 425, 1998. https://doi.org/10.1126/science.280.5362.425.
20. Ghosh, N.G., Sarkar, A., Zade, S.S., The type-II n-n inorganic/organic nanoheterojunction of Ti^{3+} self-doped TiO$_2$ nanorods and conjugated co-polymers for photoelectrochemical water splitting and photocatalytic dye degradation. *Chem. Eng. J.*, 407, 127227, 2021. https://doi.org/10.1016/j.cej.2020.127227.
21. Ibhadon, A. and Fitzpatrick, P., Heterogeneous photocatalysis: Recent advances and applications. *Catalysts*, 3, 189, 2013. https://doi.org/10.3390/catal3010189.
22. Chen, F., Huang, H., Guo, L., Zhang, Y., Ma, T., The Role of Polarization in Photocatalysis. *Angew. Chem. Int. Ed.*, 58, 10061, 2019. https://doi.org/10.1002/anie.201901361.

23. Feng, Y., Ling, L., Wang, Y., Xu, Z., Cao, F., Li, H., Bian, Z., Engineering spherical lead zirconate titanate to explore the essence of piezo-catalysis. *Nano Energy.*, 40, 481, 2017. https://doi.org/10.1016/j.nanoen.2017.08.058.
24. Liang, Z., Yan, C.-F., Rtimi, S., Bandara, J., Piezoelectric materials for catalytic/photocatalytic removal of pollutants: Recent advances and outlook. *Appl. Catal. B Environ.*, 241, 256, 2019. https://doi.org/10.1016/j.apcatb.2018.09.028.
25. Wang, Y. and Wu, J.M., Effect of controlled oxygen vacancy on H_2-production through the piezocatalysis and piezophototronics of ferroelectric R3C $ZnSnO_3$ nanowires. *Adv. Funct. Mater.*, 30, 1907619, 2020. https://doi.org/10.1002/adfm.201907619.
26. Yu, D., Liu, Z., Zhang, J., Li, S., Zhao, Z., Zhu, L., Liu, W., Lin, Y., Liu, H., Zhang, Z., Enhanced catalytic performance by multi-field coupling in $KNbO_3$ nanostructures: Piezo-photocatalytic and ferro-photoelectrochemical effects. *Nano Energy.*, 58, 695, 2019. https://doi.org/10.1016/j.nanoen.2019.01.095.
27. Lin, P., Pan, C., Wang, Z.L., Two-dimensional nanomaterials for novel piezotronics and piezophototronics. *Mater. Today Nano.*, 4, 17, 2018. https://doi.org/10.1016/j.mtnano.2018.11.006.
28. Peng, Y., Que, M., Tao, J., Wang, X., Lu, J., Hu, G., Wan, B., Xu, Q., Pan, C., Progress in piezotronic and piezo-phototronic effect of 2D materials. *2D Mater.*, 5, 042003, 2018. https://doi.org/10.1088/2053-1583/aadabb.
29. Li, S., Zhang, J., Kibria, M.G., Mi, Z., Chaker, M., Ma, D., Nechache, R., Rosei, F., Remarkably enhanced photocatalytic activity of laser ablated Au nanoparticle decorated $BiFeO_3$ nanowires under visible-light. *Chem. Commun.*, 49, 5856, 2013. https://doi.org/10.1039/c3cc40363g.
30. Joshi, U.A., Jang, J.S., Borse, P.H., Lee, J.S., Microwave synthesis of single-crystalline perovskite $BiFeO_3$ nanocubes for photoelectrode and photocatalytic applications. *Appl. Phys. Lett.*, 92, 242106, 2008. https://doi.org/10.1063/1.2946486.
31. Shi, H., Li, X., Wang, D., Yuan, Y., Zou, Z., Ye, J., $NaNbO_3$ nanostructures: Facile synthesis, characterization, and their photocatalytic properties. *Catal. Lett.*, 132, 205, 2009. https://doi.org/10.1007/s10562-009-0087-8.
32. Gu, Q., Zhu, K., Zhang, N., Sun, Q., Liu, P., Liu, J., Wang, J., Li, Z., Modified solvothermal strategy for straightforward synthesis of cubic $NaNbO_3$ nanowires with enhanced photocatalytic H_2 evolution. *J. Phys. Chem. C*, 119, 25956, 2015. https://doi.org/10.1021/acs.jpcc.5b08018.
33. Singh, S. and Khare, N., Coupling of piezoelectric, semiconducting and photoexcitation properties in $NaNbO_3$ nanostructures for controlling electrical transport: Realizing an efficient piezo-photoanode and piezo-photocatalyst. *Nano Energy*, 38, 335, 2017. https://doi.org/10.1016/j.nanoen.2017.05.029.
34. Liu, Q., Zhang, L., Chai, Y., Dai, W.-L., Facile fabrication and mechanism of single-crystal sodium niobate photocatalyst: Insight into the structure features influence on photocatalytic performance for H_2 evolution. *J. Phys. Chem. C*, 121, 25898, 2017. https://doi.org/10.1021/acs.jpcc.7b08819.

35. Singh, S. and Khare, N., Electrically tuned photoelectrochemical properties of ferroelectric nanostructure $NaNbO_3$ films. *Appl. Phys. Lett.*, 110, 152902, 2017. https://doi.org/10.1063/1.4980100.
36. Li, S., Zhang, J., Zhang, B.-P., Huang, W., Harnagea, C., Nechache, R., Zhu, L., Zhang, S., Lin, Y.-H., Ni, L., Sang, Y.-H., Liu, H., Rosei, F., Manipulation of charge transfer in vertically aligned epitaxial ferroelectric $KNbO_3$ nanowire array photoelectrodes. *Nano Energy*, 35, 92, 2017. https://doi.org/10.1016/j.nanoen.2017.03.033.
37. Ding, Q.-P., Yuan, Y.-P., Xiong, X., Li, R.-P., Huang, H.-B., Li, Z.-S., Yu, T., Zou, Z.-G., Yang, S.-G., Enhanced photocatalytic water splitting properties of $KNbO_3$ nanowires synthesized through hydrothermal method. *J. Phys. Chem. C*, 112, 18846, 2008. https://doi.org/10.1021/jp8042768.
38. Farooq, U., Chaudhary, P., Ingole, P.P., Kalam, A., Ahmad, T., Development of cuboidal $KNbO_3@α-Fe_2O_3$ hybrid nanostructures for improved photocatalytic and photoelectrocatalytic applications. *ACS Omega*, 5, 20491, 2020. https://doi.org/10.1021/acsomega.0c02646.
39. Zhang, J., Huang, Y., Jin, L., Rosei, F., Vetrone, F., Claverie, J.P., Efficient upconverting multiferroic core@shell photocatalysts: Visible-to-near-infrared photon harvesting. *ACS Appl. Mater. Interfaces*, 9, 8142, 2017. https://doi.org/10.1021/acsami.7b00158.
40. Chang, Y.T., Wang, Y.-C., Lai, S.-N., Su, C.-W., Leu, C.-M., Wu, J.M., Performance of hydrogen evolution reaction of R3C ferroelectric $ZnSnO_3$ nanowires. *Nanotechnology*, 30, 455401, 2019. https://doi.org/10.1088/1361-6528/ab35f9.
41. Lo, M.-K., Lee, S.-Y., Chang, K.-S., Study of $ZnSnO_3$-nanowire piezophotocatalyst using two-step hydrothermal synthesis. *J. Phys. Chem. C*, 119, 5218, 2015. https://doi.org/10.1021/acs.jpcc.5b00282.
42. Guo, R., Tian, R., Shi, D., Li, H., Liu, H., S-doped $ZnSnO_3$ nanoparticles with narrow band gaps for photocatalytic wastewater treatment. *ACS Appl. Nano Mater.*, 2, 7755, 2019. https://doi.org/10.1021/acsanm.9b01804.
43. Huang, X., Guo, F., Li, M., Ren, H., Shi, Y., Chen, L., Hydrothermal synthesis of $ZnSnO_3$ nanoparticles decorated on $g-C_3N_4$ nanosheets for accelerated photocatalytic degradation of tetracycline under the visible-light irradiation. *Sep. Purif. Technol.*, 230, 115854, 2020. https://doi.org/10.1016/j.seppur.2019.115854.
44. Zhen, C., Yu, J.C., Liu, G., Cheng, H.-M., Selective deposition of redox co-catalyst(s) to improve the photocatalytic activity of single-domain ferroelectric $PbTiO_3$ nanoplates. *Chem. Commun.*, 50, 10416, 2014. https://doi.org/10.1039/C4CC04999C.
45. Li, J., Zhang, G., Han, S., Cao, J., Duan, L., Zeng, T., Enhanced solar absorption and visible-light photocatalytic and photoelectrochemical properties of aluminium-reduced $BaTiO_3$ nanoparticles. *Chem. Commun.*, 54, 723, 2018. https://doi.org/10.1039/C7CC07636C.

46. Yu, X., Wang, S., Zhang, X., Qi, A., Qiao, X., Liu, Z., Wu, M., Li, L., Wang, Z.L., Heterostructured nanorod array with piezophototronic and plasmonic effect for photodynamic bacteria killing and wound healing. *Nano Energy*, 46, 29, 2018. https://doi.org/10.1016/j.nanoen.2018.01.033.
47. Yang, W., Yu, Y., Starr, M.B., Yin, X., Li, Z., Kvit, A., Wang, S., Zhao, P., Wang, X., Ferroelectric polarization-enhanced photoelectrochemical water splitting in TiO_2–$BaTiO_3$ core–shell nanowire photoanodes. *Nano Lett.*, 15, 7574, 2015. https://doi.org/10.1021/acs.nanolett.5b03988.
48. Zhang, J., Bang, J.H., Tang, C., Kamat, P.V., Tailored TiO_2–$SrTiO_3$ heterostructure nanotube arrays for improved photoelectrochemical performance. *ACS Nano.*, 4, 387, 2010. https://doi.org/10.1021/nn901087c.
49. Cheng, X., Zhang, Y., Hu, H., Shang, M., Bi, Y., High-efficiency $SrTiO_3$/TiO_2 hetero-photoanode for visible-light water splitting by charge transport design and optical absorption management. *Nanoscale*, 10, 3644, 2018. https://doi.org/10.1039/C7NR09023D.
50. Zhang, C., Li, Y., Chu, M., Rong, N., Xiao, P., Zhang, Y., Hydrogen-treated $BiFeO_3$ nanoparticles with enhanced photoelectrochemical performance. *RSC Adv.*, 6, 24760, 2016. https://doi.org/10.1039/C5RA23699A.
51. Wang, Z., Cao, D., Wen, L., Xu, R., Obergfell, M., Mi, Y., Zhan, Z., Nasori, N., Demsar, J., Lei, Y., Manipulation of charge transfer and transport in plasmonic-ferroelectric hybrids for photoelectrochemical applications. *Nat. Commun.*, 7, 10348, 2016. https://doi.org/10.1038/ncomms10348.
52. Huang, H., Li, D., Lin, Q., Shao, Y., Chen, W., Hu, Y., Chen, Y., Fu, X., Efficient photocatalytic activity of PZT/TiO_2 heterojunction under visible light irradiation. *J. Phys. Chem. C*, 113, 14264, 2009. https://doi.org/10.1021/jp902330w.
53. Tijare, S.N., Joshi, M.V., Padole, P.S., Mangrulkar, P.A., Rayalu, S.S., Labhsetwar, N.K., Photocatalytic hydrogen generation through water splitting on nano-crystalline $LaFeO_3$ perovskite. *Int. J. Hydrog. Energy*, 37, 10451, 2012. https://doi.org/10.1016/j.ijhydene.2012.01.120.
54. Peng, Q., Shan, B., Wen, Y., Chen, R., Enhanced charge transport of $LaFeO_3$ via transition metal (Mn, Co, Cu) doping for visible light photoelectrochemical water oxidation. *Int. J. Hydrog. Energy*, 40, 15423, 2015. https://doi.org/10.1016/j.ijhydene.2015.09.072.
55. Celorrio, V., Bradley, K., Weber, O.J., Hall, S.R., Fermín, D.J., Photoelectrochemical properties of $LaFeO_3$ nanoparticles. *ChemElectroChem.*, 1, 1667, 2014. https://doi.org/10.1002/celc.201402192.
56. Shen, H., Xue, T., Wang, Y., Cao, G., Lu, Y., Fang, G., Photocatalytic property of perovskite $LaFeO_3$ synthesized by sol-gel process and vacuum microwave calcination. *Mater. Res. Bull.*, 84, 15, 2016. https://doi.org/10.1016/j.materresbull.2016.07.024.
57. Saito, K. and Kudo, A., Niobium-complex-based syntheses of sodium niobate nanowires possessing superior photocatalytic properties. *Inorg. Chem.*, 49, 2017, 2010. https://doi.org/10.1021/ic902107u.

58. Guo, Y., Zhang, N., Huang, H., Li, Z., Zou, Z., A novel wide-spectrum response hexagonal YFeO$_3$ photoanode for solar water splitting. *RSC Adv.*, 7, 18418, 2017. https://doi.org/10.1039/C6RA28390J.
59. Díez-García, M.I., Celorrio, V., Calvillo, L., Tiwari, D., Gómez, R., Fermín, D.J., YFeO$_3$ photocathodes for hydrogen evolution. *Electrochim. Acta*, 246, 365, 2017. https://doi.org/10.1016/j.electacta.2017.06.025.
60. Humayun, M., Sun, N., Raziq, F., Zhang, X., Yan, R., Li, Z., Qu, Y., Jing, L., Synthesis of ZnO/Bi-doped porous LaFeO$_3$ nanocomposites as highly efficient nano-photocatalysts dependent on the enhanced utilization of visible-light-excited electrons. *Appl. Catal. B Environ.*, 231, 23, 2018. https://doi.org/10.1016/j.apcatb.2018.02.060.
61. Long, X., Wang, T., Jin, J., Zhao, X., Ma, J., The enhanced water splitting activity of a ZnO-based photoanode by modification with self-doped lanthanum ferrite. *Nanoscale*, 13, 11215, 2021. https://doi.org/10.1039/D1NR02673A.
62. Long, X., Wang, C., Wei, S., Wang, T., Jin, J., Ma, J., Layered double hydroxide onto perovskite oxide-decorated ZnO nanorods for modulation of carrier transfer behavior in photoelectrochemical water oxidation. *ACS Appl. Mater. Interfaces*, 12, 2452, 2020. https://doi.org/10.1021/acsami.9b17965.
63. Liu, Y., Quiñonero, J., Yao, L., Da Costa, X., Mensi, M., Gómez, R., Sivula, K., Guijarro, N., Defect engineered nanostructured LaFeO$_3$ photoanodes for improved activity in solar water oxidation. *J. Mater. Chem. A*, 9, 2888, 2021. https://doi.org/10.1039/D0TA11541J.
64. Gupta, M.V.N.S., Baig, H., Reddy, K.S., Mallick, T.K., Pesala, B., Tahir, A.A., Photoelectrochemical water splitting using a concentrated solar flux-assisted LaFeO$_3$ photocathode. *ACS Appl. Energy Mater.*, 3, 9002, 2020. https://doi.org/10.1021/acsaem.0c01428.
65. Pawar, G.S. and Tahir, A.A., Unbiased spontaneous solar fuel production using stable LaFeO$_3$ photoelectrode. *Sci. Rep.*, 8, 3501, 2018. https://doi.org/10.1038/s41598-018-21821-z.
66. Tang, P., Chen, H., Cao, F., Pan, G., Magnetically recoverable and visible-light-driven nanocrystalline YFeO$_3$ photocatalysts. *Catal. Sci. Technol.*, 1, 1145, 2011. https://doi.org/10.1039/c1cy00199j.
67. Saito, K., Koga, K., Kudo, A., Lithium niobate nanowires for photocatalytic water splitting. *Dalt. Trans.*, 40, 3909, 2011. https://doi.org/10.1039/c0dt01844a.
68. Dong, S., Sun, J., Li, Y., Yu, C., Li, Y., Sun, J., ZnSnO$_3$ hollow nanospheres/reduced graphene oxide nanocomposites as high-performance photocatalysts for degradation of metronidazole. *Appl. Catal. B Environ.*, 144, 386, 2014. https://doi.org/10.1016/j.apcatb.2013.07.043.
69. Papadas, I., Christodoulides, J.A., Kioseoglou, G., Armatas, G.S., A high surface area ordered mesoporous BiFeO$_3$ semiconductor with efficient water oxidation activity. *J. Mater. Chem. A*, 3, 1587, 2015. https://doi.org/10.1039/C4TA05272B.

70. Liu, Q., Zhou, Y., You, L., Wang, J., Shen, M., Fang, L., Enhanced ferroelectric photoelectrochemical properties of polycrystalline $BiFeO_3$ film by decorating with Ag nanoparticles. *Appl. Phys. Lett.*, 108, 022902, 2016. https://doi.org/10.1063/1.4939747.
71. Ahn, C.W., Borse, P.H., Kim, J.H., Kim, J.Y., Jang, J.S., Cho, C.-R., Yoon, J.-H., Lee, B., Bae, J.-S., Kim, H.G., Lee, J.S., Effective charge separation in site-isolated Pt-nanodot deposited $PbTiO_3$ nanotube arrays for enhanced photoelectrochemical water splitting. *Appl. Catal. B Environ.*, 224, 804, 2018. https://doi.org/10.1016/j.apcatb.2017.11.002.
72. Kooshki, H., Sobhani-Nasab, A., Eghbali-Arani, M., Ahmadi, F., Ameri, V., Rahimi-Nasrabadi, M., Eco-friendly synthesis of $PbTiO_3$ nanoparticles and $PbTiO_3$/carbon quantum dots binary nano-hybrids for enhanced photocatalytic performance under visible light. *Sep. Purif. Technol.*, 211, 873, 2019. https://doi.org/10.1016/j.seppur.2018.10.057.
73. Jang, J.S., Ahn, C.W., Won, S.S., Kim, J.H., Choi, W., Lee, B.-S., Yoon, J.-H., Kim, H.G., Lee, J.S., Vertically aligned core–shell $PbTiO_3$@TiO_2 heterojunction nanotube array for photoelectrochemical and photocatalytic applications. *J. Phys. Chem. C.*, 121, 15063, 2017. https://doi.org/10.1021/acs.jpcc.7b03081.
74. Ji, W., Yao, K., Lim, Y.-F., Liang, Y.C., Suwardi, A., Epitaxial ferroelectric $BiFeO_3$ thin films for unassisted photocatalytic water splitting. *Appl. Phys. Lett.*, 103, 062901, 2013. https://doi.org/10.1063/1.4817907.
75. Lee, H., Joo, H., Yoon, C., Lee, J., Lee, H., Choi, J., Park, B., Choi, T., Ferroelectric $BiFeO_3$/TiO_2 nanotube heterostructures for enhanced photoelectrochemical performance. *Curr. Appl. Phys.*, 17, 679, 2017. https://doi.org/10.1016/j.cap.2017.02.015.
76. May, K.J., Fenning, D.P., Ming, T., Hong, W.T., Lee, D., Stoerzinger, K.A., Biegalski, M.D., Kolpak, A.M., Shao-Horn, Y., Thickness-dependent photoelectrochemical water splitting on ultrathin $LaFeO_3$ films grown on Nb:$SrTiO_3$. *J. Phys. Chem. Lett.*, 6, 977, 2015. https://doi.org/10.1021/acs.jpclett.5b00169.
77. Guigoz, V., Balan, L., Aboulaich, A., Schneider, R., Gries, T., Heterostructured thin $LaFeO_3$/g-C_3N_4 films for efficient photoelectrochemical hydrogen evolution. *Int. J. Hydrog. Energy.*, 45, 17468, 2020. https://doi.org/10.1016/j.ijhydene.2020.04.267.
78. Yu, Q., Meng, X., Wang, T., Li, P., Liu, L., Chang, K., Liu, G., Ye, J., A highly durable p-$LaFeO_3$/n-Fe_2O_3 photocell for effective water splitting under visible light. *Chem. Commun.*, 51, 3630, 2015. https://doi.org/10.1039/C4CC09240F.
79. Chakrabartty, J., Barba, D., Jin, L., Benetti, D., Rosei, F., Nechache, R., Photoelectrochemical properties of $BiMnO_3$ thin films and nanostructures. *J. Power Sources.*, 365, 162, 2017. https://doi.org/10.1016/j.jpowsour.2017.08.064.

80. Kim, S., Nguyen, N., Bark, C., Ferroelectric materials: A novel pathway for efficient solar water splitting. *Appl. Sci.*, 8, 1526, 2018. https://doi.org/10.3390/app8091526. https://doi.org/10.3390/app8091526.
81. Yilmaz, P., Yeo, D., Chang, H., Loh, L., Dunn, S., Perovskite $BiFeO_3$ thin film photocathode performance with visible light activity. *Nanotechnology*, 27, 345402, 2016. https://doi.org/10.1088/0957-4484/27/34/345402.
82. Yi, H.T., Choi, T., Choi, S.G., Oh, Y.S., Cheong, S.-W., Mechanism of the switchable photovoltaic effect in ferroelectric $BiFeO_3$. *Adv. Mater.*, 23, 3403, 2011. https://doi.org/10.1002/adma.201100805.
83. Ren, Y., Nan, F., You, L., Zhou, Y., Wang, Y., Wang, J., Su, X., Shen, M., Fang, L., Enhanced photoelectrochemical performance in reduced graphene oxide/$BiFeO_3$ heterostructures. *Small.*, 13, 1603457, 2017. https://doi.org/10.1002/smll.201603457.
84. Chen, X.Y., Yu, T., Gao, F., Zhang, H.T., Liu, L.F., Wang, Y.M., Li, Z.S., Zou, Z.G., Liu, J.-M., Application of weak ferromagnetic $BiFeO_3$ films as the photoelectrode material under visible-light irradiation. *Appl. Phys. Lett.*, 91, 022114, 2007. https://doi.org/10.1063/1.2757132.
85. Gu, S., Zhou, X., Zheng, F., Fang, L., Dong, W., Shen, M., Improved photocathodic performance in Pt catalyzed ferroelectric $BiFeO_3$ films sandwiched by a porous carbon layer. *Chem. Commun.*, 53, 7052, 2017. https://doi.org/10.1039/C7CC03222F.
86. Shen, H., Zhou, X., Dong, W., Su, X., Fang, L., Wu, X., Shen, M., Dual role of TiO_2 buffer layer in Pt catalyzed $BiFeO_3$ photocathodes: Efficiency enhancement and surface protection. *Appl. Phys. Lett.*, 111, 123901, 2017. https://doi.org/10.1063/1.4999969.
87. Cheng, X., Shen, H., Dong, W., Zheng, F., Fang, L., Su, X., Shen, M., Nano-Au and ferroelectric polarization mediated Si/ITO/$BiFeO_3$ tandem photocathode for efficient H_2 production. *Adv. Mater. Interfaces*, 3, 1600485, 2016. https://doi.org/10.1002/admi.201600485.
88. Cao, D., Wang, Z., Nasori, N., Wen, L., Mi, Y., Lei, Y., Switchable charge-transfer in the photoelectrochemical energy-conversion process of ferroelectric $BiFeO_3$ photoelectrodes. *Angew. Chem. Int. Ed.*, 53, 11027, 2014. https://doi.org/10.1002/anie.201406044.
89. Comes, R. and Chambers, S., Interface structure, band alignment, and built-in potentials at $LaFeO_3$/n–$SrTiO_3$ heterojunctions. *Phys. Rev. Lett.*, 117, 226802, 2016. https://doi.org/10.1103/PhysRevLett.117.226802.
90. Mizusaki, J., Sasamoto, T., Cannon, W.R., Bowen, H.K., Electronic conductivity, seebeck coefficient, and defect structure of $La_{1-x}Sr_xFeO_3$ (x=0.1, 0.25). *J. Am. Ceram. Soc.*, 66, 247, 1983. https://doi.org/10.1111/j.1151-2916.1983.tb15707.x.
91. Wheeler, G.P. and Choi, K.-S., Photoelectrochemical properties and stability of nanoporous p-type $LaFeO_3$ photoelectrodes prepared by electrodeposition. *ACS Energy Lett.*, 2, 2378, 2017. https://doi.org/10.1021/acsenergylett.7b00642.

92. Wul, B. and Goldman, J.M., Ferroelectric switching in BaTiO$_3$ ceramics. *C.R. Acad. Sci. URSS*, 51, 21, 1946.
93. Kennedy, J.H. and Frese, K.W., Photo-oxidation of water at barium titanate electrodes. *J. Electrochem. Soc.*, 123, 1683, 1976. https://doi.org/10.1149/1.2132667.
94. Upadhyay, S., Shrivastava, J., Solanki, A., Choudhary, S., Sharma, V., Kumar, P., Singh, N., Satsangi, V.R., Shrivastav, R., Waghmare, U.V., Dass, S., Enhanced photoelectrochemical response of BaTiO$_3$ with Fe doping: Experiments and first-principles analysis. *J. Phys. Chem. C*, 115, 24373, 2011. https://doi.org/10.1021/jp202863a.
95. Huang, H.-C., Yang, C.-L., Wang, M.-S., Ma, X.-G., Chalcogens doped BaTiO$_3$ for visible light photocatalytic hydrogen production from water splitting. *Spectrochim. Acta A Mol. Biomol. Spectrosc.*, 208, 65, 2019. https://doi.org/10.1016/j.saa.2018.09.048.
96. Senthilkumar, P., Jency, D.A., Kavinkumar, T., Dhayanithi, D., Dhanuskodi, S., Umadevi, M., Manivannan, S., Giridharan, N.V., Thiagarajan, V., Sriramkumar, M., Jothivenkatachalam, K., Built-in electric field assisted photocatalytic dye degradation and photoelectrochemical water splitting of ferroelectric Ce doped BaTiO$_3$ nanoassemblies. *ACS Sustain. Chem. Eng.*, 7, 12032, 2019. https://doi.org/10.1021/acssuschemeng.9b00679.
97. Li, C., Fang, T., Hu, H., Wang, Y., Liu, X., Zhou, S., Fu, J., Wang, W., Synthesis and enhanced bias-free photoelectrochemical water-splitting activity of ferroelectric BaTiO$_3$/Cu$_2$O heterostructures under solar light irradiation. *Ceram. Int.*, 47, 11379, 2021. https://doi.org/10.1016/j.ceramint.2020.12.264.
98. Wysmulek, K., Sar, J., Osewski, P., Orlinski, K., Kolodziejak, K., Trenczek-Zajac, A., Radecka, M., Pawlak, D.A., A SrTiO$_3$-TiO$_2$ eutectic composite as a stable photoanode material for photoelectrochemical hydrogen production. *Appl. Catal. B Environ.*, 206, 538, 2017. https://doi.org/10.1016/j.apcatb.2017.01.054.
99. Wrighton, M.S., Ellis, A.B., Wolczanski, P.T., Morse, D.L., Abrahamson, H.B., Ginley, D.S., Strontium titanate photoelectrodes. Efficient photoassisted electrolysis of water at zero applied potential. *J. Am. Chem. Soc.*, 98, 2774, 1976. https://doi.org/10.1021/ja00426a017.
100. Kawasaki, S., Takahashi, R., Yamamoto, T., Kobayashi, M., Kumigashira, H., Yoshinobu, J., Komori, F., Kudo, A., Lippmaa, M., Photoelectrochemical water splitting enhanced by self-assembled metal nanopillars embedded in an oxide semiconductor photoelectrode. *Nat. Commun.*, 7, 11818, 2016. https://doi.org/10.1038/ncomms11818.
101. Liu, J., Wei, Z., Shangguan, W., Enhanced photocatalytic water splitting with surface defective SrTiO$_3$ nanocrystals. *Front. Energy*, 15, 700, 2021. https://doi.org/10.1007/s11708-021-0735-2.
102. Wang, C., Qiu, H., Inoue, T., Yao, Q., Band gap engineering of SrTiO$_3$ for water splitting under visible light irradiation. *Int. J. Hydrog. Energy.*, 39, 12507, 2014. https://doi.org/10.1016/j.ijhydene.2014.06.059.

103. Wu, F., Yu, Y., Yang, H., German, L.N., Li, Z., Chen, J., Yang, W., Huang, L., Shi, W., Wang, L., Wang, X., Simultaneous enhancement of charge separation and hole transportation in a TiO_2–$SrTiO_3$ core–shell nanowire photoelectrochemical system. *Adv. Mater.*, 29, 1701432, 2017. https://doi.org/10.1002/adma.201701432.

104. Jin, Y., Jiang, D., Li, D., Xiao, P., Ma, X., Chen, M., $SrTiO_3$ nanoparticle/ $SnNb_2O_6$ nanosheet 0D/2D heterojunctions with enhanced interfacial charge separation and photocatalytic hydrogen evolution activity. *ACS Sustain. Chem. Eng.*, 5, 9749, 2017. https://doi.org/10.1021/acssuschemeng.7b01548.

105. Vijay, A. and Vaidya, S., Tuning the morphology and exposed facets of $SrTiO_3$ nanostructures for photocatalytic dye degradation and hydrogen evolution. *ACS Appl. Nano Mater.*, 4, 3406, 2021. https://doi.org/10.1021/acsanm.0c03160.

106. Olagunju, M.O., Poole, X., Blackwelder, P., Thomas, M.P., Guiton, B.S., Shukla, D., Cohn, J.L., Surnar, B., Dhar, S., Zahran, E.M., Bachas, L.G., Knecht, M.R., Size-controlled $SrTiO_3$ nanoparticles photodecorated with Pd cocatalysts for photocatalytic organic dye degradation. *ACS Appl. Nano Mater.*, 3, 4904, 2020. https://doi.org/10.1021/acsanm.0c01086.

107. Ismael, M., Elhaddad, E., Taffa, D., Wark, M., Synthesis of phase pure hexagonal $YFeO_3$ perovskite as efficient visible light active photocatalyst. *Catalysts*, 7, 326, 2017. https://doi.org/10.3390/catal7110326.

108. Maarouf, A.A., Gogova, D., Fadlallah, M.M., Metal-doped $KNbO_3$ for visible light photocatalytic water splitting: A first principles investigation. *Appl. Phys. Lett.*, 119, 063901, 2021. https://doi.org/10.1063/5.0058065.

109. Modak, B. and Ghosh, S.K., Improving $KNbO_3$ photocatalytic activity under visible light. *RSC Adv.*, 6, 9958, 2016. https://doi.org/10.1039/C5RA26079E.

110. Zhang, T., Lei, W., Liu, P., Rodriguez, J.A., Yu, J., Qi, Y., Liu, G., Liu, M., Organic pollutant photodecomposition by $Ag/KNbO_3$ nanocomposites: A combined experimental and theoretical study. *J. Phys. Chem. C*, 120, 2777, 2016. https://doi.org/10.1021/acs.jpcc.5b11297.

111. Kumar, D., Sharma, S., Khare, N., Piezo-phototronic and plasmonic effect coupled Ag-$NaNbO_3$ nanocomposite for enhanced photocatalytic and photoelectrochemical water splitting activity. *Renew. Energy.*, 163, 1569, 2021. https://doi.org/10.1016/j.renene.2020.09.132.

112. Kumar, D., Sharma, S., Khare, N., Enhanced photoelectrochemical performance of $NaNbO_3$ nanofiber photoanodes coupled with visible light active g-C_3N_4 nanosheets for water splitting. *Nanotechnology*, 31, 135402, 2020. https://doi.org/10.1088/1361-6528/ab59a1.

113. Kumar, D., Sharma, S., Khare, N., RGO nanosheets coupled $NaNbO_3$ nanorods based nanocomposite for enhanced photocatalytic and photoelectrochemical water splitting activity. *Adv. Powder Technol.*, 32, 4754, 2021. https://doi.org/10.1016/j.apt.2021.10.030.

114. Xu, J., Zhu, J., Niu, J., Chen, M., Yue, J., Efficient and stable photocatalytic hydrogen evolution activity of multi-heterojunction composite

photocatalysts: CdS and NiS_2 Co-modified $NaNbO_3$ nanocubes. *Front. Chem.*, 7, 880, 2020. https://doi.org/10.3389/fchem.2019.00880.
115. Gallo, K., Spatial wave dynamics in 2-D periodically poled $LiNbO_3$ waveguides. *IEEE J. Quantum Electron.*, 45, 1415, 2009. https://doi.org/10.1109/JQE.2009.2027447.
116. Xia, Y., Revisiting lattice thermal transport in PbTe: The crucial role of quartic anharmonicity. *Appl. Phys. Lett.*, 113, 073901, 2018. https://doi.org/10.1063/1.5040887.
117. Liu, G., Ma, L., Yin, L.-C., Wan, G., Zhu, H., Zhen, C., Yang, Y., Liang, Y., Tan, J., Cheng, H.-M., Selective chemical epitaxial growth of TiO_2 islands on ferroelectric $PbTiO_3$ crystals to boost photocatalytic activity. *Joule*, 2, 1095, 2018. https://doi.org/10.1016/j.joule.2018.03.006.
118. Yin, S., Liu, S., Yuan, Y., Guo, S., Ren, Z., Octahedral shaped $PbTiO_3$-TiO_2 nanocomposites for high-efficiency photocatalytic hydrogen production. *Nanomaterials*, 11, 2295, 2021. https://doi.org/10.3390/nano11092295.
119. Reddy, K.H. and Parida, K., Fabrication, characterization, and photoelectrochemical properties of Cu-doped $PbTiO_3$ and its hydrogen production activity. *ChemCatChem.*, 5, 3812, 2013. https://doi.org/10.1002/cctc.201300462.
120. Hu, Y., Dong, W., Zheng, F., Fang, L., Shen, M., Fe(III) doped and grafted $PbTiO_3$ film photocathode with enhanced photoactivity for hydrogen production. *Appl. Phys. Lett.*, 105, 082903, 2014. https://doi.org/10.1063/1.4894097.
121. Wang, C., Shan, L., Song, D., Xiao, Y., Suriyaprakash, J., Hydrothermal synthesis of rGO/$PbTiO_3$ photocatalyst and its photocatalytic H_2 evolution activity. *J. Nanomater.*, 2019, 1, 2019. https://doi.org/10.1155/2019/4869728.
122. Zhang, J., Yao, K.L., Liu, Z.L., Gao, G.Y., Sun, Z.Y., Fan, S.W., First-principles study of the ferroelectric and nonlinear optical properties of the $LiNbO_3$-type $ZnSnO_3$. *Phys. Chem. Chem. Phys.*, 12, 9197, 2010. https://doi.org/10.1039/b920065g.
123. Guo, F., Huang, X., Chen, Z., Shi, Y., Sun, H., Cheng, X., Shi, W., Chen, L., Formation of unique hollow $ZnSnO_3$@$ZnIn_2S_4$ core-shell heterojunction to boost visible-light-driven photocatalytic water splitting for hydrogen production. *J. Colloid Interface Sci.*, 602, 889, 2021. https://doi.org/10.1016/j.jcis.2021.06.074.

10

Solar-Driven H_2 Production in PVE Systems

Zaki N. Zahran[1,2]*, Yuta Tsubonouchi[1] and Masayuki Yagi[1†]

[1]Department of Materials Science and Technology, Faculty of Engineering, Niigata University, Niigata, Japan
[2]Faculty of Science, Tanta University, Tanta, Egypt

Abstract

Water splitting driven by solar energy is an important technology for carbon-neutral and sustainable hydrogen (H_2) production to replace the fossil fuels. This chapter focuses on photovoltaic-electrolyzer (PVE) systems, consisting of photovoltaic (PV) cells connected by wires with electrolyzers equipped with an anode and a cathode in an electrolyte solution as one of the most promising approaches for solar-driven water splitting. The principles and the mechanisms of solar-driven water splitting in the PVE systems are exposited to understand the performance of the PVE systems. Recent advances on the PVE systems in the last decade are reviewed with classification according to types of the PV cell materials to highlight the efficient PVE systems with insights into the important factors. Finally, a future prospective is mentioned to achieve practical use of PVE systems for solar-driven water splitting.

Keywords: Hydrogen production, solar-driven water splitting, photovoltaic cells, electrolyzers-solar hydrogen production - wired photovoltaic-electrolyzer

---

*Corresponding author: znzahran@eng.niigata-u.ac.jp
†Corresponding author: yagi@eng.niigata-u.ac.jp

10.1 Introduction

Recently, the inevitable depletion of fossil fuels including oil, coal, and natural gas and the huge emissions of global warming CO_2 gas in our social activity emergently demand to develop sustainable and carbon-neutral fuels to replace the fossil ones [1, 2]. H_2 is expected as one of the most promising carbon-neutral fuels because of its high weight energy density (the energy in 1 kg H_2 gas is about the same as that in 2.8 kg of gasoline) and zero CO_2 emission in use [3, 4]. However, about 90% of H_2 is produced worldwide from methane manufactured from fossil fuels with CO_2 emission [5, 6], and in some cases with a CO_2 capture and storage technology [7]. The innovative technology on carbon-neutral and sustainable H_2 production remains challenges in current science, technology, and industry [8], and much attention has been paid to solar-driven water splitting for H_2 production as such a technology.

$$2H_2O \xrightarrow{\text{Solar light}} O_2 + 2H_2 \qquad (10.1)$$

Solar-driven water splitting technology requires development of functional materials of photoactive materials such as photocatalysts [9–14], and solar cells [15–18] for capturing the solar light energy and efficient catalysts [14, 19–27], to promote the O_2 (OER) and H_2 (HER) evolution reactions. Massive research on developing these materials has been performed to construct efficient systems for solar-driven water splitting. However, it still remains challenges to improve and conflate these materials for realizing practical abundant material-based systems with sufficiently high *STH* and durability [28, 29].

In this chapter, at first, the different approaches are overviewed to design the solar-driven water splitting systems for H_2 production including the photocatalytic (PC), photoelectrochemical (PEC), photovoltaic-electrolyzer (PVE), and PVE-PEC hybrid systems. We will focus on PVE systems due to advantage of high *STH* values. Following the principle of designing PVE systems and improvements of functional materials for their essential components of the photovoltaic (PV) cells and electrolyzers, recent advances on the PVE systems in the last decade will be reviewed with classification according to types of the PV cell materials to highlight the efficient PVE systems with insights into the important factors. Finally, a future prospective will be mentioned to achieve practical use of PVE systems for solar-driven water splitting.

10.2 Approaches for H_2 Production *via* Solar-Driven Water Splitting

Several approaches to produce H_2 *via* solar-driven water splitting have been reported to be classified into four main systems: PC [30–34], PEC [35–37], PVE [38–42], and PVE-PEC [43, 44] hybrid systems (Figure 10.1). PC systems have advantages of being the simplest and lowest cost for potential scalable H_2 production because semiconductor photocatalyst powders for water splitting are dispersed in water under solar irradiation (Figure 10.1A). Upon solar irradiation, charge separation occurs on the surface of the semiconductor photocatalyst to form electrons and holes, which travel to the surface/water interface for OER and HER, commonly with help of cocatalysts, respectively. The photocatalysts need thermodynamically possible energy levels of valence band and conduction band edges for OER and HER, respectively with keeping their narrow band gap properties to adsorb visible light effectively. Otherwise, different photocatalysts for OER and HER are separately dispersed in solution containing redox mediators, which serve electron transfer from the OER photocatalyst to the HER photocatalysts [45]. Although several excellent PC systems [30–34] have been developed, attaining a maximum *STH* of 5% [31] so far, the PC systems still face several challenges in improving both the *STH* and durability which are the main targets in this technology for practical application. Moreover, the necessity of separation of H_2 and O_2 is a disadvantage in PC systems although a great progress has been achieved in gas separation by zeolite membranes [46].

PEC systems consist of either one or both of a photoanode and a photocathode in an electrolyte solution (Figure 10.1B). The photoanode and photocathode are fabricated from n- and p-type semiconductors, respectively. Upon solar irradiation, photogenerated carriers are separated by the space-charge field formed at the semiconductor surface/electrolyte junction and the minority carriers (holes and electrons for the photoanode and photocathode) travel to the interface for OER and HER, respectively, in the same manners as semiconductor photocatalyst powders [47–49]. There is no need for gas separation in PEC systems because the H_2 and O_2 produced are spatially separated at different compartments for the (photo)anode and (photo)cathode. *STH* values in the PEC systems are much higher than those in the PC systems, attaining the maximum value of 19% so far [37]. Theoretical studies predict that the maximum attainable *STH* in PEC systems is 23–32% [35, 50, 51]. Eventually, either PC or PEC systems require further development of innovative photoactive materials for capturing the solar light with longer wavelength and efficient cocatalysts to promote OER and HER on the material surface to improve *STH* and long-term stability.

PVE systems consist of PV cells coupled with OER and HER sites including three types reported so far: PV cells connected by wires with electrolyzers equipped with an anode and a cathode in an electrolyte solution [40, 52, 53] (Figure 10.1, C.1), wireless PV cells directly attaching an anode and a cathode on either side of the PV cell [36, 54, 55] (Figure 10.1, C.2), and PV cells directly attaching an anode or a cathode on the single side of the PV cell and the other connected with a wire (Figure 10.1, C.3). The hybrid systems of PEC assisted by PV cells [43, 44] are also reported (Figure 10.1, D). The PV cells require the open-circuit voltage (V_{oc}) more than the theoretical thermodynamic potential (1.23 V) plus the overpotential ($\eta_{overall}$) for water splitting including the overpotentials of OER (η_{O2}) and HER (η_{H2}) on the anode and the cathode respectively. In case of using PV cells with insufficient V_{oc}, multi-junction module cells are fabricated, or a few cells are connected in series to increase the total output voltage of PV cells in PVE systems. Development of prominent anode and cathode materials are also a key task to fabricate efficient electrolyzers for PVE systems, as it is for PV

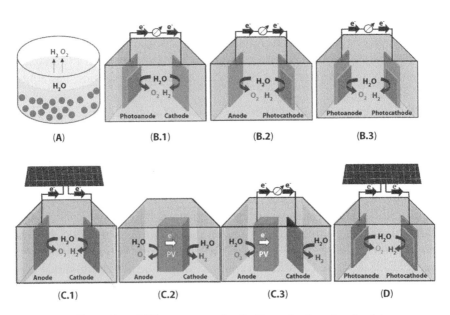

Figure 10.1 Illustration of different approaches for H_2 production via solar-driven water splitting. (A) Photocatalytic (PC), (B1-3) Photoelectrochemical (PEC), (C1-3) Photovoltaic-electrolyzer (PVE), and (D) PV-PEC hybrid systems. Detailed explanations of the figures are mentioned in the text.

Table 10.1 Summary of state-of-the-art photovoltaic-electrolyzer (PVE) systems for solar-driven water splitting.[a]

PV devices						Electrolyzer					Performances of solar water splitting						Ref.
Device	Junc.	No. of cells	V_{oc}/V	J_{sc}/mA cm^{-2}	STE (%)	Anode	Cathode	Electrolyte	No. of cells	Light Int. (Sun)	$(V_{op\,expect})$/V[b]	(ETH_{expect}) (%)[c]	(MF_{expect}) (%)[d]	J_{op}	STH (%)	Stab./h	
PVE based on Si PV cells																	
Amorphous-Si	–	4	2.75	8.26	15.0	NiFe/Ni	NiFe/Ni	1M KOH	1	1	1.75	70	90	7.7	9.5	25	[52]
Crystalline-Si	–	4	2.45	8.5	16.0	NiBi	NiMoZn	0.5 M KBi, pH 9.2	1	1	1.90	65	94	7.9	9.8	168	[40]
Crystalline-Si	–	3	1.67	13.3	14.4	Ni-Co-P/Ni-Co-S/NF	Ni-Co-P/Ni-Co-S/NF	1M NaOH	1		1.60	77	97	8.8	10.8	48	[53]
Crystalline-Si	–	3	2.08	11.9	20.6	IrO_x	Pt	PEM, H_2O	1	1	1.67	74	93	11.5	14.2	100	[104]
PVE based on group III-V compound PV cells																	
InGaP/GaAs/GaInNAs(Sb)	3	1	2.8	13.4	33.0	Pt	Pt	1M H_2SO_4	1	1	1.70	72	67	13.0	16	24	[107]
GaAs/GaAs	2	1	1.79	13.40	16.5	$FeNiWO_x$/NF	Pt/NF	1M KOH	1	1	1.5 (1.45)[e]	82 (85)[f]	~100 (99)[g]	11.3	13.9	72	[116]
GaInP/GaAs/Ge	3	1	2.6	1,770	38	NF	NF	1M NaOH	3	100	2.18	56	~100	18.2	22.4	24	[29]
GaInP/GaAs/Ge	3	3	3.0	13,200	31.2	Pt/C	Pt/C	PEM, H_2O	5	800	1.50	82	95	18.2	24.4	NR	[42]

(*Continued*)

Table 10.1 Summary of state-of-the-art photovoltaic-electrolyzer (PVE) systems for solar-driven water splitting.[a] (Continued)

PV devices						Electrolyzer				Performances of solar water splitting						Ref.
Device	Junc.	No. of cells	V_{OC}/V	J_{SC}/mA cm^{-2}	STE (%)	Anode	Cathode	Electrolyte	No. of cells	Light Int. (Sun)	$(V_{op})_{expect}$/V[b]	(ETH_{expect}) (%)[c]	(MF_{expect}) (%)[d]	J_{op}	STH (%)	Stab./h	
GaInP/GaInAs	2	1	2.3	14.5	21.0	Ir	Pt	PEM, H$_2$O	1	81.2	1.60	77	100	13.2	16.2[h]	2	[38]
GaInP/GaAs/GaInAsNSb	3	1	3.2	584	39.0	Ir	Pt	PEM, H$_2$O	2	42	2.90	42	183	24.4	30.0	48	[56]
AlGaAs/Si	2	1	1.57	23.6	21.2	RuO$_2$/Ti foil	Pt black/Pt mesh	1M HClO$_4$	1	1.4	1.40	88	98	14.9	18.3	14	[108]
PVE based on chalcogenide PV cells																	
(Ag, Cu)(In, Ga)Se$_2$	-	4	2.90	7.30	16.0	NiO/NF	NiMoV/NF	1M KOH	1	1	2.30	53	100	6.9	8.5	100	[121]
	-	3	2.20	9.50	14.5	NiO/NF	NiMoV/NF	1M KOH	1	1	1.63	75	92	8.1	10.0	NR	[121]
CIGSe	-	3	1.97	11.7	17.0	Pt/Pt foil	Pt/Pt foil	3M H$_2$SO$_4$	1	1	1.72	72	90	8.9	11.0	10 min.	[117]
(Ag, Cu)(In, Ga)Se$_2$	2	1	NR	NR	17.3	NiFeLDH	NiFeLDH	1M KOH	1	1	NR	NR	-	9.2	11.3	10	[133]
PVE based on perovskite PV cells																	
CH$_3$NH$_3$PbI$_3$	-	2	2.00	10.6	15.7	NiFe-LDH/NF	NiFe-LDH/NF	1M NaOH	1	1	1.53	80	98	10.0	12.3	2	[123]

(Continued)

Table 10.1 Summary of state-of-the-art photovoltaic-electrolyzer (PVE) systems for solar-driven water splitting.[a] (Continued)

PV devices						Electrolyzer				Performances of solar water splitting					Ref.
Device	Junc.	No. of cells	V_{OC}/ V	J_{SC}/mA cm^{-2}	STE (%)	No. of cells	Anode	Cathode	Electrolyte	Light Int. (Sun)	$(V_{op})_{expect}$/ V[b]	(ETH_{expect}) (%)[c]	(MF_{expect}) (%)[d]	J_{op}	STH (%)	Stab./h
$CH_3NH_3PbI_3$	-	2	2.29	12.3	19.8	1	NiFe	MoS_2	1M KOH	1	1.91	64	100	10.3	12.7	270s	[124]
PVE based on organic PV cells																	
PBDTTPD/ $PC_{71}BM$	2	1	1.84	6.54	8.35	1	NF	Pt	1M NaOH	1	1.50	82	89	4.9	6.1	NR	[125]
PF10TBT- $PC_{61}BM$/ PDPPTPT- $PC_{61}BM$/ PDPPTPT- $PC_{61}BM$	3	1	2.50	4.42	5.3	1	Pt	Pt	1M KOH	1	1.70	80	73	2.5	3.1	NR	[126]

[a] Junc.: Junctions; No. of cells: Number of cells; Light Int.: Light intensity; Stab.: Stability; NR: not reported; PBDTTPD: poly{benzo[1,2-b:4, 5-b'] dithiophene-thieno[3,4-c]pyrrole-4,6-dione}; $PC_{71}BM$: ([6,6]-phenyl-C_{71} butyric acid methyl ester) fullerenes; PF10TBT: (poly[2,7-(9, 9-didecylfluorene)-alt-5,5-(4',7'-di-2-thienyl-2', 1',3'-benzothiadiazole)]; $PC_{61}BM$: [6,6]-phenyl-C_{61} butyric acid methyl ester; PDPPTPT: poly{[2, 5-bis(2-hexyldecyl)-2,3,5,6-tetrahydro3,6-dioxopyrrolo[3,4-c]pyrrole-1,4-diyl]-alt-[2,2'-(1,4-phenylene) bisthiophene]-5,5'-diyl}]; LDH: layered double hydroxide; PEM: polymer electrolyte membrane; MEA: membrane electrode assembly. [b] Expected operating voltages (($V_{op})_{expect}$ / V) between the anode and cathode were provided from the intersection between J-V curves for PV cells and electrolyzers. [c] Expected ETH (($ETH)_{expect}$) calculated according to $(ETH)_{expect} = 1.23 / (V_{op})_{expect}$. [d] Expected MF (MF)$_{expect}$ calculated from (ETH) and measured STH according to eq 10.3. [e] Measured operating voltage (V_{op}); [f] Measured ETH provided from V_{op} according to $ETH = 1.23 / V_{op}$. [g] Measured MF provided according to eq 10.8. [h] STH of 18% based on V_{th} = 1.45 V was reported and recalculated as STH of 16.2% based on V_{th} = 1.23 V.

and PEC systems. Nevertheless, PVE systems have many advantages for solar-driven water splitting, compared with PC and PEC systems involving obstacles inherent to a lack of efficient solar light absorber materials (the band gap must be < 2.0 eV), the corrosion of the photoactive materials during OER and HER processes, and the difficulty in thermodynamic matching of the band edge energies of photoactive materials for OER and HER [35]. The highest STH of 30% were yielded in the PVE system using an InGaP/GaAs/GaInNAsSb triple-junction PV cells connected with two sets of electrolyzers quipped with an Ir anode and a Pt cathode under concentrated light illumination (42 Sun) [56], so far (Table 10.1). Although the STH value of the PVE systems exceeds the requirements for industrial H_2 production via solar-driven water splitting (~ 10%) [57], the fabrication of PV cells is still relatively complicated and expensive, and the cost for H_2 product is far beyond that of H_2 produced from fossil fuels presently. The high STH values of the PVE systems allow us to consider that they are one of the most feasible systems for solar-driven water splitting in the near future. Henceforward, we focus on the PVE systems in which the PV cells connected by wires with electrolyzers (Figure 10.1, C.1).

10.3 Principle of Designing of PVE Systems for Solar-Driven Water Splitting

STH is defined as the ratio of between energies (E_g / J) of H_2 produced and solar input onto PV cells, as expressed by eq (10.2).

$$STH(\%) = \frac{V_{th} \times J_{op} \times FE_{H2} \times 100}{P_s} \qquad (10.2)$$

where V_{th}, J_{op}, FE_{H2} and P_s are the theoretical voltage (V) for H_2 production via water splitting, the operating current density (mA cm^{-2}) based on the solar irradiation area of PV cells (in case the solar irradiation area is smaller than the geometric areas of the anode and the cathode) and Faraday efficiency (%) and irradiated solar power (mW cm^{-2}). The V_{th} value adopts 1.23 or 1.48 V based on standard Gibbs energy ($\Delta G°_f$ = -237.2 kJ mol^{-1}) or standard enthalpy ($\Delta H°_f$ = -285.8 kJ mol^{-1}) of formation for H_2 at 25 °C, corresponding to 1.23 and 1.45 eV, respectively, the latter including the temperature-depending entropy factor of $T\Delta S°$ (48.6 kJ mol^{-1}), where T (K) and $\Delta S°$ are absolute temperature and standard entropy. J_{op} and FE_{H2}

are experimentally provided under the solar light irradiation power of P_s. On the other hand, the STH values of the PVE systems are represented by eq (10.3):

$$STH\ (\%) = STE \times ETH \times MF \times 100 \tag{10.3}$$

where STE, ETH and MF are the solar-to-electricity efficiency for PV cells, the electricity-to-hydrogen efficiency for the electrolyzers, and the matching factor in performance between the PV cells and the electrolyzers respectively. As for STE, the typical current density-voltage (J-V, green) and output power-voltage (P-V, violet) curves of PV cells on solar light irradiation with intensity P_s are shown in Figure 10.2. The J and V values to afford the maximum output power (P_{max}) value are defined as J_{max} and V_{max}, respectively. STE is represented by eq (10.4).

$$STE(\%) = \frac{V_{max} \times J_{max} \times 100}{P_s} \tag{10.4}$$

$$= \frac{V_{oc} \times J_{sc} \times FF \times 100}{P_s} \tag{10.5}$$

Eq (10.4) can be transformed to eq (10.5) including the open circuit voltage (V_{oc}) and the short circuit current density (J_{sc}) of PV cells if fill factors (FF) of the ratio of $V_{max} \times J_{max} / V_{oc} \times J_{sc}$ are introduced. Efficient PV

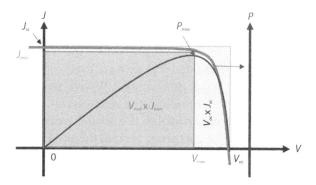

Figure 10.2 Current density-voltage (J-V, green) and output power-voltage (P-V, violet) curves for PV cells.

cells with high STE and sufficient V_{oc} values are essentially desired to be fabricated to achieve high STH in the PVE systems.

ETH is represented by eq (10.6).

$$ETH(\%) = \frac{V_{th} \times 100}{V_{op}} \tag{10.6}$$

where V_{op} is the operating voltage between the anode and the cathode in the electrolyzer during solar-driven water splitting. 1.23 or 1.48 V is employed for V_{th} based on $\Delta G°_f$ or $\Delta H°_f$ for H_2 at 25 °C in a similar manner to eq (10.2) for STE. Since in addition to V_{th}, the overall water splitting process requires an extra input voltage of $\eta_{overall}$ of the electrolyzer, which arises from the sluggish kinetics of OER and HER on the anode and the cathode, respectively, V_{op} is expressed by eq (10.7) using $\eta_{overall}$.

$$V_{op} = V_{th} + \eta_{overall} \tag{10.7}$$

The $\eta_{overall}$ value is needed to be minimized for high ETH values by developing efficient catalysts for OER and HER on the anode and the cathode, respectively.

MF is defined as the ratio of the operating output power ($V_{op} \times J_{op}$) in the PVE system versus the maximum output power ($V_{max} \times J_{max}$) of the PV cells, as expressed by eq (10.8):

$$MF = \frac{V_{op} \times J_{op}}{V_{max} \times J_{max}} \tag{10.8}$$

To have better understandability of MF, the typical relationship between the current-voltage (I-V) curves of PV cells (green) and of electrolyzers (red) are shown in in Figure 10.3, classified into the three types of Case A-C. The current (not current density) axes are used considering the different areas of solar light irradiation on the PV cells from the geometric areas of the anode and the cathode in the electrolyzers. The I values are converted to J values by being divided by the smaller area of solar light irradiation on the PV cells or the geometric areas of the anode and the cathode for the MF definition in eq 10.8 (in most case the solar irradiation area is smaller than the geometric areas of the anode and the cathode). V_{op} and J_{op} in the PEV systems can be expected as $(V_{op})_{expect}$ and $(J_{op})_{expect}$ from the intersection of I-V curves of the PV cells and the electrolyzers, respectively. In Case A with $(V_{op})_{expect} > V_{max}$, the operating output power in the PVE

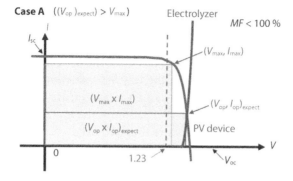

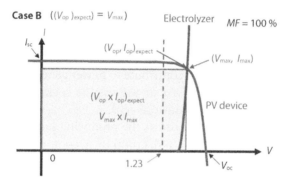

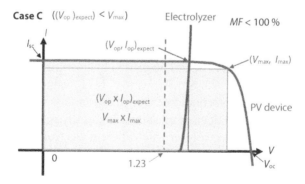

Figure 10.3 Illustration of the matching image between I-V curves of PV cells (green) and electrolyzers (red). Green and red squares represent the maximum output power ($V_{max} \times I_{max}$) of the PV cells and expected operating output power (($V_{op} \times I_{op})_{expect}$), respectively. Case A with $(V_{op})_{expect} > V_{max}$, MF < 100%; Case B with $(V_{op})_{expect} = V_{max}$, MF = 100%; Case C with $(V_{op})_{expect} < V_{max}$, MF < 100%. I values are converted to J values by being divided by the smaller area of solar light irradiation on the PV cells or the geometric areas of the anode and the cathode for $V_{op} \times J_{op}$ and $V_{max} \times J_{max}$ the MF definition in eq 10.8 (in most case the solar irradiation area is smaller than the geometric areas of the anode and the cathode).

system is lower than the maximum output power of the PV cells, expecting less *MF* values due to insufficient J_{op} compared with J_{max} of the PV cell. In Case B with $(V_{op})_{expect} = V_{max}$, the operating output power in the PVE systems matches the maximum output power of the PV cells, expecting ideal *MF* of 100%. In a similar manner, the less *MF* values are expected due to insufficient V_{op} compared with V_{max} of PV cells in Case C with $(V_{op})_{expect} < V_{max}$. To achieve high *STH* in the PVE systems, the matching task between the PV cells and the electrolyzers is thus required considering their *I-V* characters, not only developing efficient PV cells with high *STE* and sufficient V_{oc} and the efficient electrolyzers with high *ETH*.

10.4 Development of PVE Systems for Solar-Driven Water Splitting

Currently, PV cells appreciably contribute to power generation worldwide, [58] being progressively improved to address the social demand of high *STE* and reducing the cost of industrial cells.

Recent progress on PV cells shows the promising possibility of several kinds of PV cells based on Si [59–68], group III-V compound semiconductors [17, 69–73], chalcogenide [74–80], perovskite [81–83], and organic heterojunction [84, 85] for application to PVE systems for solar-driven water splitting. The single junction module of these PV cells can provide the *STE* values of ~28% [17] under one sun but with insufficient V_{oc} of 0.51-1.43V [17, 59–67, 69, 70, 74–79, 81, 82, 85–88], which is lower than the essential voltage of $V_{th} + \eta_{overall}$ for the water splitting for most case. The sufficient V_{oc} of 1.92 ~ 5.15 V can be gained by forming multi-junction of double-sextuple [68, 71–73, 83]. However, the I_{sc} values of the PV cells commonly decreases by formation of the junction.

Electrolyzers are vital devises to electrochemically split water to O_2 and H_2. Three types of the electrolyzer with different electrolyte media of alkaline solutions [90–93], proton exchange membranes (PEMs) [94–96], and alkaline anion exchange membranes (AEM) [89, 97] have been utilized for the PEV systems (Figure 10.4). Although the electrolyzers with alkaline electrlyte solutons [90–93] have the advantage of low cost, high stability, and availavility of non-precious metal catalysts [92, 98–100], relatively low current densities of 300-400 mA cm^{-2} gained at an applied potentials of ~1.7-2.4 V (60–90 °C) is an essential issue for alkaline electrolyzers [90–93]. The electrolyzers with PEM electrolytes operate at higher current densities (1–6 A cm^{-2} at ~2.0 V) than that of the alkaline electrolyzers [89–93, 95, 97]. However, the PEM electrolyzer is nessesary to use the catalysts

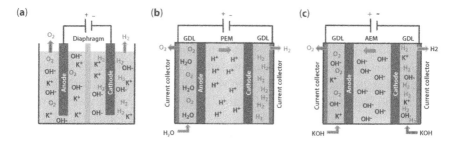

Figure 10.4 Schematic illustration of three types of the electrolyzers with different electrolyte media of (a) alkaline solutions, (b) proton exchange membranes (PEMs), and (c) alkaline anion exchange membranes (AEM). GDL expresses a gas diffusion layer.

with acid-resistance such as precious Ir and Ru-based materials [94–96]. Developing non-precious and active catalysts with acid-resistance and low cost PEM is a challenging task. The electrolyzers with AEM electrolytes can use non-precious metal-based catalysts [89, 97] and result in the highter current density of ~ 1.7 A cm^{-2} at 1.8 V in circulating of a 24 wt% KOH solution at 80 °C using nickel-based catalyst [89, 97], compared with those of alkaline electrolyzers. The current desities (~ 0.4–2.7 mA cm^{-2}) in pure water at 1.8 V have been reported to be lower that those of the PEM electrolyters [89, 101, 102].

The state-of-the art PEC systems consist of PV cells connected with electrolyzers by wires for solar-driven water splitting are classified by the used PV cells, as listed in Table 10.1. The performances of the PV cells and the electrolyzer are discussed in terms of *STE* and *ETH*, respectively to understand overall performance of the PVE systems.

10.4.1 PVE Systems Based on Si PV Cells

The PVE systems based on Si PV cells are good candidates toward the industrialization and deployment of cost effective solar-driven water splitting with long durability and scalability due to the commercial availability of Si PV cells. So far, the Si PV cells have been eagerly developed to reach a *STE* of 26.7% [59], and have demonstrated stabilities in excess of 25 years [103]; however the V_{oc} values of single (0.55-0.74 V) [59–67] or double (~1.34) [67] junction modules of Si cells are still lower than common values of $V_{th} + \eta_{overall}$ required for water splitting in the electrolyzers. Consequently, the state-of-the art PVE systems based on Si PV cells have to use at least 3 cells connected in series (Table 10.1). Several Si PV cells with STE of 14.4-20.6 % have been employed in PVE systems for the solar-driven water splitting, demonstrating STH of 9.5-14.2% [40, 52, 53, 104]. (Table 10.1). The STH value of 14.2% with the 100 h durability test has been reported using the three-interconnected crystalline Si PV cells (in series) with STE of 20.6% coupled with a PEM electrolyzer equipped with the IrO_x anode and the Pt cathode [104] (Table 10.1). The V_{op} value was not measured for the actual ETH and MF values. However, the expected ETH and MF ((ETH)$_{expect}$ and (MF)$_{expect}$) values of this system can be calculated to be 74 and 93%, respectively based on the expected J_{op} (J_{op})$_{expect}$) value given from the J-V characters of the PV cells and the electrolyzer. The STH value was calculated from the measured STE value and (ETH)$_{expect}$ and (MF)$_{expect}$ values according eq 10.3 to be 13.8%, which is nearly consistent with the measured STH of 14.2% from the J_{op} value according to eq 10.2, assuming FE_{H2} of 100%.

10.4.2 PVE Systems Based on Group III-V Compound PV Cells

Unlike Si PV cells with the predicted maximum STE of ~30% [58], very high STE values of 38.8% [71] and ~46% [105] were reported under either one sun or a concentrated solar irradiation (508 Sun), respectively for the multi-junction III–V group compound-based PV cells of GaInP/GaAs;GaInAsP/GaInAs. The theoretical studies predict the possibility of the STE values of ~57% for a triple-junction module and of ~62% for a quadruple or quintuple-junction module for III–V group compound-based PV cells [106]. These values indicate that there is significant room for further improvement in the performance of III–V group compound-based PV cells. A number of PVE systems based on III–V group compound PV cells have been reported to achieve STH values of 13.0-30.0 under one sun or concentrated solar irradiation conditions (Table 10.1) [29, 38, 42, 56, 107, 108].

A PVE system based on a triple-junction InGaP/GaAs/GaInNAs(Sb) PV cell with V_{oc} of 2.8 V and very high STE of 33% attained STH of 16% under one sun irradiation conditions by connecting with an electrolyzer equipped with the Pt plate anode and cathode in a 1M H_2SO_4 solution [107]. The V_{op} value in this system was not measured directly. However, the (ETH)$_{expect}$ value is calculated from the (V_{op})$_{expect}$ value to be 72%

being relatively low, implying that *STH* of this system could be further improved by employing the more efficient electrolyzer with a superior anode and a cathode. Recently, the FeNiWO$_x$ film was reported to work as a highly efficient and robust OER anode, attaining the very low overpotentials of η_{O2}^{10} = 167 mV (for 10 mA cm^{-2} current generation) with at least 100 h stability for OER in a 1.0 M KOH solution [19], which compares advantageously with the excellent performances of only a few state-of-the-art OER anodes with excellent η_{O2}^{10} < 200 mV [19, 27, 109–115]. A PVE system based on double-junction GaAs/GaAs of an efficient alkaline electrolyzer with a FeNiWO$_x$/Nickel form (NF) anode and a Pt/NF cathode was fabricated to attain the *STH* of 13.9% under one sun irradiation conditions [116]. The V_{op} value in this system was directly measured to be 1.45V, which is very low to provide the high *ETH* value of 85%. This demonstrates the importance of further development of electrolyzers for efficient PVE systems.

Excellent *STH* values of 22.4~30% were reported under the concentrated solar irradiation (1.4 ~ 800 suns) conditions through condenser lens in PVE systems with multi-electrolyzers (2 ~ 5 cells) in series [29, 42, 56]. As illustrated in Figure 10.5, a PVE system was fabricated using a triple-junction InGaP/GaAs/GaInNAsSb PV cell (very high *STE* of 39%) connected in series with two PEM electrolyzers with a Nafion membrane, an Ir black anode and a Pt black cathode [56]. The PV cell is a commercially available and composed of three subcells of InGaP (E_g = 1.895 eV), GaAs (E_g = 1.414 eV) and GaInNAs(Sb) (E_g = 0.965 eV). The PVE system achieved an average *STH* of 30%, which is the highest value reported to date among all PVE systems. The V_{op} value in this system was not measured directly. The $(ETH)_{expect}$ and $(MF)_{expect}$ values were calculated to be 42 and 183%, respectively from $(V_{op})_{expect}$ of 2.90 V given from the *J-V* characters of the PV cells and the electrolyzer. The $(MF)_{expect}$ value of 183% is aberrantly high, which might be caused by some complicated systems of concentrated solar light irradiation conditions. The direct measurement of V_{op} values is very important to understand the performance of PVE systems in terms of *ETH* and *MF* values. For the stability test, the performance of the PV cell did not change appreciably before and after the 48 h operation under concentrated 42 sun irradiation conditions, whereas the *I–V* curves of the dual electrolyzer changed significantly with the current (at 2.91 V) decreased by 10% from 177 to 160 mA. This shows that the very robust electrolyzer is desired for the practical PVE system.

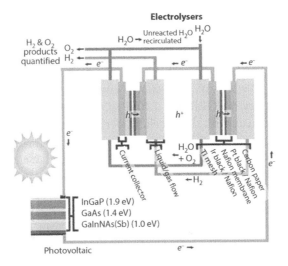

Figure 10.5 Schematic illustration of a PVE system fabricated using a triple-junction InGaP/GaAs/GaInNAsSb PV cell connected in series with two PEM electrolyzers with a Nafion membrane, an Ir black anode and a Pt black cathode. (Reproduced from reference [56] with open access.)

10.4.3 PVE Systems Based on Chalcogenide PV Cells

$CuIn_xGa_{1-x}Se_2$ (CIGS) is well-known as a classic chalcogenide PV cell with STE of 10.0-23.35 % [74–79]. CIGS has advantages of not only a high efficiency and low fabrication cost but also a tunable band gap (1.0 – 1.7 V) through changing the indium to gallium ratio, which is very beneficial in terms of designing multi-junctions and a series-interconnected cells with sufficient V_{oc} for water splitting. However, only a few PVE systems based on CIGS PV cell [117–121] are reported because of relatively low V_{oc} value (0.51-0.88 V) [74–79] for single cells. The series-interconnected three CIGS PV cells with V_{oc} of 1.97 V and STE of 17.0% attained STH of 11.0% by combining electrolyzes equipped with Pt/Pt foil anode and cathode in 3 M H_2SO_4 solution [117]. This means that the term of ETH x MF is 65% according to eq 10.3; however, the actual V_{op} value was not measured. From the intersection of the J-V curves of the PV cell and the electrolyzer, the $(V_{op})_{expect}$ value can be calculated to 1.72 V, corresponding to 72% of $(ETH)_{expect}$ calculated with V_{th} of 1.23 V according to eq 10.6. Consequently, the

MF value can be estimated as 90%, suggesting that there is still room for improvement in terms of *MF*.

A PVE systems based on three- or four-interconnected PV cells of (Ag, Cu)(In, Ga)Se$_2$ ((A)CIGS)) coupled with an alkaline electrolyzer equipped with a NiO/NF anode and a NiMoV/NF cathode have been reported [121], as shown in Figures 10.6a and b. The three- and four-interconnected PV cells (82 and 78 cm^2 irradiation area) provided the *I-V* curves with V_{oc} of 2.2 and 2.9 V and I_{sc} of 814 and 572 mA under one sun irradiation at 25 °C, respectively. The *I-V* curve of the electrolyzer exhibited the intersection with the *I-V* curve for the three-cell near its V_{max} of 1.8 V at 25 °C (compared with V_{max} of 2.3 V for the four-cell). The PVE system with the three-cell provided higher *STH* of 11% than that (9 %) with the four-cell at 25 °C due to low $(ETH)_{expect}$ value. The *STH* value with the three-cell decreased from 11 to 10% with temperate increase from 25 to 50 °C due to the *I-V* performance change with the V_{oc} decrease, whereas the *STH* value with the four-cell hardly changed as the intersections between the *I-V* curves the PV cell and the electrolyzer located on the plateau-current region of the *J-V* curve of the PV cell.

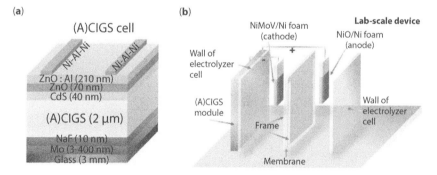

Figure 10.6 (a) Configuration of an (Ag,Cu)(In,Ga)Se2 ((A)CIGS)) PV cell. (b) Schematic illustration of PVE system fabricated by the (A)CIGS) PV cells connected in series with the electrolyzers equipped with a NiO/Ni foam (NF) anode and a NiMoV/NF cathode. (Reproduced from reference [121] with open access.)

10.4.4 PVE Systems Based on Perovskite PV Cells

Recently, much attention has been paid to perovskite PV cells, consisting of organic cations ($CH_3NH_3^+$), metal cations (Pb^{2+} or Sn^{2+}), and halides (I^-, Br^-, Cl^- or mixtures) due to their high *STE* and facile preparation [122]. The *STE* values of perovskite PV cells have dramatically jumped from 4% to 24.8%% [83], making these PV cells highly promising for practical use, although the toxicity of lead, instability, and high cost remain as challenges for its commercialization. The perovskite PV cells have commonly achieved V_{oc} values of ~ 0.9 V, which are lower than the essential voltage of $V_{th} + \eta_{overall}$ for the water splitting. However, the band gaps of the perovskites can be altered by changing the elements of metal cations and halides [122], and the PV cells based on $CsPbI_2Br$ and $CH_3NH_3PbBr_3$ perovskites was reported to achieve relatively high V_{oc} values of 1.43 [88] and 1.40 V [15].

A PVE system was fabricated by two-interconnected $CH_3NH_3PbI_3$ perovskite PV cells in series and placed side by side, which are coupled with an electrolyzer equipped with a bifunctional NiFe-layered double hydroxide (LDH) anode and cathode, as shown in Figure 10.7 [123]. The *J–V* curve of the PV cells under one sun irradiation conditions provided V_{oc} of 2.00 V, J_{sc} of 10.6 mA cm^{-2}, and *STE* of 15.7% (Table 10.1). The J_{op} value of 10.0 mA cm^{-2} corresponding to *STH* of 12.3% was observed under one

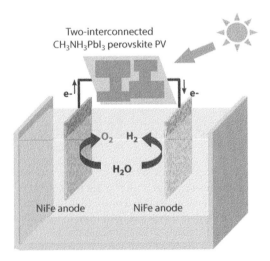

Figure 10.7 Schematic diagram of a PVE system fabricated by two-interconnected $CH_3NH_3PbI_3$ perovskite PV cells coupled with an electrolyzer equipped with a bifunctional NiFe-layered double hydroxide (LDH) anode and cathode.

sun irradiation conditions, be consistent with the $(J_{op})_{expect}$ value from the intersection of the J-V curves of the PV cells and the electrolyzer. Moreover, the $(V_{op})_{expect}$ value was calculated from the intersection to be ~ 1.53 V, corresponding to ETH of 80% with V_{th} of 1.23 V (Table 10.1). In this case, the $(MF)_{expect}$ value was calculated to be 98%, suggesting the better munching between the J-V curves of the PV cells and the electrolyzer. The stability of the PVE system was examined under AM 1.5G chopped light illumination, showing a 30 % decrease in the current density after 2 h, attributed to the instability of the perovskite PV cells. Another PVE system using an two-interconnected $CH_3NH_3PbI_3$ perovskite PV cells were also reported to provide STH of 12.7% using an electrolyzer with a NiFe anode and a MoS_2 cathode; however, this system was examined for only a short time of 270 s [124]. Improvement of stability of the PV cells using protecting techniques such as passivation and encapsulation is an important task for development of efficient PVE system based on perovskite PV cells.

10.4.5 PVE Systems Based on Organic Heterojunction PV Cells

Organic heterojunction PV cells have advantage of their lightweight, mechanical flexibility, and facile solution-processability that is beneficial for manufacture in large-area, high-throughput, and cost-efficient ways [125]. To date, neither single-junction nor tandem organic PV cells yield sufficient V_{oc} for essential voltage of $V_{th} + \eta_{overall}$ for the water splitting, and accordingly multi-junction module is required for the fabrication of the PVE systems based on organic heterojunction PV cells. A homo-tandem organic heterojunction PV cell was fabricated from two identical subcells of a blend of aPBDTTPD polymer (see note in Table 10.1) with a $PC_{71}BM$ fullerene derivative to afford V_{oc} of 1.84 V, J_{sc} of 6.54 mA cm^{-2} and STE of 8.35% (Figures 10.8a and b, and Table 10.1). For the PVE system using the homo-tandem organic heterojunction PV cell coupled with an electrolyzer equipped with a NF anode and a Pt cathode (Figure 10.8c), the solar-driven water splitting was demonstrated with STH of 6.1% under one sun irradiation conditions [125].

A triple junction module of organic heterojunction PV cell consisting of a high-bandgap subcell of a blend of a PF10TBT polymer (see note in Table 10.1) with a $PC_{61}BM$ fullerene derivative and the middle and back cells of a blend of a PDPPTPT polymer and $PC_{61}BM$ was fabricated to provide STE

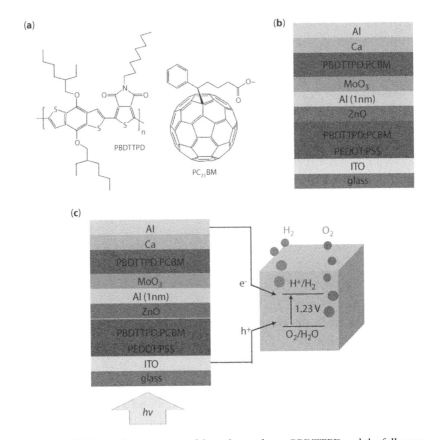

Figure 10.8 (a) Molecular structures of the polymer donor PBDTTPD and the fullerene acceptor PC71BM used in the fabrication of MJ polymer solar cells. (b) Schematic illustration of the construction of the homo-tandem polymer solar cell. (c) Schematic illustration of the PVE system fabricated from a homo-tandem polymer solar cell and electrolyzer.

of 5.3%. The PVE system based on this organic heterojunction PV cell was fabricated by connecting with an electrolyzer equipped with a Pt anode and a Pt cathode demonstrated solar-driven water splitting with STH of 3.1% [126]. $(ETH)_{expect}$ and $(MF)_{expect}$ were estimated to 80 and 73%, respectively from the $(V_{op})_{expect}$ of 1.70 V calculated from the intersection of the J-V curves of the PV cells and the electrolyzer. The STH value could be further improved if superior electrolyzers with efficient anodes and cathodes are employed. The long-term stability of the PVE system is still hindered by the instability issues of the polymer solar cells [125, 126].

10.5 Conclusions and Future Perspective

H_2 production *via* solar-driven water splitting is an indispensable technology for realizing a carbon-neutral and sustainable society in future. PVE systems consist of PV cells connected by wires with electrolyzers equipped with an anode and a cathode in an electrolyte solution are promising systems for solar-driven water splitting due to their advantages of higher *STE* values compared with those of other system of PC and PEC. The principle of designing efficient PVE systems were carefully discussed to suggest that *MF* in performance between the PV cells and the electrolyzers is important in addition to development of efficient PV cells with high *STE* and efficient electrolyzers with high *ETH*. The state-of-the-art photovoltaic-electrolyzer (PVE) systems for solar-driven water splitting were summarized with classification according to types of the PV cell materials in Table 10.1. This indicates possible applications of a various type of PV cells for efficient PVE systems. However, the conventional anodes and cathodes are employed in the electrolyzer except some new Ni-based electrodes [19, 20, 27, 115, 127], although a large variety of active and stable anodes and cathodes have been reported recently [128–132]. The further collaboration between researchers in the fields of PV cells and electrocatalysis is expected for improved PVE systems. For most PVE systems the J_{op} values were measured for acquirement of *STH* values, but the J_{op} values were hardly measured though they are necessary to acquire *ETH* and *MF* values with respect to the principle of designing efficient PVE systems. The consistency of the *ETH* and *MF* values with the $(MF)_{expect}$ and $(MF)_{expect}$ values calculated from the intersection between *J-V* curves of PV cells and electrolyzers ensures that the PVE systems work with propre coupling of PV cells and the electrolyzers. Moreover, the *ETH* and *MF* values tell us the points of improvement on the performance of PVE systems in addition to *STE* of PV cells. Advance of PV cells will provide higher *STE* than ever under one sun irradiation conditions. This will be of service for advance of PVE systems. Collaterally, further improvement of electrolyzers with active and robust anodes and cathodes prepared easily with low-cost is desired for efficient and practical PVE systems.

References

1. Dalle, K.E., Warnan, J., Leung, J.J., Reuillard, B., Karmel, I.S., Reisner, E., Electro- and solar-driven fuel synthesis with first row transition metal

complexes. *Chem. Rev.*, 119, 2752–2875, 2019. https://doi.org/10.1021/acs. chemrev.8b00392.
2. Zahran, Z.N., Tsubonouchi, Y., Mohamed, E.A., Yagi, M., Recent advances in development of molecular catalyst-based anodes for water oxidation toward artificial photosynthesis. *ChemSusChem*, 12, 1775–1793, 2019.
3. Turner, J.A., Sustainable hydrogen production. *Science*, 305, 972–974, 2004. https://doi.org/10.1126/science.1103197.
4. Meyer, T.J., The art of splitting water. *Nature*, 451, 778–779, 2008. https://doi.org/10.1038/emboj.2008.1.
5. Simpson, A.P. and Lutz, A.E., Exergy analysis of hydrogen production via steam methane reforming. *Int. J. Hydrogen Energy*, 32, 4811–4820, 2007. https://doi.org/10.1016/j.ijhydene.2007.08.025.
6. Qian, J.X., Chen, T.W., Enakonda, L.R., Liu, D.B., Basset, J.M., Zhou, L., Methane decomposition to pure hydrogen and carbon nano materials: State-of-the-art and future perspectives. *Int. J. Hydrogen Energy*, 45, 15721–15743, 2020. https://doi.org/10.1016/j.ijhydene.2020.04.100.
7. Bard, A.J. and Fox, M.A., Artificial photosynthesis: Solar splitting of water to hydrogen and oxygen water splitting. *Acc. Chem. Res.*, 28, 141–145, 1995. https://doi.org/10.1021/ar00051a007.
8. Dresselhaus, M.S. and Thomas, I.L., Alternative energy technologies. *Nature*, 332–337, 2001. https://doi.org/10.1038/35104599.
9. Mohamed, E.A., Zahran, Z.N., Naruta, Y., Simple preparation of highly active water splitting FTO/BiVO4 photoanode modified with tri-layers water oxidation catalysts. *J. Mater. Chem. A*, 5, 6825–6831, 2017. https://doi.org/10.1039/C7TA00156H.
10. Katsuki, T., Zahran, Z.N., Tanaka, K., Eo, T., Mohamed, E.A., Tsubonouchi, Y., Berber, M.R., Yagi, M., Facile fabrication of a highly crystalline and well-interconnected hematite nanoparticle photoanode for efficient visible-light-driven water oxidation. *ACS Appl. Mater. Interfaces*, 13, 39282–39290, 2021. https://doi.org/10.1021/acsami.1c08949.
11. Chandra, D., Li, D., Sato, T., Tanahashi, Y., Togashi, T., Ishizaki, M., Kurihara, M., Mohamed, E.A., Tsubonouchi, Y., Zahran, Z.N., et al., Characterization and mechanism of efficient visible-light-driven water oxidation on an *in situ* N2-intercalated WO3 nanorod photoanode. *ACS Sustain. Chem. Eng.*, 7, 17896–17906, 2019. https://doi.org/10.1021/acssuschemeng.9b04467.
12. Eo, T., Katsuki, T., Berber, M.R., Zahran, Z.N., Mohamed, E.A., Tsubonouchi, Y., Alenad, A.M., Althubiti, N.A., Yagi, M., Handy protocol of nitrogen-doped BiVO4 photoanode for visible light-driven water oxidation. *ACS Appl. Energy Mater.*, 4, 2983–2989, 2021. https://doi.org/10.1021/acsaem.1c00261.
13. Abdel Haleem, A., Majumder, S., Perumandla, N., Zahran, Z.N., Naruta, Y., Enhanced performance of pristine Ta3N5 photoanodes for solar water splitting by modification with Fe-Ni-Co mixed-metal oxide cocatalysts. *J. Phys. Chem. C*, 121, 20093–20100, 2017. https://doi.org/10.1021/acs.jpcc.7b04403.

14. Zahran, Z.N., Mohamed, E.A., Haleem, A.A., Naruta, Y., Efficient photoelectrochemical O2 and CO production using BiVO4 water oxidation photoanode and CO2 reduction Au nanoparticle cathode prepared by *in situ* deposition from Au3+ containing solution. *Adv. Sustain. Syst.*, 1, 1700111, 2017. https://doi.org/10.1002/adsu.201700111.
15. Ryu, S., Noh, J.H., Jeon, N.J., Chan Kim, Y., Yang, W.S., Seo, J., Seok, S., II. Voltage output of efficient perovskite solar cells with high open-circuit voltage and fill factor. *Energy Environ. Sci.*, 7, 8, 2614–2618, 2014. https://doi.org/10.1039/c4ee00762j.
16. Cotal, H., Fetzer, C., Boisvert, J., Kinsey, G., King, R., Hebert, P., Yoon, H., Karam, N., III-V multijunction solar cells for concentrating photovoltaics. *Energy Environ. Sci.*, 2, 174–192, 2009. https://doi.org/10.1039/b809257e.
17. Kayes, B.M., Nie, H., Twist, R., Spruytte, S.G., Reinhardt, F., Kizilyalli, I.C., Higashi, G.S., 27.6% conversion efficiency, a new record for single-junction solar cells under 1 sun illumination. *Conf. Rec. IEEE Photovolt. Spec. Conf.*, 000004–000008, 2011. https://doi.org/10.1109/PVSC.2011.6185831.
18. Hill, S.P. and Hanson, K., Harnessing molecular photon upconversion in a solar cell at sub-solar irradiance: Role of the redox mediator. *J. Am. Chem. Soc.*, 139, 10988–10991, 2017. https://doi.org/10.1021/jacs.7b05462.
19. Zahran, Z.N., Mohamed, E.A., Tsubonouchi, Y., Ishizaki, M., Togashi, T., Kurihara, M., Saito, K., Yui, T., Yagi, M., Concisely synthesized FeNiWox film as a highly efficient and robust catalyst for electrochemical water oxidation. *ACS Appl. Energy Mater.*, 4, 1410–1420, 2021. https://doi.org/10.1021/acsaem.0c02628.
20. Zahran, Z.N., Mohamed, E.A., Katsuki, T., Tsubonouchi, Y., Yagi, M., Nickel sulfate as an influential precursor of amorphous high-valent Ni(III) oxides for efficient water oxidation in preparation via a mixed metal-imidazole casting method. *ACS Appl. Energy Mater.*, 5, 1894–1904, 2022. https://doi.org/10.1021/acsaem.1c03379.
21. Mohamed, E.A., Zahran, Z.N., Naruta, Y.A., Flexible cofacial Fe porphyrin dimer as an extremely efficient and selective electrocatalyst for the CO2 to CO conversion in non-aqueous and aqueous media. *J. Mater. Chem. A*, 9, 18213–18221, 2021. https://doi.org/10.1039/d1ta04176b.
22. Mohamed, E.A., Zahran, Z.N., Tsubonouchi, Y., Saito, K., Yui, T., Yagi, M., Highly efficient and selective electrocatalytic CO_2-to-CO conversion by a non-heme iron complex with an in-plane N4 ligand. *ACS Appl. Energ. Mater.*, 3, 4114–4120, 2020. https://doi.org/10.1021/acsaem.9b02548.
23. Zahran, Z.N., Mohamed, E.A., Naruta, Y., Kinetics and mechanism of heterogeneous water oxidation by α-Mn_2O_3 sintered on an FTO electrode. *ACS Catal.*, 6, 4470–4476, 2016. https://doi.org/10.1021/acscatal.6b00413.
24. Zahran, Z.N., Mohamed, E.A., Naruta, Y., Bio-inspired cofacial Fe porphyrin dimers for efficient electrocatalytic CO_2 to CO conversion: Overpotential tuning by substituents at the porphyrin rings. *Sci. Rep.*, 6, 24533, 2016. https://doi.org/10.1038/srep24533.

25. Zahran, Z.N., Mohamed, E.A., Ohta, T., Naruta, Y., Electrocatalytic water oxidation by a highly active and robust α-Mn2O3 thin film sintered on a fluorine-doped tin oxide electrode. *ChemCatChem*, 8, 532–535, 2016. https://doi.org/10.1002/cctc.201501073.
26. Zahran, Z.N., Mohamed, E.A., Naruta, Y., Electrocatalytic water oxidation at low energy cost by a highly active and robust calcium–manganese oxide thin film sintered on an FTO electrode with ethyl methyl imidazolium triflate ionic liquid. *J. Mater. Chem. A*, 5, 15167–15174, 2017. https://doi.org/10.1039/c7ta03665e.
27. Zahran, Z.N., Mohamed, E.A., Tsubonouchi, Y., Ishizaki, M., Togashi, T., Kurihara, M., Saito, K., Yui, T., Yagi, M., Electrocatalytic water splitting with unprecedentedly low overpotentials by nickel sulfide nanowires stuffed into carbon nitride scabbards. *Energy Environ. Sci.*, 14, 5358–5365, 2021.
28. Pinaud, B.A., Benck, J.D., Seitz, L.C., Forman, A.J., Chen, Z., Deutsch, T.G., James, B.D., Baum, K.N., Baum, G.N., Ardo, S., *et al.*, Technical and economic feasibility of centralized facilities for solar hydrogen production via photocatalysis and photoelectrochemistry. *Energy Environ. Sci.*, 6, 1983–2002, 2013. https://doi.org/10.1039/c3ee40831k.
29. Bonke, S.A., Wiechen, M., Macfarlane, D.R., Spiccia, L., Renewable fuels from concentrated solar power: Towards practical artificial photosynthesis. *Energy Environ. Sci.*, 8, 2791–2796, 2015. https://doi.org/10.1039/x0xx00000x.
30. Liao, L., Zhang, Q., Su, Z., Zhao, Z., Wang, Y., Li, Y., Lu, X., Wei, D., Feng, G., Yu, Q., *et al.*, Efficient solar water-splitting using a nanocrystalline CoO photocatalyst. *Nat Nano*, 69–73, 2014. https://doi.org/10.1038/nnano.2013.272.
31. Wang, Q., Hisatomi, T., Jia, Q., Tokudome, H., Zhong, M., Wang, C., Pan, Z., Takata, T., Nakabayashi, M., Shibata, N., *et al.*, Scalable water splitting on particulate photocatalyst sheets with a solar-to-hydrogen energy conversion e ciency exceeding 1 %. *Nat. Mater.*, 15, 611–615, 2016. https://doi.org/10.1038/NMAT4589.
32. Gao, L., Li, Y., Ren, J., Wang, S., Wang, R., Fu, G., Hu, Y., Passivation of defect states in anatase tio$_2$ hollow spheres with mg doping: Realizing efficient photocatalytic overall water splitting. *Appl. Catal. B Environ.*, 202, 127–133, 2017. https://doi.org/10.1016/j.apcatb.2016.09.018.
33. Li, Y., Peng, Y.K., Hu, L., Zheng, J., Prabhakaran, D., Wu, S., Puchtler, T.J., Li, M., Wong, K.Y., Taylor, R.A., *et al.*, Photocatalytic water splitting by N-TiO$_2$ on MgO (111) with exceptional quantum efficiencies at elevated temperatures. *Nat. Commun.*, 10, 4421, 2019. https://doi.org/10.1038/s41467-019-12385-1.
34. Liu, J., Liu, Y., Liu, N., Han, Y., Zhang, X., Huang, H., Lifshitz, Y., Lee, S.T., Zhong, J., Kang, Z., Metal-free efficient photocatalyst for stable visible water splitting via a two-electron pathway. *Science*, 347, 970–974, 2015. https://doi.org/10.1126/science.aaa3145.
35. Khaselev, O. and Turner, J.A., A monolithic photovoltaic-photoelectrochemical device for hydrogen production via water splitting. *Science*, 280, 425–427, 1998. https://doi.org/10.1126/science.280.5362.425.

36. Reece, S.Y., Hamel, J.A., Sung, K., Jarvi, T.D., Esswein, A.J., Pijpers, J.J.H., Nocera, D.G., Wireless solar water splitting using silicon-based semiconductors and earth-abundant catalysts. *Science*, 334, 645–648, 2011. https://doi.org/10.1126/science.1209816.
37. Cheng, W.H., Richter, M.H., May, M.M., Ohlmann, J., Lackner, D., Dimroth, F., Hannappel, T., Atwater, H.A., Lewerenz, H.J., Monolithic photoelectrochemical device for direct water splitting with 19% efficiency. *ACS Energy Lett.*, 3, 1795–1800, 2018. https://doi.org/10.1021/acsenergylett.8b00920.
38. Peharz, G., Dimroth, F., Wittstadt, U., Solar hydrogen production by water splitting with a conversion efficiency of 18 %. *Int. J. Hydrogen Energy*, 32, 3248–3252, 2007. https://doi.org/10.1016/j.ijhydene.2007.04.036.
39. Kelly, N.A., Gibson, T.L., Ouwerkerk, D.B., A solar-powered, high-efficiency hydrogen fueling system using high-pressure electrolysis of water: Design and initial results. *Int. J. Hydrogen Energy*, 33, 2747–2764, 2008. https://doi.org/10.1016/j.ijhydene.2008.03.036.
40. Cox, C.R., Lee, J.Z., Nocera, D.G., Buonassisi, T., Ten-percent solar-to-fuel conversion with nonprecious materials. *Proc. Natl. Acad. Sci. U. S. A.*, 111, 14057–14061, 2014. https://doi.org/10.1073/pnas.1414290111.
41. Fujii, K., Nakamura, S., Sugiyama, M., Watanabe, K., Bagheri, B., Nakano, Y., Characteristics of hydrogen generation from water splitting by polymer electrolyte electrochemical cell directly connected with concentrated photovoltaic cell. *Int. J. Hydrogen Energy*, 38, 14424–14432, 2013. https://doi.org/10.1016/j.ijhydene.2013.07.010.
42. Nakamura, A., Ota, Y., Koike, K., Hidaka, Y., Nishioka, K., Sugiyama, M., Fujii, K., A 24.4% Solar to hydrogen energy conversion efficiency by combining concentrator photovoltaic modules and electrochemical cells. *Appl. Phys. Express*, 8, 107101, 2015. https://doi.org/10.7567/APEX.8.107101.
43. Pihosh, Y., Turkevych, I., Mawatari, K., Uemura, J., Kazoe, Y., Kosar, S., Makita, K., Sugaya, T., Matsui, T., Fujita, D., et al., Photocatalytic generation of hydrogen by core-shell WO3/BiVO4 nanorods with ultimate water splitting efficiency. *Sci. Rep.*, 5, 11141, 2015. https://doi.org/10.1038/srep11141.
44. Fan, R., Cheng, S., Huang, G., Wang, Y., Zhang, Y., Vanka, S., Botton, G.A., Mi, Z., Shen, M., Unassisted solar water splitting with 9.8% Efficiency and over 100 h stability based on Si solar cells and photoelectrodes catalyzed by bifunctional Ni-Mo/Ni. *J. Mater. Chem. A*, 7, 5, 2200–2209, 2019. https://doi.org/10.1039/c8ta10165e.
45. Sayama, K., Mukasa, K., Abe, R., Abe, Y., Arakawa, H., Stoichiometric water splitting into H_2 and O_2 using a mixture of two different photocatalysts and an IO_3^-/I^- shuttle redox mediator under visible light irradiation. *Chem. Commun.*, 1, 2416–2417, 2001. https://doi.org/10.1039/b107673f.
46. Kosinov, N., Gascon, J., Kapteijn, F., Hensen, E.J.M., Recent developments in zeolite membranes for gas separation. *J. Memb. Sci.*, 499, 65–79, 2016. https://doi.org/10.1016/j.memsci.2015.10.049.

47. Yu, Q., Meng, X., Wang, T., Li, P., Ye, J., Hematite films decorated with nanostructured ferric oxyhydroxide as photoanodes for efficient and stable photoelectrochemical water splitting. *Adv. Funct. Mater.*, 25, 2686–2692, 2015. https://doi.org/10.1002/adfm.201500383.
48. Weng, B., Grice, C.R., Ge, J., Poudel, T., Deng, X., Yan, Y., Barium bismuth niobate double perovskite/tungsten oxide nanosheet photoanode for high-performance photoelectrochemical water splitting. *Adv. Energy Mater.*, 8, 10, 1701655, 2018. https://doi.org/10.1002/aenm.201701655.
49. Du, C., Yang, X., Mayer, M.T., Hoyt, H., Xie, J., McMahon, G., Bischoping, G., Wang, D., Hematite-based water splitting with low turn-on voltages. *Angew. Chem. Int. Ed.*, 52, 12692–12695, 2013. https://doi.org/10.1002/anie.201306263.
50 Bolton, J.R., Strickler, S.J., Connolly, J.S., Limiting and realizable efficiencies of solar photolysis of water. *Nature*, 316, 6028, 495–500, 1985. https://doi.org/10.1038/316495a0.
51. Ager, J.W., Shaner, M.R., Walczak, K.A., Sharp, I.D., Ardo, S., Experimental demonstrations of spontaneous, solar-driven photoelectrochemical water splitting. *Energy Environ. Sci.*, 8, 10, 2811–2824, 2015. https://doi.org/10.1039/c5ee00457h.
52. Song, H., Oh, S., Yoon, H., Kim, K.H., Ryu, S., Oh, J., Bifunctional NiFe inverse opal electrocatalysts with heterojunction Si solar cells for 9.54%-efficient unassisted solar water splitting. *Nano Energy*, 42, 1–7, 2017. https://doi.org/10.1016/j.nanoen.2017.10.028.
53. Zhou, X., Zhou, J., Huang, G., Fan, R., Ju, S., Mi, Z., Shen, M., A bifunctional and stable Ni-Co-S/Ni-Co-P bistratal electrocatalyst for 10.8%-efficient overall solar water splitting. *J. Mater. Chem. A*, 6, 20297–20303, 2018. https://doi.org/10.1039/c8ta07197g.
54. Kim, J.H., Jo, Y., Kim, J.H., Jang, J.W., Kang, H.J., Lee, Y.H., Kim, D.S., Jun, Y., Lee, J.S., Wireless solar water splitting device with robust cobalt-catalyzed, dual-doped $BiVO_4$ photoanode and perovskite solar cell in tandem: A dual absorber artificial leaf. *ACS Nano*, 9, 12, 11820–11829, 2015. https://doi.org/10.1021/acsnano.5b03859.
55. Kang, D., Young, J.L., Lim, H., Klein, W.E., Chen, H., Xi, Y., Gai, B., Deutsch, T.G., Yoon, J., Printed assemblies of GaAs photoelectrodes with decoupled optical and reactive interfaces for unassisted solar water splitting. *Nat. Energy*, 2, 5, 17043, 2017. https://doi.org/10.1038/nenergy.2017.43.
56. Jia, J., Seitz, L.C., Benck, J.D., Huo, Y., Chen, Y., Ng, J.W.D., Bilir, T., Harris, J.S., Jaramillo, T.F., Solar water splitting by photovoltaic-electrolysis with a solar-to-hydrogen efficiency over 30%. *Nat. Commun.*, 7, 13237, 2016. https://doi.org/10.1038/ncomms13237.
57. Nocera, D.G., The artificial leaf. *Acc. Chem. Res.*, 45, 767–776, 2012. https://doi.org/10.1021/ar2003013.

58. Andreani, L.C., Bozzola, A., Kowalczewski, P., Liscidini, M., Redorici, L., Silicon solar cells: Toward the efficiency limits. *Adv. Phys. X*, 4, 1548305, 2019. https://doi.org/10.1080/23746149.2018.1548305.
59. Yoshikawa, K., Kawasaki, H., Yoshida, W., Irie, T., Konishi, K., Nakano, K., Uto, T., Adachi, D., Kanematsu, M., Uzu, H., et al., Silicon heterojunction solar cell with interdigitated back contacts for a photoconversion efficiency over 26%. *Nat. Energy*, 2, 17032, 2017. https://doi.org/10.1038/nenergy.2017.32.
60. Haase, F., Hollemann, C., Schäfer, S., Merkle, A., Rienäcker, M., Krügener, J., Brendel, R., Peibst, R., Laser contact openings for local poly-si-metal contacts enabling 26.1%-efficient POLO-IBC solar cells. *Sol. Energy Mater. Sol. Cells*, 186, 184–193, 2018. https://doi.org/10.1016/j.solmat.2018.06.020.
61. Richter, A., Benick, J., Feldmann, F., Fell, A., Hermle, M., Glunz, S.W., N-type Si solar cells with passivating electron contact: Identifying sources for efficiency limitations by wafer thickness and resistivity variation. *Sol. Energy Mater. Sol. Cells*, 173, 96–105, 2017. https://doi.org/10.1016/j.solmat.2017.05.042.
62. Yamaguchi, T., Ichihashi, Y., Mishima, T., Matsubara, N., Yamanishi, T., Achievement of more than 25 % conversion heterojunction solar cell. *IEEE J. Photovolt*, 15–17, 2014.
63. Green, M.A., The passivated emitter and rear cell (PERC): From conception to mass production. *Sol. Energy Mater. Sol. Cells*, 143, 190–197, 2015. https://doi.org/10.1016/j.solmat.2015.06.055.
64. Zhao, J., Wang, A., Green, M.A., 24.5% efficiency silicon pert cells on mcz substrates and 24.7% efficiency PERL cells on FZ substrates. *Prog. Photovoltaics Res. Appl.*, 7, 471–474, 1999. https://doi.org/10.1002/(SICI)1099-159X(199911/12)7:6<471::AID-PIP298>3.0.CO,2-7.
65. Kobayashi, E., Watabe, Y., Hao, R., Ravi, T.S., High efficiency heterojunction solar cells on N-type Kerfless Mono crystalline silicon wafers by epitaxial growth. *Appl. Phys. Lett.*, 106, 223504, 2015. https://doi.org/10.1063/1.4922196.
66. Benick, J., Richter, A., Müller, R., Hauser, H., Feldmann, F., Krenckel, P., Riepe, S., Schindler, F., Schubert, M.C., Hermle, M., et al., High-efficiency n-Type HP Mc silicon solar cells. *IEEE J. Photovoltaics*, 7, 1171–1175, 2017. https://doi.org/10.1109/JPHOTOV.2017.2714139.
67. Sai, H., Maejima, K., Matsui, T., Koida, T., Kondo, M., Nakao, S., Takeuchi, Y., Katayama, H., Yoshida, I., High-efficiency microcrystalline silicon solar cells on honeycomb textured substrates grown with high-rate VHF plasma-enhanced chemical vapor deposition. *Jpn. J. Appl. Phys.*, 54, 08KB05, 2015. https://doi.org/10.7567/JJAP.54.08KB05.
68. Sai, H., Matsui, T., Koida, T., Matsubara, K., Kondo, M., Sugiyama, S., Katayama, H., Takeuchi, Y., Yoshida, I., Triple-junction thin-film silicon solar cell fabricated on periodically textured substrate with a stabilized efficiency of 13.6%. *Appl. Phys. Lett.*, 106, 213902, 2015. https://doi.org/10.1063/1.4921794.

69. Wanlass, W., Systems and methods for advanced ultra-high-performance InP solar cells, US9590131B2, 2017.
70. Venkatasubramanian, R., O'Quinn, B.C., Hills, J.S., Sharps, P.R., Timmons, M.L., Hutchby, J.A., Field, H., Ahrenkiel, R., Keyes, B., 18.2% (AM1.5) efficient GaAs solar cell on optical-grade polycrystalline Ge substrate. *Conf. Rec. IEEE Photovolt. Spec. Conf.*, 31–36, 1996. https://doi.org/10.1109/pvsc.1996.563940.
71. Chiu, P.T., Law, D.C., Woo, R.L., Singer, S.B., Bhusari, D., Hong, W.D., Zakaria, A., Boisvert, J., Mesropian, S., King, R.R., et al., 35.8% space and 38.8% terrestrial 5J direct bonded cells. In *2014 IEEE 40th Photovoltaic Specialist Conference, PVSC 2014*, pp 11–13, 2014. https://doi.org/10.1109/PVSC.2014.6924957.
72. Sasaki, K., Agui, T., Nakaido, K., Takahashi, N., Onitsuka, R., Takamoto, T., Development of in gaP/GaAs/InGaAs inverted triple junction concentrator solar cells. In *AIP Conference Proceedings*, Vol. 1556, pp 22–25, 2013. https://doi.org/10.1063/1.4822190.
73. Geisz, J.F., Steiner, M.A., Jain, N., Schulte, K.L., France, R.M., McMahon, W.E., Perl, E.E., Friedman, D.J., Building a six-junction inverted metamorphic concentrator solar cell. *IEEE J. Photovoltaics*, 8, 626–632, 2018. https://doi.org/10.1109/JPHOTOV.2017.2778567.
74. Yan, C., Huang, J., Sun, K., Johnston, S., Zhang, Y., Sun, H., Pu, A., He, M., Liu, F., Eder, K., et al., Cu_2ZnSnS_4 solar cells with over 10% power conversion efficiency enabled by heterojunction heat treatment. *Nat. Energy*, 3, 9, 764–772, 2018. https://doi.org/10.1038/s41560-018-0206-0.
75. Wang, W., Winkler, M.T., Gunawan, O., Gokmen, T., Todorov, T.K., Zhu, Y., Mitzi, D.B., Device characteristics of CZTSSe thin-film solar cells with 12.6% efficiency. *Adv. Energy Mater.*, 4, 7, 1301465, 2014. https://doi.org/10.1002/aenm.201301465.
76. Diermann, R., Avancis claims 19.64% efficiency for CIGS module, PV magazine international, March 4, 2021. (https://www.pv-magazine.com/2021/03/04/avancis-claims-19-64-efficiency-for-cigs-module/.
77. Nakamura, M., Yamaguchi, K., Kimoto, Y., Yasaki, Y., Kato, T., Sugimoto, H., Cd-free $Cu(In,Ga)(Se,S)_2$ thin-film solar cell with record efficiency of 23.35%. *IEEE J. Photovoltaics*, 9, 6, 1863–1867, 2019. https://doi.org/10.1109/JPHOTOV.2019.2937218.
78. First solar press release, First solar achieves yet another cell conversion efficiency world record, 24 February 2016.
79. First solar press release, First solar builds the highest efficiency thin film PV cell on record, 5 August 2014.
80. Sun, K., Yan, C., Liu, F., Huang, J., Zhou, F., Stride, J.A., Green, M., Hao, X., Over 9% efficient kesterite Cu_2ZnSnS_4 solar cell fabricated by using Zn1-XCdxS buffer layer. *Adv. Energy Mater.*, 6, 12, 1600046, 2016. https://doi.org/10.1002/aenm.201600046.

81. Jeong, M., Choi, I.W., Go, E.M., Cho, Y., Kim, M., Lee, B., Jeong, S., Jo, Y., Choi, H.W., Lee, J., et al., Stable perovskite solar cells with efficiency exceeding 24.8% and 0.3-V voltage loss. *Science*, 369, 6511, 1615–1620, 2020. https://doi.org/10.1126/science.abb7167.
82. Peng, J., Walter, D., Ren, Y., Tebyetekerwa, M., Wu, Y., Duong, T., Lin, Q., Li, J., Lu, T., Mahmud, M. A., et al., Nanoscale localized contacts for high fill factors in polymer-passivated perovskite solar cells. *Science*, 371, 6527, 390–395, 2021. https://doi.org/10.1126/science.abb8687.
83. Lin, R., Xiao, K., Qin, Z., Han, Q., Zhang, C., Wei, M., Saidaminov, M.I., Gao, Y., Xu, J., Xiao, M., et al., Monolithic all-perovskite tandem solar cells with 24.8% efficiency exploiting comproportionation to suppress Sn(Ii) oxidation in precursor ink. *Nat. Energy*, 4, 10, 864–873, 2019. https://doi.org/10.1038/s41560-019-0466-3.
84 NREL transforming energy, https://www. nrel. gov/pv/cellefficiency. htm. (accessed: F. 2020). NREL Transforming ENERGY, Https://Www.Nrel.Gov/Pv/Cellefficiency.Html (Accessed: February 2020).
85. New world record efficiency for organic solar modules, https://www. encn. de/fileadmin/user_upload/PR_opv-record__. pd. (accessed 11 N. 2019). New World Record Efficiency for Organic Solar Modules, Https://Www.Encn.de/Fileadmin/User_upload/PR_opv-Record__.Pdf (Accessed 11 November 2019).
86. Kato, N., Moribe, S., Shiozawa, M., Suzuki, R., Higuchi, K., Suzuki, A., Sreenivasu, M., Tsuchimoto, K., Tatematsu, K., Mizumoto, K., et al., Improved conversion efficiency of 10% for solid-state dye-sensitized solar cells utilizing P-type semiconducting CuI and multi-dye consisting of novel porphyrin dimer and organic dyes. *J. Mater. Chem. A*, 6, 45, 22508–22512, 2018. https://doi.org/10.1039/c8ta06418k.
87. Han, L., Fukui, A., Chiba, Y., Islam, A., Komiya, R., Fuke, N., Koide, N., Yamanaka, R., Shimizu, M., Integrated dye-sensitized solar cell module with conversion efficiency of 8.2%. *Appl. Phys. Lett.*, 94, 1, 013305, 2009. https://doi.org/10.1063/1.3054160.
88. Guo, Z., Jena, A.K., Takei, I., Kim, G.M., Kamarudin, M.A., Sanehira, Y., Ishii, A., Numata, Y., Hayase, S., Miyasaka, T., V_{OC} over 1.4 V for amorphous tin-oxide-based dopant-free $CsPbI_2Br$ Perovskite solar cells. *J. Am. Chem. Soc.*, 142, 9725–9734, 2020. https://doi.org/10.1021/jacs.0c02227.
89. Zeng, K. and Zhang, D., Recent progress in alkaline water electrolysis for hydrogen production and applications (progress in energy and combustion science (2010) 36:3 (307-326)). *Prog. Energy Combust. Sci.*, 37, 631, 2011. https://doi.org/10.1016/j.pecs.2011.02.002.
90. Ganley, J.C., High temperature and pressure alkaline electrolysis. *Int. J. Hydrogen Energy*, 34, 3604–3611, 2009. https://doi.org/10.1016/j.ijhydene.2009.02.083.

91. Haug, P., Kreitz, B., Koj, M., Turek, T., Process modelling of an alkaline water electrolyzer. *Int. J. Hydrogen Energy*, 42, 15689–15707, 2017. https://doi.org/10.1016/j.ijhydene.2017.05.031.
92. LeRoy, R.L., Industrial water electrolysis: Present and future. *Int. J. Hydrogen Energy*, 8, 401–417, 1983. https://doi.org/10.1016/0360-3199(83)90162-3.
93. Millet, P., Mbemba, N., Grigoriev, S.A., Fateev, V.N., Aukauloo, A., Etiévant, C., Electrochemical performances of PEM water electrolysis cells and perspectives. *Int. J. Hydrogen Energy*, 36, 6, 4134–4142, 2011. https://doi.org/10.1016/j.ijhydene.2010.06.105.
94. Bernt, M. and Gasteiger, H.A., Influence of ionomer content in IrO_2/TiO_2 electrodes on PEM water electrolyzer performance. *J. Electrochem. Soc.*, 163, F3179–F3189, 2016. https://doi.org/10.1149/2.0231611jes.
95. Sapountzi, F.M., Divane, S.C., Papaioannou, E.I., Souentie, S., Vayenas, C.G., The role of Nafion content in sputtered IrO_2 based anodes for low temperature PEM water electrolysis. *J. Electroanal. Chem.*, 662, 1, 116–122, 2011. https://doi.org/10.1016/j.jelechem.2011.04.005.
96. Li, D., Park, E.J., Zhu, W., Shi, Q., Zhou, Y., Tian, H., Lin, Y., Serov, A., Zulevi, B., Baca, E.D., et al., Highly quaternized polystyrene ionomers for high performance anion exchange membrane water electrolysers. *Nat. Energy*, 5, 378–385, 2020. https://doi.org/10.1038/s41560-020-0577-x.
97. Zeng, K. and Zhang, D., Recent progress in alkaline water electrolysis for hydrogen production and applications. *Prog. Energy Combust. Sci.*, 36, 307–326, 2010. https://doi.org/10.1016/j.pecs.2009.11.002.
98. Ogawa, T., Takeuchi, M., Kajikawa, Y., Analysis of trends and emerging technologies in water electrolysis research based on a computational method: A comparison with fuel cell research. *Sustain.*, 10, 478–501, 2018. https://doi.org/10.3390/su10020478.
99. Laursen, A.B., Patraju, K.R., Whitaker, M.J., Retuerto, M., Sarkar, T., Yao, N., Ramanujachary, K.V., Greenblatt, M., Dismukes, G.C., Nanocrystalline Ni_5P_4: A hydrogen evolution electrocatalyst of exceptional efficiency in both alkaline and acidic media. *Energy Environ. Sci.*, 8, 1027–1034, 2015. https://doi.org/10.1039/c4ee02940b.
100. Kraglund, M.R., Carmo, M., Schiller, G., Ansar, S.A., Aili, D., Christensen, E., Jensen, J.O., Ion-solvating membranes as a new approach towards high rate alkaline electrolyzers. *Energy Environ. Sci.*, 12, 3313–3318, 2019. https://doi.org/10.1039/c9ee00832b.
101. Leng, Y., Chen, G., Mendoza, A.J., Tighe, T.B., Hickner, M.A., Wang, C.Y., Solid-state water electrolysis with an alkaline membrane. *J. Am. Chem. Soc.*, 134, 22, 9054–9057, 2012. https://doi.org/10.1021/ja302439z.
102. Fan, J., Willdorf-Cohen, S., Schibli, E.M., Paula, Z., Li, W., Skalski, T.J.G., Sergeenko, A.T., Hohenadel, A., Frisken, B.J., Magliocca, E., et al., Poly(bisarylimidazoliums) possessing high hydroxide ion exchange capacity and high alkaline stability. *Nat. Commun.*, 10, 2306, 2019. https://doi.org/10.1038/s41467-019-10292-z.

103. Green, M.A., Dunlop, E.D., Hohl-Ebinger, J., Yoshita, M., Kopidakis, N., Hao, X., Solar cell efficiency tables (Version 58). *Prog. Photovoltaics Res. Appl.*, 29, 657–667, 2021. https://doi.org/10.1002/pip.3444.
104. Schüttauf, J.-W., Modestino, M.A., Chinello, E., Lambelet, D., Delfino, A., Dominé, D., Faes, A., Despeisse, M., Bailat, J., Psaltis, D., et al., Solar-to-hydrogen production at 14.2% efficiency with silicon photovoltaics and earth-abundant electrocatalysts. *J. Electrochem. Soc.*, 163, F1177–F1181, 2016. https://doi.org/10.1149/2.0541610jes.
105. Green, M.A., Emery, K., Hishikawa, Y., Warata, W., Dunlop, E.D., Solar cell efficiency tables (Version 40). *Prog. Photovolt Res. Appl.*, 24, 3–11, 2016. https://doi.org/10.1002/pip.
106. Vos, A.De., Detailed balance limit of the efficiency of tandem solar cells. *J. Phys. D. Appl. Phys.*, 13, 839–846, 1980. https://doi.org/10.1088/0022-3727/13/5/018.
107. Nadeem, M.A. and Idriss, H., Effect of PH, temperature, and low light flux on the performance (16% STH) of coupled triple junction solar cell to water electrolysis. *J. Power Sources*, 459, 228074, 2020. https://doi.org/10.1016/j.jpowsour.2020.228074.
108. Licht, S., Wang, B., Mukerji, S., Soga, T., Umeno, M., Tributsch, H., Efficient solar water splitting, exemplified by RuO_2-Catalyzed AlGaAs/Si photoelectrolysis. *J. Phys. Chem. B*, 104, 8920–8924, 2000. https://doi.org/10.1021/jp002083b.
109. Yao, Y., Xu, Z., Cheng, F., Li, W., Cui, P., Xu, G., Xu, S., Wang, P., Sheng, G., Yan, Y., et al., Unlocking the potential of graphene for water oxidation using an orbital hybridization strategy. *Energy Environ. Sci.*, 11, 407–416, 2018. https://doi.org/10.1039/C7EE02972A.
110. Zhang, P., Li, L., Nordlund, D., Chen, H., Fan, L., Zhang, B., Sheng, X., Daniel, Q., Sun, L., Dendritic core-shell nickel-iron-copper metal/metal oxide electrode for efficient electrocatalytic water oxidation. *Nat. Commun.*, 9, 381, 2018. https://doi.org/10.1038/s41467-017-02429-9.
111. Zhang, B., Zhang, B., Zheng, X., Voznyy, O., Comin, R., Bajdich, M., García-Melchor, M., Han, L., Xu, J., Liu, M., et al., Homogeneously dispersed multimetal oxygen-evolving catalysts. *Science*, 352, 333–337, 2016. https://doi.org/10.1126/science.aaf1525.
112. Xu, X., Song, F., Hu, X., A nickel iron diselenide-derived efficient oxygen-evolution catalyst. *Nat. Commun.*, 7, 12324, 2016. https://doi.org/10.1038/ncomms12324.
113. Liu, J., Wang, J., Zhang, B., Ruan, Y., Lv, L., Ji, X., Xu, K., Miao, L., Jiang, J., Hierarchical $NiCo_2S_4$@NiFe LDH heterostructures supported on nickel foam for enhanced overall-water-splitting activity hierarchical $NiCo_2S_4$@NiFe LDH heterostructures supported on nickel foam for enhanced overall-water-splitting activity. *ACS Appl. Mater. Interfaces*, 9, 15364–15372, 2017.

114. Cai, Z., Zhou, D., Wang, M., Bak, S.M., Wu, Y., Wu, Z., Tian, Y., Xiong, X., Li, Y., Liu, W., et al., Introducing Fe2+ into nickel–iron layered double hydroxide: Local structure modulated water oxidation activity. *Angew. Chem. Int. Ed.*, 57, 9392–9396, 2018. https://doi.org/10.1002/anie.201804881.
115. Wang, X.P., Wu, H.J., Xi, S.B., Lee, W.S.V., Zhang, J., Wu, Z.H., Wang, J.O., Hu, T.D., Liu, L.M., Han, Y., et al., Strain stabilized nickel hydroxide nanoribbons for efficient water splitting. *Energy Environ. Sci.*, 13, 229–237, 2020. https://doi.org/10.1039/c9ee02565k.
116. Zahran, Z., Miseki, Y., Mohamed, E., Tsubonouchi, Y., Makita, K., Sugaya, T., Sayama, K., Yagi, M., Perfect matching factor between a customized double-junction GaAs photovoltaic device and an electrolyzer for efficient solar water splitting. *ACS Appl. Energ. Mater.*, 5, 8241–8253, 2022. https://doi.org/doi.org/10.1021/acsaem.2c00768.
117. Jacobsson, T.J., Fjällström, V., Sahlberg, M., Edoff, M., Edvinsson, T., A monolithic device for solar water splitting based on series interconnected thin film absorbers reaching over 10% solar-to-hydrogen efficiency. *Energy Environ. Sci.*, 6, 3676–3683, 2013. https://doi.org/10.1039/c3ee42519c.
118. Marsen, B., Cole, B., Miller, E.L., Photoelectrolysis of water using thin copper gallium diselenide electrodes. *Sol. Energy Mater. Sol. Cells*, 92, 9, 1054–1058, 2008. https://doi.org/10.1016/j.solmat.2008.03.009.
119. Yokoyama, D., Minegishi, T., Maeda, K., Katayama, M., Kubota, J., Yamada, A., Konagai, M., Domen, K., Photoelectrochemical water splitting using a Cu(In,Ga)Se$_2$ thin film. *Electrochem. commun.*, 12, 6, 851–853, 2010. https://doi.org/10.1016/j.elecom.2010.04.004.
120. Moriya, M., Minegishi, T., Kumagai, H., Katayama, M., Kubota, J., Domen, K., Stable hydrogen evolution from Cds-Modified CuGaSe2 photoelectrode under visible-light irradiation. *J. Am. Chem. Soc.*, 135, 10, 3733–3735, 2013. https://doi.org/10.1021/ja312653y.
121. Pehlivan, İ.B., Oscarsson, J., Qiu, Z., Stolt, L., Edoff, M., Edvinsson, T., NiMoV and NiO-based catalysts for efficient solar-driven water splitting using thermally integrated photovoltaics in a scalable approach. *iScience*, 24, 101910, 2021. https://doi.org/10.1016/j.isci.2020.101910.
122. Moniruddin, M., Ilyassov, B., Zhao, X., Smith, E., Serikov, T., Ibrayev, N., Asmatulu, R., Nuraje, N., Recent progress on perovskite materials in photovoltaic and water splitting applications. *Mater. Today Energy*, 7, 246–259, 2018. https://doi.org/10.1016/j.mtener.2017.10.005.
123. Luo, JingshanJeong-Hyeok, I., Mayer, M.T., Schreier, M., Nazeeruddfn, M.K., Nam-Gyu, P., Tilley, S.D., Hong Jin, F., Gratzel, M., Water photolysis at 12.3% efficiency via perovskite photovoltaics and earth-abundant catalysts. *Science*, 345, 1593–1596, 2014. https://doi.org/10.1126/science.1258307.
124. Asiri, A.M., Ren, D., Zhang, H., Khan, S.B., Alamry, K.A., Marwani, H.M., Khan, M.S.J., Adeosun, W.A., Zakeeruddin, S.M., Gratzel, M., Solar water splitting using earth-abundant electrocatalysts driven by high-efficiency

perovskite solar cells. *ChemSusChem*, 15, e202102471, 2022. https://doi.org/10.1002/cssc.202102471.
125. Gao, Y., Le Corre, V.M., Gaïtis, A., Neophytou, M., Hamid, M.A., Takanabe, K., Beaujuge, P.M., Homo-tandem polymer solar cells with V_{OC} 1.8 V for efficient PV-driven water splitting. *Adv. Mater.*, 28, 3366–3373, 2016. https://doi.org/10.1002/adma.201504633.
126. Esiner, S., Van Eersel, H., Wienk, M.M., Janssen, R.A.J., Triple junction polymer solar cells for photoelectrochemical water splitting. *Adv. Mater.*, 25, 21, 2932–2936, 2013. https://doi.org/10.1002/adma.201300439.
127. Menezes, P.W., Yao, S., Beltrán-Suito, R., Hausmann, J.N., Menezes, P.V., Driess, M., Facile access to active g - NiOOH electrocatalyst for durable water oxidation derived from an intermetallic nickel germanide precursor. *Angew. Chem., Int. Ed.*, 60, 4640–4647, 2021. https://doi.org/10.1002/ange.202014331.
128. Jiao, L., Liu, E., Hwang, S., Mukerjee, S., Jia, Q., Compressive strain reduces the hydrogen evolution and oxidation reaction activity of platinum in alkaline solution. *ACS Catal.*, 11, 8165–8173, 2021. https://doi.org/10.1021/acscatal.1c01723.
129. Cheng, X., Lu, Y., Zheng, L., Pupucevski, M., Li, H., Chen, G., Sun, S., Wu, G., Engineering local coordination environment of atomically dispersed platinum catalyst via lattice distortion of support for efficient hydrogen evolution reaction. *Mater. Today Energy*, 20, 100653, 2021. https://doi.org/10.1016/j.mtener.2021.100653.
130. Yang, J., Li, W., Wang, D., Li, Y., Electronic metal–support interaction of single-atom catalysts and applications in electrocatalysis. *Adv. Mater.*, 32, e2003300, 2020. https://doi.org/10.1002/adma.202003300.
131. Zhou, K.L., Wang, C., Wang, Z., Han, C.B., Zhang, Q., Ke, X., Liu, J., Wang, H., seamlessly conductive $Co(OH)_2$ tailored atomically dispersed Pt electrocatalyst with a hierarchical nanostructure for an efficient hydrogen evolution reaction. *Energy Environ. Sci.*, 13, 3082–3092, 2020. https://doi.org/10.1039/d0ee01347a.
132. Wang, X., Zheng, Y., Sheng, W., Xu, Z.J., Jaroniec, M., Qiao, S.Z., Strategies for design of electrocatalysts for hydrogen evolution under alkaline conditions. *Mater. Today*, 36, 125–138, 2020. https://doi.org/10.1016/j.mattod.2019.12.003.
133. Calnan, S., Bagacki, R., Bao, F., Dorbandt, I., Kemppainen, E., Schary, C., Schlatmann, R., Leonardi, M., Lombardo, S.A., Milazzo, R.G., *et al.*, Development of various photovoltaic-driven water electrolysis technologies for green solar hydrogen generation. *Sol. RRL*, 2100479, 2021. https://doi.org/10.1002/solr.202100479.

11
Impactful Role of Earth-Abundant Cocatalysts in Photocatalytic Water Splitting

Yubin Chen[1]*, Xu Guo[2†], Zhichao Ge[1], Ya Liu[1] and Maochang Liu[1]

[1]*International Research Center for Renewable Energy, State Key Laboratory of Multiphase Flow in Power Engineering, Xi'an Jiaotong University, Shaanxi, China*
[2]*Shaanxi Key Laboratory for Advanced Energy Devices, School of Materials Science and Engineering, Shaanxi Normal University, Shaanxi, China*

Abstract

Photocatalytic water splitting represents an attractive strategy for storing intermittent solar energy in the form of chemical bonds. Considerable efforts have been made to design efficient and stable semiconductor photocatalysts. However, the efficiencies of most developed photocatalysts are still unsatisfactory due to the severe charge recombination and sluggish surface reaction kinetics. Loading suitable cocatalysts can efficiently capture the photogenerated carriers to reduce charge recombination and provide surface reactive sites to accelerate the chemical reaction. Typically, developing earth-abundant cocatalysts to replace noble-metal-based ones is of great significance for the practical application. To understand the impactful role of cocatalysts in photocatalytic water splitting, this chapter first introduces the basic working principle of cocatalysts and then summarizes extensively studied earth-abundant cocatalysts. The effects of intrinsic characters of cocatalysts, such as morphology, crystallization, and composition are subsequently discussed. The influence of interfacial coupling of cocatalysts with host photocatalysts, such as cocatalyst location, built-in electric field, and contact manner, is further analyzed. To demonstrate the real active species of cocatalysts and clarify the reaction mechanism, some results using advanced characterization techniques are examined. Hopefully, this chapter can offer meaningful guidance to design cost-effective and highly efficient cocatalysts for photocatalytic water splitting.

*Corresponding author: ybchen@mail.xjtu.edu.cn
†Corresponding author: guoxu@snnu.edu.cn

Keywords: Photocatalysis, cocatalyst, water splitting, solar energy, hydrogen

11.1 Introduction

Photocatalytic water splitting into H_2 and O_2 has been widely studied as a promising way to produce clean and renewable energy for the sustainable development of human beings [1–3]. It is well known that the major processes of photocatalytic water splitting reaction are considered as follows: i) light absorption by photocatalysts and electron-hole pair formation, ii) photo-induced charge carrier separation and transfer, and iii) water reduction/oxidation on active sites of catalysts (Figure 11.1a). The overall efficiency is dependent on the efficiencies of three major steps, specifically, the photon absorption efficiency, the charge separation efficiency, and the redox reaction efficiency. Though many efforts have been carried out and great breakthroughs have been achieved since the first discovery of photoelectrocatalytic water splitting by Fujishima and Honda [4–9], the efficiency is still the major factor that hind the practical application of this technology. The single photocatalysts are either wide gap semiconductors absorbing photons in the UV region or narrow bandgap semiconductors with severe electron-hole recombination. Therefore, modification strategies, such as doping, alloying, dye sensitization, and constructing heterostructures, have been applied to extend the visible light absorption and decrease the photo-induced carrier recombination [10–15]. Among them, loading cocatalysts has been demonstrated as an effective way to improve the efficiency of photocatalytic water splitting [16–21]. Rational design of efficient and stable cocatalysts is required for the practical application of solar water splitting.

Cocatalysts can influence photocatalytic water splitting by different means (Figure 11.1b). Firstly, by loading cocatalysts, the overpotential of surface redox reaction will be lowered. Photocatalytic water splitting into hydrogen and oxygen is an uphill chemical reaction with an increased Gibbs free energy (ΔG = 237 kJ/mol). Loading cocatalysts can decrease the activation energy and overpotential effectively (Figure 11.1c) [22]. For instance, $Zn_{0.5}Cd_{0.5}S$ nanoparticles are imbedded on 2D $Ni(OH)_2$ nanosheets, and the presence of $Ni(OH)_2$ nanosheets improves photocatalytic hydrogen production by reducing the overpotentials, accelerating charge separation, and offering abundant active sites [23]. Second, loading cocatalysts can accelerate the photo-induced charge carrier separation and transfer by extracting the electron or hole from the host catalyst. Then, the separated electron/hole will drive the water splitting to H_2/O_2 on the active sites provided by the cocatalysts. Third, photostability is another important

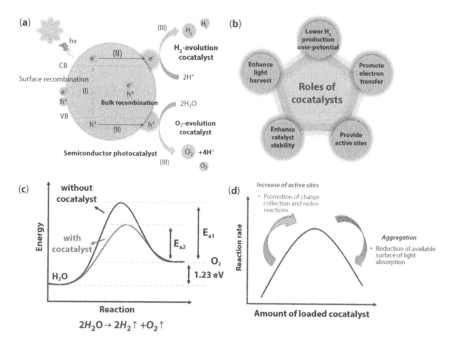

Figure 11.1 Illustration of (a) photocatalytic water splitting over semiconductor-cocatalysts system, (b) roles of loading cocatalysts, (c) reducing the overpotential by loading cocatalysts, and (d) relationship between the amount of cocatalysts and photocatalytic activity.

parameter for the practical application of photocatalytic water splitting. Loading a proper cocatalyst can consume the excessive electrons/holes timely to inhibit the photocorrosion and improve the photostability. For example, Cu_2O semiconductor is broadly used as a photocathode material but it can be easily reduced during the photoelectrochemical water splitting. Coating NiO_x layer on Cu_2O/ZnO p-n junctions as cocatalysts can not only increase the photocurrent density of the Cu_2O/ZnO photocathode but also prevent the Cu_2O film from photocorrosion [24]. In addition, light absorption can also be enhanced by cocatalysts. Metal sulfide cocatalysts have attracted tremendous attention in boosting photocatalytic efficiency owing to their low cost and high performance. Chen et al. prepared 2D MgAl layered double hydroxide (MgAl-LDH) and loaded ultrafine NiS nanoparticles to build efficient photosystem [25]. It is discovered that the loading of NiS cocatalysts can not only improve the electron-hole separation but also increase visible light absorption. As a result, NiS/MgAl-LDH displayed promising photocatalytic activity. In order to design and develop efficient and stable photocatalysts, the above principles should be considered during the preparation of cocatalysts.

Based on the above analysis, an efficient cocatalyst should have the basic features as follows: 1) The hydrogen absorption-free energy should be approximated to zero. In another word, the cocatalysts are easy to absorb the water molecules and then release hydrogen. 2) The band structure of the cocatalyst should be suitable for allowing the carrier to flow from host photocatalysts to cocatalysts. 3) The active sites are rich to accelerate the water splitting reaction. In addition, the amount of cocatalysts loading should be carefully controlled even though the photocatalytic efficiency can be remarkably improved by cocatalyst modification [26–28]. As shown in Figure 11.1d, excessive or fewer loadings are undesirable. Loading cocatalysts with an excessive amount will lead to the agglomeration on the surface of semiconductors and hinder the light absorption efficiency of host photocatalysts. Meanwhile, excessive loading of cocatalysts will increase the possibility of recombination as well. In addition, the active sites on host photocatalysts could be covered if the loading amount of cocatalysts is too high. Contrarily, few loadings of cocatalysts can't provide rich active sites for redox reaction and offer abundant charge transfer channels.

Besides the loading amount of cocatalysts, some other factors, such as categories of cocatalysts, morphology, composition, loading position, and contact manners, should be considered during the design and construction of the light absorber-cocatalysts photosystem. In the following sections, we will discuss the categories of cocatalysts utilized in photocatalytic water splitting.

11.2 Categories of Cocatalysts Utilized in Photocatalytic Water Splitting

Massive materials, including metal [12, 29–33], nonmetal [34–36], oxides [37, 38] and sulfides [39–41], are developed as cocatalysts for hydrogen and oxygen evolution reactions. They can be classified into different categories based on different physiochemical properties. For instance, they can be divided into hydrogen evolution cocatalysts and oxygen evolution cocatalysts according to the role of cocatalysts during the redox reaction [31, 42, 43]. They can be also divided into electron acceptor cocatalysts and hole acceptor cocatalysts from the view of extracting different charge carriers in charge separation and transfer [16, 44, 45]. Herein, we divided the cocatalysts into metal and nonmetal cocatalysts, metal oxides cocatalysts, metal sulfides cocatalysts, metal phosphide cocatalysts, metal carbide cocatalysts, and molecular cocatalysts according to the instinct composition of cocatalysts (Figure 11.2). Some typical examples were then introduced to understand the specific role and function of cocatalysts.

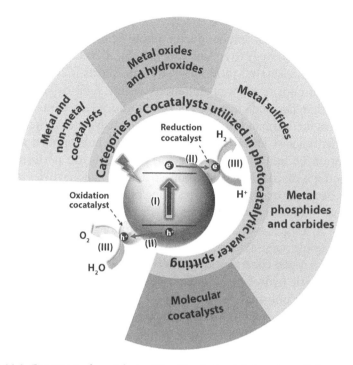

Figure 11.2 Categories of cocatalysts utilized in photocatalytic water splitting.

11.2.1 Metal and Non-Metal Cocatalysts

Noble metals (e.g., Pt and Pd) are extensively used as hydrogen evolution cocatalysts owing to their low overpotentials, stability, and superior conductivity [46–49]. Considering the scale-up application of solar water splitting, we focus on noble metal-free cocatalysts such as Fe, Co, and Ni metal as well as non-metal cocatalysts, specifically carbon-based cocatalysts. Transition metals emerged as ideal cocatalysts because of the low cost, earth-abundant, and stability though their overpotentials are slightly higher than noble metals [50–56]. Carbon-based materials are applied as cocatalysts owing to their superior electronic conductivity and excellent light absorption [29, 57, 58]. Usually, photochemical deposition of the transient metal on host photocatalysts is of great difficulty. Zhao *et al.* developed a complexing agent-assisted way to photochemically deposit transient metal (Co) on the g-C_3N_4 surface [55]. As expected, the Co/g-C_3N_4 showed the highest hydrogen generation rate of 2.30 mmol·g^{-1}·h^{-1}, which was 75 times higher than that of single g-C_3N_4, and the apparent quantum yield was calculated to be 6.2% at 400 nm. It was discovered that the emission quantum yield was reduced from 5.34% to 1.93% by Co loading, indicating the decreased electron-hole recombination and boosted photocatalytic activity.

Metallic nickel is proved to be effective for improving the photocatalytic efficiency of numerous semiconductor photocatalysts [59–61]. Di et al. developed a facile strategy to encapsulate metallic nickel with graphene and applied it as the cocatalyst to modify CdS nanosheets [62]. The prepared CdS-Ni@C composite showed enhanced visible light-driven hydrogen production than bare CdS nanosheets. Photoinduced electrons transferred from the conduction band of CdS to Ni@C cocatalysts, and then reduced water to hydrogen. Carbon-coated Cu nanoparticle is applied as hydrogen evolution cocatalyst to improve the photocatalytic performance of g-C_3N_4, and it is discovered that carbon-coated Cu nanoparticles increased the charge separation efficiency by trapping photo-induced electrons from g-C_3N_4 host [50]. As a result, the rational designed Cu@C/g-C_3N_4 showed the highest hydrogen evolution rate of 265.1 $\mu mol \cdot g^{-1} \cdot h^{-1}$, which was even superior to noble metal Pt decorated g-C_3N_4.

Apart from metal cocatalysts, nonmetal can also be used as cocatalysts for photocatalytic hydrogen evolution. High purity carbon dots were prepared via a mild one-step electrochemical method and anchored on the surface of the octahedral CoO nanoparticle [63]. The prepared C/CoO catalyst displayed enhanced visible-light overall water splitting property. The anchored C dots increased the visible-light absorption and suppressed the electron-hole recombination on CoO nanoparticles. Besides, the photostability of CoO was improved because the surface heating caused by the photothermal effect was released timely by the C dots as a heat conductor.

11.2.2 Metal Oxides and Hydroxides

Metal oxides and hydroxides, such as NiO_x, CoO_x, CeO_2, and $Ni(OH)_2$, have been developed and used as photocatalytic water splitting cocatalysts owing to the earth-abundance, low cost, nontoxicity, and good stability [42, 57, 64–69]. It has been demonstrated that the photocatalytic oxygen evolution reaction is the efficiency limiting step for overall water splitting because of the four-electron redox reaction. Developing efficient oxygen evolution cocatalysts is highly desired for realizing the overall water splitting. Fortunately, manganese oxides and cobalt oxides display excellent and stable water oxidation ability and are widely applied as oxygen evolution cocatalysts [70, 71]. For instance, g-C_3N_4 is a π-conjugated semiconductor with a bandgap of 2.7 eV, and it shows poor photocatalytic water splitting for severe electron-hole recombination. Coupling Co_3O_4 nanoparticles onto g-C_3N_4 could achieve an improved oxygen evolution rate of 25.1 μmol/h with an apparent yield of 1.1% at 420 nm [72]. The loading of Co_3O_4 promoted the charge separation and decreased the overpotential of water oxidation based on detail characterization. It was reported that photoexcited holes could be extracted by MnO_x cocatalyst from $SrTiO_3$ semiconductors and photoelectrochemical water splitting features were apparently enhanced by loading MnO_x cocatalyst [73].

In addition to the oxygen evolution cocatalysts, some metal oxides and hydroxides were used as hydrogen evolution cocatalysts as well. Zhang et al. adopted a bottom-up strategy to load MoO_x cocatalyst on CdS semiconductors [74]. Characterization and theory calculation suggest that the ultra-small MoO_x clusters could not only extract photoinduced electrons from CdS but also activate the absorbed water molecules. All these features were essential for improving the photocatalytic water splitting of CdS nanowires. Earth-abundant and nontoxic WO_3 is another metal oxide cocatalyst to strengthen the carrier separation and accelerate the charge transport. A WO_3 semiconductor decorated $Zn_xCd_{1-x}S$ photocatalyst was fabricated by Song's group [75]. By detailly studying, the effect of WO_3 cocatalyst, the highest hydrogen generation rate of 98.68 μmol/mg was obtained with excellent photostability. Photocatalytic reduction of CO_2 to chemical fuel is another strategy to solve the energy crisis and global warming. $ZnIn_2S_4$ semiconductor displays photocatalytic CO_2 reduction to CH_3OH owing to its suitable band structure and cheap nontoxic features. Yang et al. adopted a microwave-assisted hydrothermal method to build $ZnIn_2S_4$-CeO_2 catalyst [76]. The designed heterostructure exhibited the highest CH_3OH evolution rate of 0.542 μmol·$g^{-1}\cdot h^{-1}$ by optimizing the ratio of CeO_2 cocatalyst and $ZnIn_2S_4$ semiconductor. The boosted photocatalytic activity was attributed to the efficient electron trapping by CeO_2 from $ZnIn_2S_4$. The loading of CeO_2 provided more active sites for CO_2 absorption and diffusion as well.

11.2.3 Metal Sulfides

Apart from the metal oxides and hydroxides, metal sulfides play an important role in the development of efficient and stable cocatalysts [39, 77–79]. The exploration of metal sulfide as cocatalysts is highly desired for designing and building efficient and stable photocatalysts. Nickle sulfide, known as narrow bandgap material, is a good semiconductor used for light harvest. Yu's group applied a simple $S_2O_3^{2-}$-mediated photo-deposition strategy to anchor sulfur-rich NiS_{1+x} cocatalysts on TiO_2 [80]. The sulfur-rich NiS_{1+x} served as a cocatalyst for extracting photoinduced electrons from TiO_2 and providing active sites to reduce H^+ to hydrogen. Therefore, an outstanding hydrogen evolution rate of 264.42 μmol/h with good stability and repeatability was achieved over NiS_{1+x}/TiO_2. Owning to the superior electrical conductivity, rich redox features, and remarkable electrocatalytic activity, ternary transition metal sulfide $NiCo_2S_4$ is widely applied as a cocatalyst. $ZnIn_2S_4$ nanosheets were *in situ* grown on $NiCo_2S_4$ hollow spheres [81]. The prepared hierarchical $NiCo_2S_4$@$ZnIn_2S_4$ owned tight contact at the heterointerfaces and extended light absorption. The well-designed hybrid showed a 9 times higher hydrogen evolution rate than $ZnIn_2S_4$ alone, which was even superior to Pt decorated $ZnIn_2S_4$.

2D transition metal dichalcogenides (TMDs), especially MoS_2 and WS_2, have been extensively applied as efficient cocatalysts for hydrogen production owing to their low cost, graphene-like structure, and tunable band structure [82–85]. Bai's group introduced graphene and designed a novel stacked nanostructure, in which CdS nanoparticles were embedded at the interfaces of graphene and MoS_2 layers [86]. The unique structure achieved a remarkable photocatalytic hydrogen generation rate of 14.4 mmol·g^{-1}·h^{-1} and reached an apparent quantum yield of 23.7% at 420 nm with an improved stability. The increased photocatalytic performance was attributed to the intimate contact and large interfacial area between CdS and MoS_2 resulting from stacking design and inhibited photocorrosion from core-shell structure. WSe_2 nanosheet, as a representative member of TMDs, has been developed as an efficient cocatalyst [87, 88]. Owing to the excellent electrical conductivity and abundant active site, WSe_2 decorated $Zn_xCd_{1-x}S$ nanorods displayed a promising photocatalytic hydrogen evolution rate of 147.32 mmol·g^{-1}·h^{-1} and achieved an apparent quantum yield of 39.5% at 425 nm.

11.2.4 Metal Phosphides and Carbides

In addition to the above-mentioned cocatalysts, metal phosphides and carbides have attracted increasing attention in the field of solar energy conversion [19, 89–92]. A ternary photocatalyst N-TiO_2/g-C_3N_4@Ni_xP was rationally designed and built by Wang's group [93]. The TiO_2 and g-C_3N_4 were sonicated in ammonia solution and then hydrothermally treated to obtain a 3D N-TiO_2/g-C_3N_4 heterostructure. After loading Ni_xP as a cocatalyst by the photo-reduction way, the architecture showed a promising photocatalytic hydrogen generation rate of 5.43 mmol·g^{-1}·h^{-1}. The loading of Ni_xP cocatalyst as well as the intimate contact between g-C_3N_4 and TiO_2 played an important role in superior photocatalytic performance. Novel Co_2P nanosheet was applied as the cocatalyst to boost the hydrogen production over $Zn_{0.5}Cd_{0.5}S$ semiconductor. It was discovered that 0D $Zn_{0.5}Cd_{0.5}S$ were homogeneously grown on 2D Co_2P nanosheet, proving numerous heterointerfaces for charge transfer [94]. The 2D structure Co_2P could extract photo-induced electrons easily and then reduce the H^+ to H_2. As a result, Co_2P-$Zn_{0.5}Cd_{0.5}S$ displayed an outstanding hydrogen generation rate of 68.02 mmol·g^{-1}·h^{-1} with an apparent quantum yield of 30.1% at the wavelength of 420 nm.

Various metal carbides are also used as cocatalysts owing to the superior physiochemical features and chemical stability [95, 96]. For example, Zhao's group applied Co$_3$C as the cocatalyst to decorate the CdS nanorods. Expectedly, a hydrogen generation rate of 315 μmol/h was obtained [91]. Furthermore, Co$_3$C semiconductor was used to modify other photocatalysts, namely g-C$_3$N$_4$ and ZnIn$_2$S$_4$, and the enhanced photoactivity demonstrated that Co$_3$C was a universal cocatalyst for host photocatalysts.

11.2.5 Molecular Cocatalysts

For photocatalytic water splitting, molecules are widely used to modify the host photocatalysts to improve the charge separation and accelerate the photo redox reaction [97–100]. Xie's group developed trifluoroacetic acid, a water-soluble molecule, as a cocatalyst to improve the photocatalytic activity of K$_4$Nb$_6$O$_{17}$ nanosheet [101]. Four different cobalt-salen complexes were designed and adopted to decorate ZnO/CdS heterostructure [102]. The obtained photocatalytic system displayed excellent photocatalytic activity with a hydrogen generation rate of 725 μmol·h^{-1}·mg^{-1}, which was about 29 times higher than that of pure CdS NRs. The excited photocatalytic performance was attributed to the multielectron transfer pathway provided by the cocatalysts and the unique core-shell structure. Wang et al. achieved an earth-abundant photocatalyst composed of a Ni-Tu-TETN complex and graphitic C$_3$N$_4$ [103]. The photocatalyst showed a high hydrogen generation rate of 51 μmol/h, which was much better than that of other metal complexes (Co^{2+}, Fe^{2+}, and Cu^{2+}). The enhanced photocatalytic performance was attributed to the accelerated electron transfer from g-C$_3$N$_4$ to Ni molecule.

Hexagonal cadmium sulfide not only is a good candidate for visible-light-driven photocatalyst owing to its narrow band gap and suitable band position but also displays pyroelectric-catalytic hydrogen evolution owing to its noncentral-symmetric structure. When an organic molecule 2-mercaptobenzimidazole (2-MBI) was coupled onto CdS nanorods, CdS-2MBI displayed enhanced pyroelctric-catalytic hydrogen generation because the pyroelectric-induced charge carrier separation was apparently improved [104]. A trinuclear iron-oxo complex has been demonstrated to be a regeneratable cocatalyst for solar-driven hydrogen generation. The nanosized trinuclear iron-oxo cluster not only improved the visible absorption of polyoxometalate catalyst but also accelerated the charge carrier separation. As a result, the hybrid photosystem showed a hydrogen generation rate of 21.6 μmol/h [105]. A noble-metal-free hybrid photocatalyst, in which

ZnS was used as a light absorber and [Fe_2S_2] hydrogenase mimic [(μ-SPh-4-NH_2)$_2$$Fe_2$$(CO)_6$] worked as hydrogen evolution cocatalyst, was designed by Li's group [106]. The hybrids showed a hydrogen generation rate of 39 μmol/h, which was 12 times higher than the single ZnS, with excellent photocatalytic stability. Photo-induced electron transfer from ZnS absorber to [Fe_2S_2] hydrogenase mimic was also proved.

11.3 Factors Determining the Cocatalyst Activity

According to previous research, cocatalysts play an important role in improving photocatalytic efficiency. The performance of cocatalysts is largely depending on two aspects, including the instinct properties of cocatalysts and the interfacial coupling between the host semiconductors and loaded cocatalysts. In the following sections, we will detailly discuss the two aspects to provide a solid understanding of cocatalysts.

11.3.1 Intrinsic Properties of Cocatalysts

First of all, intrinsic features of cocatalysts, including the morphology, crystallization, and composition, should be considered during the rational design and development of efficient and stable cocatalysts.

a) Morphology of cocatalysts

The morphology of cocatalysts, especially the size and shape of cocatalysts, plays an important role in photocatalytic activity [107–109]. In general, efficient cocatalysts-semiconductor heterostructures can be achieved by reducing the size of cocatalysts because of the large specific area and short charge transfer distance. Besides, smaller cocatalysts can avoid the shield of host photocatalysts, which is important for light absorption and active site exposure. MoS_2 quantum dots prepared via a liquid exfoliation method were used to decorate commercial CdS and the hybrids displayed better performance than bulk MoS_2 decorated CdS [110]. The highest hydrogen generation rate of 1.032 mmol/h was achieved owing to the efficient charge separation and abundant active sites (Figure 11.3a). Recently, single-atom cocatalysts are very attractive in improving the photocatalytic efficiency because of the efficient utilization of the element [111–113]. The Ni-based Janus single-atom sites on red phosphorus (Ni-HRP) were carefully synthesized by a reactive-group guided strategy [114]. The hydrothermal treatment led to the synthesis of P-H and P-OH reactive groups, which further induced the generation of the Janus structure with two different single sites: P-Ni and P-O-Ni. The Janus sites performed different water reduction sites and oxidative sites, respectively. Because the water reduction and oxidation features in Ni-HRP were enhanced simultaneously, efficient separation of electron-hole pairs and accelerated photo-carriers transfer were achieved (Figure 11.3b).

Shape control is also important during the preparation of cocatalysts [40, 62, 115, 116]. For instance, coralline-like Ni_2P was prepared via a solvothermal method and coupled onto tetrapod bundle $Zn_{0.1}Cd_{0.9}S$ semiconductor using a calcination method [117]. It was discovered that Ni_2P loading successfully improved the photocatalytic performance of $Zn_{0.1}Cd_{0.9}S$ tetrapod by accelerating the photoinduced charge carrier separation and transport. Besides, Coralline-like Ni_2P could provide active sites for water reduction to hydrogen (Figure 11.3c). MoS_2 nanosheet was demonstrated as an efficient and low-cost cocatalyst for solar-driven water splitting. Su et al. fabricated a lamellar flower-like porous MoS_2 to decorate

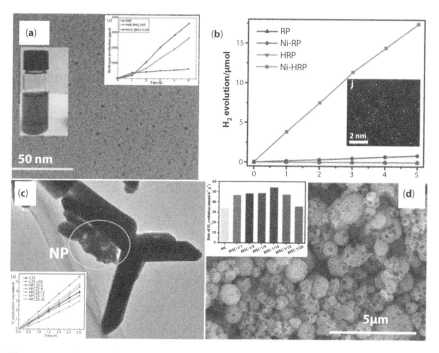

Figure 11.3 Photocatalytic hydrogen generation over different catalysts. (a) MoS_2 quantum dots decorated CdS. Reproduced from [110]. Copyright 2017 Elsevier. (b) Atomically dispersed Janus Ni decorated red phosphorus. Reproduced from [114]. Copyright 2022 Wiley. (c) Coralline-like Ni_2P decorated tetrapod-bundle $Cd_{0.9}Zn_{0.1}S$. Reproduced from [117]. Copyright 2021 Elsevier, and (d) lamellar flower-like porous MoS_2 decorated CdS. Reproduced from [118]. Copyright 2021 Royal Society of Chemistry.

the CdS semiconductor [118]. After loading MoS_2 cocatalyst, the photocatalytic hydrogen evolution rate reached 54.1 mmol·g^{-1}·h^{-1}, which was 36 times higher than CdS alone. Efficient charge separation and active sites were expected in CdS/MoS_2 owing to the spherical porous and lamellar structure of MoS_2 as well as the intimate contact between CdS and MoS_2 (Figure 11.3d).

b) Crystallization of cocatalysts

Usually, well-crystalized cocatalysts have a positive effect on photocatalytic performance due to the fewer defects [90, 119, 120]. The crystallinity and surface hydrophobicity of CoP cocatalyst were detailly investigated by Wang's group [121]. The prepared CoP cocatalysts were calcinated at different temperatures for 2 h in the H_2/Ar atmosphere. The CoP thermally post-treated at 600°C exhibited enhanced crystallinity and tailored surface hydrophobicity. The promoted charge separation and reinforced CO_2 chemisorption were achieved in CoP-600 with the best photocatalytic CO_2 reduction activity (Figure 11.4a).

Interestingly, it was reported that the amorphous cocatalyst could provide an enhanced performance as well [122–125]. Ultra-small noncrystalline Cu_xP nanodots were used as the cocatalyst to boost the hydrogen production of TiO_2 [126]. The prepared Cu_xP with an average size of 2 to 5 nm were homogeneously decorated onto TiO_2 surface via a triethanolamine-assisted photo deposition method. The loading of noncrystalline Cu_xP introduced a built-in electric field and inhibited the charge recombination.

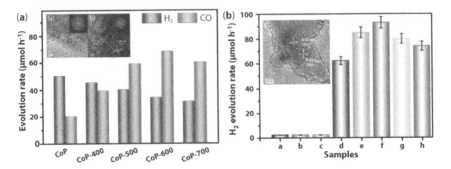

Figure 11.4 (a) Photocatalytic CO_2 reduction over CoP/[Ru(bpy)$_3$]$^{2+}$. Reproduced from [121]. Copyright 2021 Wiley. (b) Photocatalytic hydrogen generation from Cu_xP-TiO_2. Reproduced from [126]. Copyright 2020 Royal Society of Chemistry.

As expected, Cu_xP-TiO_2 displayed a boosted photocatalytic activity compared with pure TiO_2 (Figure 11.4b).

c) Composition of cocatalysts

The photocatalytic performance of cocatalyst-semiconductor hybrids can be further improved by tuning the composition of cocatalysts [127]. Taking transition-metal-based semiconductors as an example, owing to the multiple valence states, M_xN_y (M: Fe, Co, Ni; N: S, O, P) cocatalysts can be adjusted by varying the values of x and y [128]. The efficiencies certainly differ based on the composition of cocatalysts. Ni_xS_y cocatalyst, composed of NiS and Ni_3S_4, has been demonstrated to be an efficient cocatalyst for trapping electrons and improving photocatalytic activity by optimizing the NiS/Ni_3S_4 molar ratio [41]. The composition of cocatalysts can be tuned by introducing other elements [81, 122, 129–131]. For example, NiCoP, a new type of ternary metal phosphide, emerged as an efficient cocatalyst for solar energy conversion owing to good electronic conductivity and tunable electronic structure. Song et al. adjusted the proportions of Ni and Co in NiCoP and investigated the cocatalytic activity [132]. The optimized Ni-Co ratio was determined to be 7:3 and the cocatalyst significantly improved the activity and stability of CdS photocatalyst, which was even superior to the Pt/CdS system.

Both CoSe and NiSe are promising cocatalysts for solar water splitting. To improve the photocatalytic activity of $ZnIn_2S_4$, $Ni_{1-x}Co_xSe_2$-C cocatalysts were well designed and prepared [133]. The constructed 2D/2D heterostructure exhibited intimate 2D heterointerface with rich heterojunctions. According to the spectroscopic study and theoretical calculations, the charge transfer property from $ZnIn_2S_4$ to $Ni_{1-x}Co_xSe_2$ was greatly enhanced. Under visible light irradiation, the hierarchical nanocages of $Ni_{1-x}Co_xSe_2$-C/$ZnIn_2S_4$ showed excellent photoactivity toward hydrogen generation with a rate of 5.10 mmol·g^{-1}·h^{-1}. Multicomponent cocatalyst, polyaniline-derived N-doped carbon encapsulated copper/copper phosphide (Cu/Cu_xP@NC) nanoparticles, was well designed and prepared using a facile one-step phosphorization-pyrolysis way [134]. The Cu/Cu_xP@NC cocatalysts could effectively improve the photocatalytic activity of TiO_2 and g-C_3N_4. The improved catalytic activity was attributed to efficient charge transfer from TiO_2 or g-C_3N_4 to Cu/Cu_xP@NC and sufficient sites for reduction on Cu/Cu_xP@NC cocatalyst.

Noble metal Pt and Au were demonstrated to be good cocatalysts for hydrogen generation. Recently, Li's group adopted Au atoms to optimize the hydrogen absorption energy of Pt cocatalyst, and the obtained Pt-Au alloy successfully improved the photocatalytic activity of TiO_2. Typically, the cocatalytic performance was much better than Pt metal alone [135].

11.3.2 Interfacial Coupling of Cocatalysts With Host Semiconductors

Charge transfer efficiency across the interface of semiconductor/cocatalyst plays a vital role in determining the photocatalytic performance and charge separation efficiency. In general, the poor interfacial contact and inadequate coupling could increase the interfacial charge transfer resistance, which is undesirable for photocatalytic water splitting. Interfacial engineering of semiconductor/cocatalyst heterostructures has attracted tremendous attention and received remarkable achievements [59, 136, 137]. For example, Qin *et al.* adopted a one-step hydrothermal method to construct NiS/CdS hybrids. CdS semiconductor and NiS_x cocatalysts underwent intergrowth with each other during the hydrothermal reaction [138]. Owing to the two-phase intergrowth effect, NiS_x cocatalyst was tightly coupled to CdS semiconductor. The hybrids showed an exciting hydrogen generation rate of 2.86 mmol/h with an apparent quantum yield of 60.4% at 420 nm. The high performance was attributed to the intimate contact at the interface of NiS_x/CdS and monodispersed NiS_x nanoparticles on the surface of CdS.

Loading cocatalyst on specific locations could control the carrier transfer direction and reduce the electron-hole recombination [7, 139–142]. Hydrogen evolution cocatalysts (Pt) and oxygen evolution cocatalysts (MnO_x) were selectively deposited on the (100) and (001) plane of $BiVO_4$ semiconductor, and photoinduced charge carriers could transfer to desired active sites for the redox reaction. Tantalum nitride is regarded as a promising photocatalyst or photoanode for visible-light driven overall water splitting owing to its narrow band gap (E_g = 2.1 eV) and suitable band edge position. CoO_x has been proved to be a good water oxidation cocatalyst. Coupling them together to construct the heterostructure could achieve boosted photocatalytic activity. By introducing a magnesia monolayer, the overall water splitting activity was improved by 23 times [143]. The magnesia monolayer favored the loading of CoO_x cocatalysts and expanded the interfacial contact area. Meanwhile, the monolayer magnesia served as a passivation layer to prevent the oxidation of Ta_3N_5. All these merits led to a recorded apparent quantum yield of 11.3% at 500 to 600 nm for Ta_3N_5 photocatalysts.

a) Cocatalyst location
Though metal sulfides display excellent photocatalytic hydrogen generation, it is difficult to realize the overall water splitting owing to the severe electron-hole recombination and limited active site for water oxidation. Chai's group developed RuO_2/CdS/MoS_2 hybrids with spatially separated catalytic sites, in which hydrogen evolution cocatalyst MoS_2

was anchored on the ends of CdS nanorods and oxygen evolution cocatalyst RuO_2 was deposited on the sidewall [144]. The prepared CdS-based photosystem showed overall water splitting activity owing to the efficient charge separation. Hole trapping by the sidewall RuO_2 inhibited the oxidation of sulfide ions and provided the active sites for oxidation of OH^- (Figure 11.5a).

Wei et al. adopted a selectively deposition strategy to achieve a unique structure, in which Pt cocatalyst was loaded on the inner surface of $ZnTiO_{3-x}N_y$ hollow nanospheres, while RhO_x cocatalyst was loaded onto the outer surface [145]. On one hand, the selective deposition of Pt and RhO_x inhibited the back reaction of H_2 with O_2. On the other hand, the structure introduced a strong potential gradient promoting charge separation and migration. Consequently, the rational designed $Pt-ZnTiO_{3-x}N_y-RhO_x$ exhibited an outstanding performance for photocatalytic overall water splitting (Figure 11.5b).

Li's group synthesized a p-type semiconductor $Cu_{1.94}S$ nanosphere as seeds and $Zn_xCd_{1-x}S$ nuclei were attached to the (800) plane of $Cu_{1.94}S$ nanosphere coherently. Then $Zn_xCd_{1-x}S$ grew to nanorods along the [001] direction [146]. The well-designed $Cu_{1.94}S-Zn_xCd_{1-x}S$ hybrids displayed an efficient photocatalytic hydrogen evolution rate of 7.735 mmol·g^{-1}·h^{-1}, which was 59 times higher than single CdS nanorods. The p-type $Cu_{1.94}S$ could extend the light absorption of $Zn_xCd_{1-x}S$ owing to its narrow bandgap and surface plasmon resonances (Figure 11.5c). Meanwhile, the coherently epitaxial growth formed a tight contact between $Cu_{1.94}S$ and $Zn_xCd_{1-x}S$, leading to efficient charge transfer at the heterointerfaces. The photocatalytic activity could be further improved by depositing Pt cocatalysts on the $Cu_{1.94}S-Zn_xCd_{1-x}S$ nanorods.

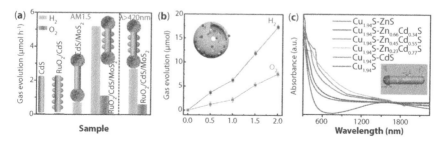

Figure 11.5 Photocatalytic hydrogen generation over different catalysts. (a) RuO_2/CdS/MoS_2 nanodumbbells. Reproduced from [144]. Copyright 2020 Wiley. (b) Pt@$ZnTiO_{3-x}Ny$@RhO. Reproduced from [145]. Copyright 2021 Wiley. (c) $Cu_{1.94}S-Zn_xCd_{1-x}S$. Reproduced from [146]. Copyright 2016 American Chemical Society.

b) Built-in electric field

When the cocatalyst was coupled to the host photocatalysts, a built-in electric field such as p-n junction or van der Waals junction would form, which could greatly improve the charge transfer ability [35, 147–150]. Xu et al. applied 1T-MoS_2 to decorate monolayer g-C_3N_4 (O-g-C_3N_4) [151]. The binary composite displayed a photocatalytic hydrogen evolution rate of 1.84 mmol·g^{-1}·h^{-1}, four times higher than that of the Pt/g-C_3N_4 hybrids. Owing to the *in-situ* growth strategy, the close contact between MoS_2 and g-C_3N_4 was achieved. Thus, charge carrier transfer from g-C_3N_4 to MoS_2 was facilitated. Meanwhile, 1T-MoS_2 could provide abundant active sites for photocatalytic reaction, which was another reason for the remarkable performance of 1T-MoS_2/O-g-C_3N_4 system. Last but not least, the two-dimensional heterostructure formed by van der Waals interaction between 2D MoS_2 and 2D g-C_3N_4 reduced the Schottky barrier and accelerated the efficiency of charge transfer (Figure 11.6a).

A p-n heterojunction photosystem composed of Cu_3P/g-C_3N_4 was constructed by Chen's group [152]. According to the electronic structure of

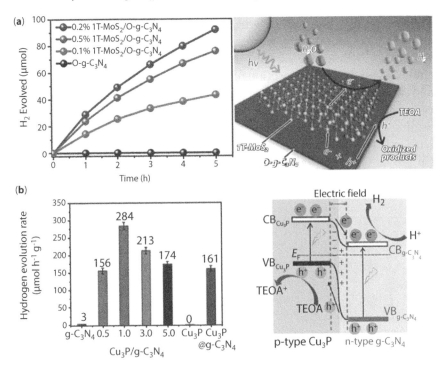

Figure 11.6 Photocatalytic hydrogen generation over different catalysts. (a) Metallic 1T-MoS_2/monolayer O-g-C_3N_4. Reproduced from [151]. Copyright 2018 Elsevier. (b) Cu_3P/g-C_3N_4. Reproduced from [152]. Copyright 2018 Springer.

Cu_3P and g-C_3N_4, the Fermi level of Cu_3P was lower than that of g-C_3N_4. After loading Cu_3P cocatalysts on g-C_3N_4 to form the heterojunction, a built-in electric field was formed at the heterointerface owing to the balance of the Fermi level. Under visible light irradiation, the constructed p-n heterojunction showed enhanced photocatalytic hydrogen production because of the efficient charge separation from the conduction band of Cu_3P to that of g-C_3N_4 (Figure 11.6b).

c) Contact manners

As shown in Figure 11.7, the cocatalyst-semiconductor photocatalysts can be classified into four different types, namely point (0D), line (1D), face (2D), and non-contact (bridge and free), from a geometrical view [94, 153, 154]. Generally, the tight and large contact between cocatalysts and host photocatalysts is favorable for designing cocatalysts because of rich carrier transfer channel and small transfer resistance. 2D tandem heterojunction composed of $Cd_xZn_{1-x}In_2S_4$, CdS, and MoS_2 was obtained by the hydrothermal growth of CdS on $Cd_xZn_{1-x}In_2S_4$ with the subsequent loading MoS_2 on $Cd_xZn_{1-x}In_2S_4$-CdS [155]. By optimizing the weight ratio of components, $Cd_xZn_{1-x}In_2S_4$-CdS-MoS_2 with 5wt% CdS and 3wt% MoS_2 exhibited the highest hydrogen generation rate of 27.14 mmol·g^{-1}·h^{-1}, 47 times higher than that of $ZnIn_2S_4$, with an apparent quantum yield of 19.97% at 400 nm. The tandem 2D interface provided an efficient charge transfer path for electron-hole separation. Furthermore, the 2D structure of MoS_2 offered abundant active sites for water reduction.

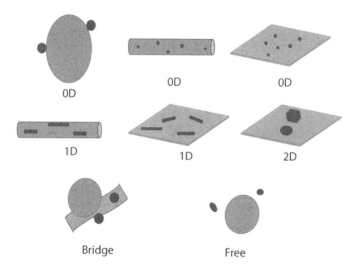

Figure 11.7 Contact manners in cocatalysts-semiconductor photosystems.

The contact manner can be adjusted by various strategies [19, 156]. For instance, the effects of annealing conditions were investigated by Domen's group [157]. Under different air conditions, the annealed CoO_x-$BaTaO_2N$ displayed varied photocatalytic oxygen evolution properties. Namely, after annealing in the H_2 flow, the photocatalytic activity of post-annealed CoO_x-$BaTaO_2N$ could be further increased. The superior catalytic performance is attributed to the intimate contact and abundance of exposed surface area.

Generally, the tight contact between cocatalysts and host semiconductor is beneficial for charge transfer at the heterointerface. For instance, coupling Pt-MoS_2 cocatalysts with environment-friendly semiconductors with a tight interaction could reduce charge recombination and lead to continuous charge transfer [158]. However, it is interesting to find that charge transfer from the host to cocatalyst can also occur with an indirect contact, even in separate cocatalyst-semiconductor system without direct contact [159, 160]. Commonly, a mediate is introduced in the

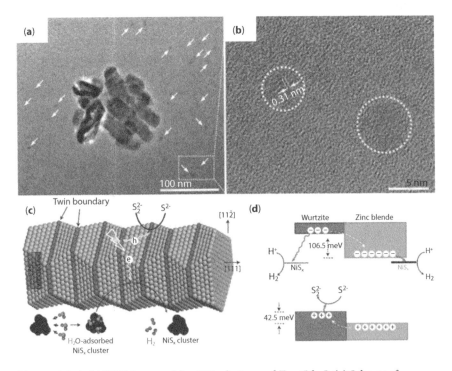

Figure 11.8 (a, b) TEM images of free NiS_x clusters and $Zn_{0.5}Cd_{0.5}S$. (c) Scheme of collisions between NiS_x clusters and $Zn_{0.5}Cd_{0.5}S$. (d) Band-alignment and charge transfer between NiS_x clusters and $Zn_{0.5}Cd_{0.5}S$. Reproduced from [163]. Copyright 2016 Springer Nature.

cocatalyst-semiconductor system to bridge the two components and assists the charge separation in these photosystems. Carbon-based 2D materials, such as graphene and reduced graphene oxide, are used as bridge materials owing to their superior electronic conductivity [161, 162]. A novel collision contact mechanism was proposed between NiS_x cocatalyst and $Zn_{0.5}Cd_{0.5}S$ photocatalyst, which yielded an internal quantum efficiency approaching 100% [163]. The prepared NiS_x cocatalyst was not anchored on $Zn_{0.5}Cd_{0.5}S$ and instead existed in the form of freestanding clusters (Figure 11.8a and b). As shown in Figure 11.8c, free NiS_x clusters absorbed H_2O and formed a Ni-H bond. During the collision process, photogenerated electrons could transfer to NiS_x clusters and reduce water to hydrogen (Figure 11.8d). The collision could significantly facilitate charge separation and inhibit the back reaction.

11.4 Advanced Characterization Techniques for Cocatalytic Process

Semiconductor-cocatalyst heterostructures have been demonstrated as efficient photocatalysts for hydrogen production compared with semiconductors alone. Efficient charge separation between semiconductors and cocatalysts is expected commonly. With the rapid development of time-resolved spectra technology, the process of photoinduced charge carrier formation, transition, separation, and recombination could be observed and characterized directly [164, 165]. WSe_2 cocatalysts with a wide range of semiconductors (CdS nanorods, CdSe/CdS dot-in-rods, TiO_2 nanoparticles, g-C_3N_4 nanosheet, and $Zn_xCd_{1-x}S$ nanorods) was constructed, and the charge separation states of $WSe_2^{(-)}$-semiconductor$^{(+)}$ were characterized by transient absorption spectra study (Figure 11.9a and b) [87, 88]. According to the photocarrier dynamic kinetics, electron transfer from CdS and CdSe/CdS nanorods to Pt tip was approximated to unity, while holes trapping to the sacrificial agency was positively correlated to the photocatalytic hydrogen generation efficiency [166]. All these results indicated that in a quantum semiconductor-metal system, the efficiency-limiting step is the hole trapping rather than electron transfer (Figure 11.9c).

A stratified $CdS/Cu_{2-x}S$-MoS_2 heteronanoarchitecture was carefully designed by Liu et al. $CdS/Cu_{2-x}S$ core-shell nanorods were firstly synthesized using a cation exchange way and then the epitaxial vertical growth of MoS_2 layer on $CdS/Cu_{1-x}S$ nanorods was achieved by a seed growth method [167]. The MoS_2 coating layer could enhance photocatalytic activity and photostability. Due to the stabilized interfacial electronic state, $CdS/Cu_{2-x}S$

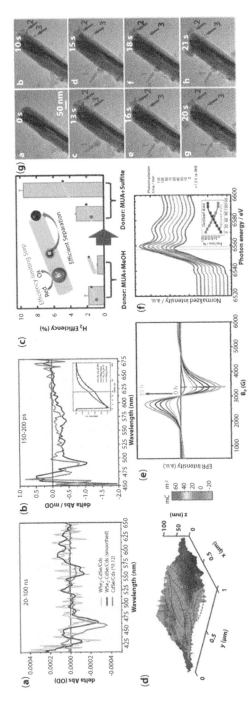

Figure 11.9 (a) Charge separation state in WSe$_2$-CdSe/CdS. Reproduced from [87]. Copyright 2020 American Chemical Society. (b) Charge separation state in WSe$_2$-Zn$_{0.1}$Cd$_{0.9}$S. Reproduced from [88]. Copyright 2020 Elsevier. (c) Photocatalytic activity in Pt decorated CdS and CdSe/CdS. Reproduced from [166]. Copyright 2014 American Chemical Society. (d) Surface charge distribution in stratified CdS-Cu$_{2-x}$S/MoS$_2$ photocatalysts. Reproduced from [167]. 2020 American Chemical Society. (e) *In situ* EPR spectroscopy study of Ni/g-C$_3$N$_4$. Reproduced from [168]. Copyright 2016 Royal Society of Chemistry. (f) XAFS spectroscopy study of MnO$_x$-SrTiO$_3$. Reproduced from [73]. Copyright 2014 American Chemical Society. (g) Time-lapse screenshots of a Cu nanowire in HAuCl$_4$ solution at initial stages. Reproduced from [172]. Copyright 2021 Royal Society of Chemistry.

owned the longest carrier lifetime. After coating the stratified MoS_2 layer, photo-induced carrier lifetime was reduced to 51 ps, indicating the efficient charge transfer from $CdS/Cu_{2-x}S$ nanorods to MoS_2 cocatalyst. The charge separation behaviors were further studied using a scanning ion conductance microscope and the surface charge distributions were mapped. As shown in Figure 11.9d, efficient charge separation was achieved on the rationally designed $CdS/Cu_{2-x}S$-MoS_2 heteronanoarchitecture.

To identify the real active species during the water splitting process, an *in situ* electron paramagnetic resonance (EPR) spectroscopy was adopted to investigate the active species in Ni decorated g-C_3N_4 [168]. During the photocatalytic reaction, Ni^{2+} could be reduced to Ni particle and the reduced Ni acted as cocatalysts to extract the electrons for water reduction (Figure 11.9e). Besides, X-ray absorption fine structure (XAFS) spectroscopy is a powerful tool to study the local structure of photocatalysts. It was discovered that photo induced hole transfer from the $SrTiO_3$ photoelectrode to the MnO_x cocatalyst was dependent on both the electrode potential and the UV light photon intensity using an *in situ* XAFS spectroscopy (Figure 11.9f) [73].

In situ observation and characterization of cocatalysts during photocatalytic water splitting is of great importance for efficient catalyst design [169–171]. Gao *et al.* prepared large-sized cocatalysts (Cu-Au) to decorate ultra-small TiO_2 semiconductors and applied *in situ* liquid cell transmission electron microscopy to observe the microstructure evolution of Au nanoparticles on Cu nanowires [172]. After the injection of $HAuCl_4$ solution, free Au particles could adhere onto Cu nanowires, and simultaneously the adhered large Au particle at the initial stage would be collapsed and reshaped (Figure 11.9g). It was found that Cu-Au cocatalysts formed at the initial state, which was composed of solid Cu nanowire and a smaller amount of Au particles, displayed the best photocatalytic performance.

11.5 Conclusion

Photocatalytic water splitting will be cost-competitive when the solar-to-hydrogen conversion efficiency reaches about 10%. Cocatalysts modified host photocatalysts display enhanced photocatalytic efficiency and stability, and hold the promise for the practical application of this technology. Based on this chapter, we have a fundamental understanding of the cocatalyst-semiconductor system. In order to achieve practically used photosystems, efficiency and stability are required to be further improved. Developing low-cost, nontoxic, earth-abundant, and efficient cocatalysts is critically required in the future. The following points should be considered in exploring ideal cocatalysts: 1) deep insight into the photocatalytic process and mechanism; 2) precise control of intrinsic physiochemical properties of cocatalysts and the coupling manner between cocatalysts and host photocatalysts; 3) adjustment of photo-induced charge transfer kinetics by interfacial engineering and band alignment engineering. It is expected that this chapter can provide meaningful guidance to develop efficient and durable cocatalysts for diverse photocatalytic reactions.

Acknowledgments

The authors thank support from the National Natural Science Foundation of China (52076177, 52106259), China National Key Research and Development Plan Project (2021YFF0500503), Sichuan Science and Technology Program (2021YFSY0047), the Postdoctoral Science Foundation (2021M692005), and the China Fundamental Research Funds for the Central Universities.

References

1. Wang, Q. and Domen, K., Particulate photocatalysts for light-driven water splitting: Mechanisms, challenges, and design strategies. *Chem. Rev.*, 120, 2, 919–985, 2020.
2. Zhu, J., Hu, L., Zhao, P., Lee, L., Wong, K., Recent advances in electrocatalytic hydrogen evolution using nanoparticles. *Chem. Rev.*, 120, 2, 851–918, 2019.
3. Wang, Z., Li, C., Domen, K., Recent developments in heterogeneous photocatalysts for solar-driven overall water splitting. *Chem. Soc. Rev.*, 48, 7, 2109–2125, 2019.
4. Fujishima, A. and Honda, K., Electrochemical photolysis of water at a semiconductor electrode. *Nature*, 238, 5358, 37–38, 1972.
5. Wang, L., Zhang, Y., Chen, L., Xu, H., Xiong, Y., 2D polymers as emerging materials for photocatalytic overall water splitting. *Adv. Mater.*, 120, 2, 1801955, 2018.
6. Kato, H., Asakura, K., Kudo, A., Highly efficient water splitting into H_2 and O_2 over lanthanum-doped $NaTaO_3$ photocatalysts with high crystallinity and surface nanostructure. *J. Am. Chem. Soc.*, 125, 10, 3082–3089, 2003.
7. Chen, M., Liu, Y., Li, C., Li, A., Chang, X., Liu, W., Sun, Y., Wang, T., Gong, J., Spatial control of cocatalysts and elimination of interfacial defects towards efficient and robust CIGS photocathodes for solar water splitting. *Energy Environ. Sci.*, 11, 8, 2025–2034, 2018.

8. Wan, L., Xiong, F., Zhang, B., Che, R., Li, Y., Yang, M., Achieving photocatalytic water oxidation on LaNbON$_2$ under visible light irradiation. *J. Energy Chem.*, 27, 2, 367–371, 2018.
9. Naldoni, A., Altomare, M., Zoppellaro, G., Liu, N., Kment, S., Zboril, R., Schmuki, P., Photocatalysis with reduced TiO$_2$: From black TiO$_2$ to cocatalyst-free hydrogen production. *ACS Catal.*, 9, 1, 345–364, 2019.
10. Guo, X., Chen, Y., Qin, Z., Wang, M., Guo, L., One-step hydrothermal synthesis of Zn$_x$Cd$_{1-x}$S/ZnO heterostructures for efficient photocatalytic hydrogen production. *Int. J. Hydrogen Energy*, 41, 34, 15208–15217, 2016.
11. Guo, X., Chen, Y., Qin, Z., Su, J., Guo, L., Facet-selective growth of cadmium sulfide nanorods on zinc oxide microrods: Intergrowth effect for improved photocatalytic performance. *ChemCatChem*, 10, 1, 153–158, 2018.
12. Gao, Y., Li, X., Wu, H., Meng, S., Fan, X., Huang, M., Guo, Q., Tung, C., Wu, L., Exceptional catalytic nature of quantum dots for photocatalytic hydrogen evolution without external cocatalysts. *Adv. Funct. Mater.*, 28, 33, 1801769, 2018.
13. Garcia-Mulero, A., Rendon-Patino, A., Asiri, A.M., Primo, A., Garcia, H., Band engineering of semiconducting microporous graphitic carbons by phosphorous doping: Enhancing of photocatalytic overall water splitting. *ACS Appl. Mater. Interfaces*, 13, 41, 48753–48763, 2021.
14. Lv, Z., Li, W., Cheng, X., Liu, B., Guo, Z., Zhuang, T., Zhang, C., Constructing internal electric field in CdS via Bi, Ni co-doping strategy for enhanced visible-light H$_2$ production. *Appl. Surf. Sci.*, 556, 149758, 2021.
15. Li, Y. and Tsang, S.C.E., Recent progress and strategies for enhancing photocatalytic water splitting. *Mater. Today Sustain.*, 9, 100032, 2020.
16. Song, H., Meng, X., Wang, S., Zhou, W., Song, S., Kako, T., Ye, J., Selective photo-oxidation of methane to methanol with oxygen over dual-cocatalyst-modified titanium dioxide. *ACS Catal.*, 10, 23, 14318–14326, 2020.
17. Chen, S., Takata, T., Domen, K., Particulate photocatalysts for overall water splitting. *Nat. Rev. Mater.*, 2, 10, 17050, 2017.
18. Peng, W., Li, Y., Zhang, F., Zhang, G., Fan, X., Roles of two-dimensional transition metal dichalcogenides as cocatalysts in photocatalytic hydrogen evolution and environmental remediation. *Ind. Eng. Chem. Res.*, 56, 16, 4611–4626, 2017.
19. Li, T., Wang, K., Fang, Q., Zhang, Y., Wang, B., Li, R., Lin, Y., Liu, K., Xie, H., Li, K., Conductive polymer supported and confined iron phosphide nanocrystals for boosting the photocatalytic hydrogen production of graphitic carbon nitride. *J. Mater. Chem. C*, 8, 41, 14540–14547, 2020.
20. Guo, J., Liang, Y., Wang, H., Liu, L., Cui, W., The cocatalyst in photocatalytic hydrogen evolution. *Prog. Chem.*, 33, 7, 1100–1114, 2021.
21. Gong, S., Jiang, Z., Shi, P., Fan, J., Xu, Q., Min, Y., Noble-metal-free heterostructure for efficient hydrogen evolution in visible region: Molybdenum nitride/ultrathin graphitic carbon nitride. *Appl. Catal. B Environ.*, 238, 318–327, 2018.

22. Deng, J., Zhang, Q., Feng, K., Lan, H., Zhong, J., Chaker, M., Ma, D., Efficient photoelectrochemical water oxidation on hematite with fluorine-doped FeOOH and FeNiOOH as dual cocatalysts. *ChemSusChem*, 11, 21, 3783–3789, 2018.
23. Gao, X., Zeng, D., Yang, J., Ong, W., Fujita, T., He, X., Liu, J., Wei, Y., Ultrathin Ni(OH)$_2$ nanosheets decorated with Zn$_{0.5}$Cd$_{0.5}$S nanoparticles as 2D/0D heterojunctions for highly enhanced visible light-driven photocatalytic hydrogen evolution. *Chin. J. Catal.*, 42, 7, 1137–1146, 2021.
24. Jian, J., Kumar, R., Sun, J., Cu$_2$O/ZnO p–n junction decorated with NiO$_x$ as a protective layer and cocatalyst for enhanced photoelectrochemical water splitting. *ACS Appl. Energy Mater.*, 3, 11, 10408–10414, 2020.
25. Chen, J., Wang, C., Zhang, Y., Guo, Z., Luo, Y., Mao, C., Engineering ultrafine NiS cocatalysts as active sites to boost photocatalytic hydrogen production of MgAl layered double hydroxide. *Appl. Surf. Sci.*, 506, 144999, 2020.
26. Zhong, W., Wu, X., Liu, Y., Wang, X., Fan, J., Yu, H., Simultaneous realization of sulfur-rich surface and amorphous nanocluster of NiS$_{1+x}$ cocatalyst for efficient photocatalytic H$_2$ evolution. *Appl. Catal. B Environ.*, 280, 119455, 2021.
27. Bhavani, P., Praveen Kumar, D., Jeong, S., Kim, E.H., Park, H., Hong, S., Gopannagari, M., Amaranatha Reddy, D., Song, J.K., Kim, T.K., Multidirectional-charge-transfer urchin-type Mo-doped W$_{18}$O$_{49}$ nanostructures on CdS nanorods for enhanced photocatalytic hydrogen evolution. *Catal. Sci. Technol.*, 8, 7, 1880–1891, 2018.
28. Gao, D., Wu, X., Wang, P., Yu, H., Zhu, B., Fan, J., Yu, J., Selenium-enriched amorphous NiSe$_{1+x}$ nanoclusters as a highly efficient cocatalyst for photocatalytic H$_2$ evolution. *Chem. Eng. J.*, 408, 127230, 2021.
29. Qiu, S., Shen, Y., Wei, G., Yao, S., Xi, W., Shu, M., Si, R., Zhang, M., Zhu, J., An, C., Carbon dots decorated ultrathin CdS nanosheets enabling *in-situ* anchored Pt single atoms: A highly efficient solar-driven photocatalyst for hydrogen evolution. *Appl. Catal. B Environ.*, 259, 118036, 2019.
30. Li, H., Lu, D., Chen, S., Hisatomi, T., Vequizo, J.J.M., Xiao, J., Wang, Z., Lin, L., Xiao, Q., Sun, Y., Miseki, Y., Sayama, K., Yamakata, A., Takata, T., Domen, K., A Na-containing Pt cocatalyst for efficient visible-light-induced hydrogen evolution on BaTaO$_2$N. *J. Mater. Chem. A*, 9, 24, 13851–13854, 2021.
31. Sun, K., Shen, J., Liu, Q., Tang, H., Zhang, M., Zulfiqar, S., Lei, C., Synergistic effect of Co(II)-hole and Pt-electron cocatalysts for enhanced photocatalytic hydrogen evolution performance of P-doped g-C$_3$N$_4$. *Chin. J. Catal.*, 41, 1, 72–81, 2020.
32. Chen, H., Zhang, F., Sun, X., Zhang, W., Li, G., Effect of reaction atmosphere on photodeposition of Pt nanoparticles and photocatalytic hydrogen evolution from SrTiO$_3$ suspension system. *Int. J. Hydrogen Energy*, 43, 10, 5331–5336, 2018.
33. Gao, D., Wu, X., Wang, P., Xu, Y., Yu, H., Yu, J., Simultaneous realization of direct photoinduced deposition and improved H$_2$-evolution performance of

Sn-nanoparticle-modified TiO_2 photocatalyst. *ACS Sustain. Chem. Eng.*, 7, 11, 10084–10094, 2019.

34. Kong, L., Ji, Y., Dang, Z., Yan, J., Li, P., Li, Y., Liu, S.F., g-C_3N_4 loading black phosphorus quantum dot for efficient and stable photocatalytic H_2 generation under visible light. *Adv. Funct. Mater.*, 28, 22, 1800668, 2018.
35. Wei, R., Huang, Z., Gu, G., Wang, Z., Zeng, L., Chen, Y., Liu, Z., Dual-cocatalysts decorated rimous CdS spheres advancing highly-efficient visible-light photocatalytic hydrogen production. *Appl. Catal. B Environ.*, 231, 101–107, 2018.
36. Xu, Q., Zhu, B., Cheng, B., Yu, J., Zhou, M., Ho, W., Photocatalytic H_2 evolution on graphdiyne/g-C_3N_4 hybrid nanocomposites. *Appl. Catal. B Environ.*, 255, 117770, 2019.
37. Wang, M., Ma, Y., Fo, Y., Lyu, Y., Zhou, X., Theoretical insights into the origin of highly efficient photocatalyst NiO/$NaTaO_3$ for overall water splitting. *Int. J. Hydrogen Energy*, 45, 38, 19357–19369, 2020.
38. Zhao, G. and Xu, X., Cocatalysts from types, preparation to applications in the field of photocatalysis. *Nanoscale*, 13, 24, 10649–10667, 2021.
39. Fu, J., Bie, C., Cheng, B., Jiang, C., Yu, J., Hollow CoS_x polyhedrons act as high-efficiency cocatalyst for enhancing the photocatalytic hydrogen generation of g-C_3N_4. *ACS Sustain. Chem. Eng.*, 6, 2, 2767–2779, 2018.
40. Ren, S., Wei, W., Han, C., Li, Y., Tian, Q., Fan, M., Zhuang, J., Morphology and phase engineering of MoS_2 cocatalyst for high-efficiency hydrogen evolution: One-step clean synthesis and comparative studies. *J. Phys. Chem. C*, 125, 44, 24451–24462, 2021.
41. Han, L., Peng, C., Huang, J., Sun, L., Wang, S., Zhang, X., Yang, Y., Noble-metal-free Ni_xS_y-C_3N_5 hybrid nanosheet with highly efficient photocatalytic performance. *Catalysts*, 11, 9, 1089, 2021.
42. Li, Z., Qi, Y., Wang, W., Li, D., Li, Z., Xiao, Y., Han, G., Shen, J., Li, C., Blocking backward reaction on hydrogen evolution cocatalyst in a photosystem II hybrid Z-scheme water splitting system. *Chin. J. Catal.*, 40, 4, 486–494, 2019.
43. Ran, J., Jaroniec, M., Qiao, S., Cocatalysts in semiconductor-based photocatalytic CO_2 reduction: Achievements, challenges, and opportunities. *Adv. Mater.*, 30, 7, 1704649, 2018.
44. Wang, L., Xu, N., Pan, X., He, Y., Wang, X., Su, W., Cobalt lactate complex as a hole cocatalyst for significantly enhanced photocatalytic H_2 production activity over CdS nanorods. *Catal. Sci. Technol.*, 8, 6, 1599–1605, 2018.
45. Qiu, B., Du, M., Ma, Y., Zhu, Q., Xing, M., Zhang, J., Integration of redox cocatalysts for artificial photosynthesis. *Energy Environ. Sci.*, 14, 10, 5260–5288, 2021.
46. Bajpai, H., Patra, K.K., Ranjan, R., Nalajala, N., Reddy, K.P., Gopinath, C.S., Can half-a-monolayer of Pt simulate activity like that of bulk Pt? Solar hydrogen activity demonstration with quasi-artificial leaf device. *ACS Appl. Mater. Interfaces*, 12, 27, 30420–30430, 2020.

47. Wu, Q., Ye, J., Qiao, W., Li, Y., Niemantsverdriet, J.W., Richards, E., Pan, F., Su, R., Inhibit the formation of toxic methylphenolic by-products in photodecomposition of formaldehyde–toluene/xylene mixtures by Pd cocatalyst on TiO_2. *Appl. Catal. B Environ.*, 291, 120118, 2021.
48. Tong, T., Zhu, B., Jiang, C., Cheng, B., Yu, J., Mechanistic insight into the enhanced photocatalytic activity of single-atom Pt, Pd or Au-embedded g-C_3N_4. *Appl. Surf. Sci.*, 433, 1175–1183, 2018.
49. Liu, Q., Wang, Z., Chen, H., Wang, H., Song, H., Ye, J., Weng, Y., Rules for selecting metal cocatalyst based on charge transfer and separation efficiency between ZnO nanoparticles and noble metal cocatalyst Ag/Au/Pt. *ChemCatChem*, 12, 15, 3838–3842, 2020.
50. Chen, S., Yang, S., Sun, X., He, K., Ng, Y.H., Cai, X., Zhou, W., Fang, Y., Zhang, S., Carbon-coated Cu nanoparticles as a cocatalyst of g-C_3N_4 for enhanced photocatalytic H_2 evolution activity under visible-light irradiation. *Energy Technol.*, 7, 8, 1800846, 2019.
51. Xiang, X., Zhu, B., Cheng, B., Yu, J., Lv, H., Enhanced photocatalytic H_2 production activity of CdS quantum dots using Sn^{2+} as cocatalyst under visible light irradiation. *Small*, 16, 26, 2001024, 2020.
52. Yang, Y., Chen, H., Zou, X., Shi, X., Liu, W., Feng, L., Suo, G., Hou, X., Ye, X., Zhang, L., Sun, C., Li, H., Wang, C., Chen, Z., Flexible carbon-fiber/semimetal Bi nanosheet arrays as separable and recyclable plasmonic photocatalysts and photoelectrocatalysts. *ACS Appl. Mater Interfaces*, 12, 22, 24845–24854, 2020.
53. Xu, F., Meng, K., Zhu, B., Liu, H., Xu, J., Yu, J., Graphdiyne: A new photocatalytic CO_2 reduction cocatalyst. *Adv. Funct. Mater.*, 29, 43, 1904256, 2019.
54. Peng, C., Wei, P., Li, X., Liu, Y., Cao, Y., Wang, H., Yu, H., Peng, F., Zhang, L., Zhang, B., Lv, K., High efficiency photocatalytic hydrogen production over ternary Cu/TiO_2@$Ti_3C_2T_x$ enabled by low-work-function 2D titanium carbide. *Nano Energy*, 53, 97–107, 2018.
55. Zhao, N., Kong, L., Dong, Y., Wang, G., Wu, X., Jiang, P., Insight into the crucial factors for photochemical deposition of cobalt cocatalysts on g-C_3N_4 photocatalysts. *ACS Appl. Mater Interfaces*, 10, 11, 9522–9531, 2018.
56. Zhu, J., Cheng, G., Xiong, J., Li, W., Dou, S., Recent advances in Cu-based cocatalysts toward solar-to-hydrogen evolution: Categories and roles. *Sol. RRL*, 3, 10, 1900256, 2019.
57. Barik, B., Maji, B., Bag, J., Mishra, M., Singh, J., Dash, P., Design of a non-cytotoxic $ZnFe_2O_4$-CeO_2/BRGO direct Z-scheme photocatalyst with bioreduced graphene oxide as cocatalyst. *ChemistrySelect*, 6, 1, 101–112, 2021.
58. Bie, C., Zhu, B., Xu, F., Zhang, L., Yu, J., *In situ* grown monolayer N-doped graphene on CdS hollow spheres with seamless contact for photocatalytic CO_2 reduction. *Adv. Mater.*, 31, 42, 1902868, 2019.
59. Vu, M., Sakar, M., Nguyen, C., Do, T., Chemically bonded Ni cocatalyst onto the S doped g-C_3N_4 nanosheets and their synergistic enhancement in H_2

production under sunlight irradiation. *ACS Sustain. Chem. Eng.*, 6, 3, 4194–4203, 2018.
60. Zhang, K., Ran, J., Zhu, B., Ju, H., Yu, J., Song, L., Qiao, S., Nanoconfined nickel@carbon core-shell cocatalyst promoting highly efficient visible-light photocatalytic H_2 production. *Small*, 14, 38, 1801705, 2018.
61. Liu, X. and Zhuang, H., Recent progresses in photocatalytic hydrogen production: Design and construction of Ni-based cocatalysts. *Int. J. Energy Res.*, 45, 2, 1480–1495, 2020.
62. Di, T., Zhang, L., Cheng, B., Yu, J., Fan, J., CdS nanosheets decorated with Ni@graphene core-shell cocatalyst for superior photocatalytic H2 production. *J. Mater. Sci. Technol.*, 56, 170–178, 2020.
63. Shi, W., Guo, F., Zhu, C., Wang, H., Li, H., Huang, H., Liu, Y., Kang, Z., Carbon dots anchored on octahedral CoO as a stable visible-light-responsive composite photocatalyst for overall water splitting. *J. Mater. Chem. A*, 5, 37, 19800–19807, 2017.
64. Liu, Y., Xu, X., Lv, S., Li, H., Si, Z., Wu, X., Ran, R., Weng, D., Nitrogen doped graphene quantum dots as a cocatalyst of $SrTiO_3(Al)/CoO_x$ for photocatalytic overall water splitting. *Catal. Sci. Technol.*, 11, 9, 3039–3046, 2021.
65. Albukhari, S.M. and Shawky, A., Ag/Ag_2O-decorated sol-gel-processed TeO_2 nanojunctions for enhanced H_2 production under visible light. *J. Mol. Liq.*, 336, 116870, 2021.
66. Ramadan, W., Feldhoff, A., Bahnemann, D., Assessing the photocatalytic oxygen evolution reaction of $BiFeO_3$ loaded with IrO_2 nanoparticles as cocatalyst. *Sol. Energy Mater. Sol. Cells*, 232, 11349–11349, 2021.
67. Li, D., Dong, Y., Wang, G., Jiang, P., Zhang, F., Zhang, H., Li, J., Lyu, J., Wang, Y., Liu, Q., Controllable photochemical synthesis of amorphous $Ni(OH)_2$ as hydrogen production cocatalyst using inorganic phosphorous acid as sacrificial agent. *Chin. J. Catal.*, 41, 5, 889–897, 2020.
68. Reddy, N.L., Rao, V.N., Kumari, M.M., Ravi, P., Sathish, M., Shankar, M.V., Effective shuttling of photoexcitons on CdS/NiO core/shell photocatalysts for enhanced photocatalytic hydrogen production. *Mater. Res. Bull.*, 101, 223–231, 2018.
69. Qi, Y., Zhao, Y., Gao, Y., Li, D., Li, Z., Zhang, F., Li, C., Redox-based visible-light-driven Z-scheme overall water splitting with apparent quantum efficiency exceeding 10%. *Joule*, 2, 11, 2393–2402, 2018.
70. Ketwong, P., Yoshihara, S., Takeuchi, S., Takashima, M., Ohtani, B., Light intensity-dependence studies on the role of surface deposits for titania-photocatalyzed oxygen evolution: Are they really cocatalysts? *J. Chem. Phys.*, 153, 12, 124709, 2020.
71. Zhang, J., Bai, T., Huang, H., Yu, M., Fan, X., Chang, Z., Bu, X., Metal-organic-framework-based photocatalysts optimized by spatially separated cocatalysts for overall water splitting. *Adv. Mater.*, 32, 49, 2004747, 2020.

72. Zhang, J., Grzelczak, M., Hou, Y., Maeda, K., Domen, K., Fu, X., Antonietti, M., Wang, X., Photocatalytic oxidation of water by polymeric carbon nitride nanohybrids made of sustainable elements. *Chem. Sci.*, 3, 2, 443–446, 2012.
73. Yoshida, M., Yomogida, T., Mineo, T., Nitta, K., Kato, K., Masuda, T., Nitani, H., Abe, H., Takakusagi, S., Uruga, T., Asakura, K., Uosaki, K., Kondoh, H., Photoexcited hole transfer to a MnO_x cocatalyst on a $SrTiO_3$ photoelectrode during oxygen evolution studied by *in situ* X-ray absorption spectroscopy. *J. Phys. Chem. C*, 118, 42, 24302–24309, 2014.
74. Zhang, H., Zhang, P., Qiu, M., Dong, J., Zhang, Y., Lou, X.W.D., Ultrasmall MoO_x clusters as a novel cocatalyst for photocatalytic hydrogen evolution. *Adv. Mater.*, 31, 6, 1804883, 2019.
75. Song, L., Liu, D., Zhang, S., Wei, J., WO_3 cocatalyst improves hydrogen evolution capacity of ZnCdS under visible light irradiation. *Int. J. Hydrogen Energy*, 44, 31, 16327–16335, 2019.
76. Yang, C., Li, Q., Xia, Y., Lv, K., Li, M., Enhanced visible-light photocatalytic CO_2 reduction performance of $ZnIn_2S_4$ microspheres by using CeO_2 as cocatalyst. *Appl. Surf. Sci.*, 464, 388–395, 2019.
77. Chang, C. and Tsai, W., CuS ZnS decorated Fe_3O_4 nanoparticles as magnetically separable composite photocatalysts with excellent hydrogen production activity. *Int. J. Hydrogen Energy*, 44, 37, 20872–20880, 2019.
78. Zhang, L., Hao, X., Li, Y., Jin, Z., Performance of WO_3/g-C_3N_4 heterojunction composite boosting with NiS for photocatalytic hydrogen evolution. *Appl. Surf. Sci.*, 499, 143862, 2020.
79. Nguyen, T.N., Kampouri, S., Valizadeh, B., Luo, W., Ongari, D., Planes, O.M., Zuttel, A., Smit, B., Stylianou, K.C., Photocatalytic hydrogen generation from a visible-light-responsive metal-organic framework system: Stability versus activity of molybdenum sulfide cocatalysts. *ACS Appl. Mater. Interfaces*, 10, 36, 30035–30039, 2018.
80. Gao, D., Xu, J., Wang, L., Zhu, B., Yu, H., Yu, J., Optimizing atomic hydrogen desorption of sulfur-rich NiS_{1+x} cocatalyst for boosting photocatalytic H_2 evolution. *Adv. Mater.*, 34, 6, e2108475, 2022.
81. Xiong, Z., Hou, Y., Yuan, R., Ding, Z., Ong, W., Wang, S., Hollow $NiCo_2S_4$ nanospheres as a cocatalyst to support $ZnIn_2S_4$ nanosheets for visible-light-driven hydrogen production. *Acta Phys. Chim. Sin.*, 38, 2111021, 2021.
82. Jena, A., Chen, C., Chang, H., Hu, S., Liu, R., Comprehensive view on recent developments in hydrogen evolution using MoS_2 on a Si photocathode: From electronic to electrochemical aspects. *J. Mater. Chem. A*, 9, 7, 3767–3785, 2021.
83. Nagajyothi, P.C., Devarayapalli, K.C., Shim, J., Prabhakar Vattikuti, S.V., Highly efficient white-LED-light-driven photocatalytic hydrogen production using highly crystalline $ZnFe_2O_4$/MoS_2 nanocomposites. *Int. J. Hydrogen Energy*, 45, 57, 32756–32769, 2020.

84. Li, S., Xiong, J., Zhu, X., Li, W., Chen, R., Cheng, G., Recent advances in synthesis strategies and solar-to-hydrogen evolution of 1T phase MS_2 (M = W, Mo) co-catalysts. *J. Mater. Sci. Technol.*, 101, 242–263, 2022.
85. Sun, B., Zhou, W., Li, H., Ren, L., Qiao, P., Li, W., Fu, H., Synthesis of particulate hierarchical tandem heterojunctions toward optimized photocatalytic hydrogen production. *Adv. Mater.*, 30, 43, e1804282, 2018.
86. Liu, Q., Wang, S., Ren, Q., Li, T., Tu, G., Zhong, S., Zhao, Y., Bai, S., Stacking design in photocatalysis: Synergizing cocatalyst roles and anti-corrosion functions of metallic MoS_2 and graphene for remarkable hydrogen evolution over CdS. *J. Mater. Chem. A*, 9, 3, 1552–1562, 2021.
87. Guo, X., Li, Q., Liu, Y., Jin, T., Chen, Y., Guo, L., Lian, T., Enhanced light-driven charge separation and H_2 generation efficiency in WSe_2 nanosheet-semiconductor nanocrystal heterostructures. *ACS Appl. Mater. Interfaces*, 12, 40, 44769–44776, 2020.
88. Guo, X., Guo, P., Wang, C., Chen, Y., Guo, L., Few-layer WSe_2 nanosheets as an efficient cocatalyst for improved photocatalytic hydrogen evolution over $Zn_{0.1}Cd_{0.9}S$ nanorods. *Chem. Eng. J.*, 383, 123183–12391, 2020.
89. Ye, M., Wang, X., Liu, E., Ye, J., Wang, D., Boosting the photocatalytic activity of P25 for carbon dioxide reduction by using a surface-alkalinized titanium carbide MXene as cocatalyst. *ChemSusChem*, 11, 10, 1606–1611, 2018.
90. Wang, T., Yang, L., Jiang, D., Cao, H., Minja, A.C., Du, P., CdS nanorods anchored with crystalline FeP nanoparticles for efficient photocatalytic formic acid dehydrogenation. *ACS Appl. Mater. Interfaces*, 13, 20, 23751–23759, 2021.
91. Irfan, R.M., Tahir, M.H., Iqbal, S., Nadeem, M., Bashir, T., Maqsood, M., Zhao, J., Gao, L., Co_3C as a promising cocatalyst for superior photocatalytic H_2 production based on swift electron transfer processes. *J. Mater. Chem. C*, 9, 9, 3145–3154, 2021.
92. Hu, T., Dai, K., Zhang, J., Chen, S., Noble-metal-free Ni_2P modified step-scheme $SnNb_2O_6$/CdS-diethylenetriamine for photocatalytic hydrogen production under broadband light irradiation. *Appl. Catal. B Environ.*, 269, 118844, 2020.
93. Wu, M., Zhang, J., Liu, C., Gong, Y., Wang, R., He, B., Wang, H., Rational design and fabrication of noble-metal-free Ni_xP cocatalyst embedded 3D $N-TiO_2/g-C_3N_4$ heterojunctions with enhanced photocatalytic hydrogen evolution. *ChemCatChem*, 10, 14, 3069–3077, 2018.
94. Liang, Z. and Dong, X., Co_2P nanosheet cocatalyst-modified $Cd_{0.5}Zn_{0.5}S$ nanoparticles as 2D-0D heterojunction photocatalysts toward high photocatalytic activity. *J. Photochem. Photobiol. A Chem.*, 407, 113081, 2021.
95. Shen, R., Ding, Y., Li, S., Zhang, P., Xiang, Q., Ng, Y.H., Li, X., Constructing low-cost Ni_3C/twin-crystal $Zn_{0.5}Cd_{0.5}S$ heterojunction/homojunction nanohybrids for efficient photocatalytic H_2 evolution. *Chin. J. Catal.*, 42, 1, 25–36, 2021.

96. Yue, X., Yi, S., Wang, R., Zhang, Z., Qiu, S., Well-controlled $SrTiO_3$@Mo_2C core-shell nanofiber photocatalyst: Boosted photo-generated charge carriers transportation and enhanced catalytic performance for water reduction. *Nano Energy*, 47, 463–473, 2018.
97. Huang, Y. and Zhang, B., Active cocatalysts for photocatalytic hydrogen evolution derived from nickel or cobalt amine complexes. *Angew. Chem. Int. Ed.*, 56, 47, 14804–14806, 2017.
98. Dong, H., Meng, X., Zhang, X., Tang, H., Liu, J., Wang, J., Wei, J., Zhang, F., Bai, L., Sun, X., Boosting visible-light hydrogen evolution of covalent-organic frameworks by introducing Ni-based noble metal-free co-catalyst. *Chem. Eng. J.*, 379, 122342, 2020.
99. Zhao, G., Hu, B., Busser, G.W., Peng, B., Muhler, M., Photocatalytic oxidation of alpha-C-H bonds in unsaturated hydrocarbons through a radical pathway induced by a molecular cocatalyst. *ChemSusChem*, 12, 12, 2795–2801, 2019.
100. Yang, J., Wang, D., Han, H., Li, C., Roles of cocatalysts in photocatalysis and photoelectrocatalysis. *Acc. Chem. Res.*, 46, 8, 1900–1909, 2013.
101. Bi, W., Li, X., Zhang, L., Jin, T., Zhang, L., Zhang, Q., Luo, Y., Wu, C., Xie, Y., Molecular co-catalyst accelerating hole transfer for enhanced photocatalytic H_2 evolution. *Nat. Commun.*, 6, 8647, 2015.
102. Irfan, R.M., Jiang, D., Sun, Z., Zhang, L., Cui, S., Du, P., Incorporating a molecular co-catalyst with a heterogeneous semiconductor heterojunction photocatalyst: Novel mechanism with two electron-transfer pathways for enhanced solar hydrogen production. *J. Catal.*, 353, 274–285, 2017.
103. Wang, D., Zhang, Y., Chen, W., A novel nickel-thiourea-triethylamine complex adsorbed on graphitic C_3N_4 for low-cost solar hydrogen production. *Chem. Commun.*, 50, 14, 1754–1756, 2014.
104. Zhang, M., Hu, Q., Ma, K., Ding, Y., Li, C., Pyroelectric effect in CdS nanorods decorated with a molecular Co-catalyst for hydrogen evolution. *Nano Energy*, 73, 104810, 2020.
105. Wang, Y., Shi, L., Qian, B., Huang, X., Sun, X., Wang, B., Song, X., Huang, H., Zhang, Y., Ma, T., Regeneratable trinuclear iron-oxo cocatalyst for photocatalytic hydrogen evolution. *Chem. Eng. J.*, 413, 127551, 2021.
106. Wen, F., Wang, X., Huang, L., Ma, G., Yang, J., Li, C., A hybrid photocatalytic system comprising ZnS as light harvester and an [Fe_2S_2] hydrogenase mimic as hydrogen evolution catalyst. *ChemSusChem*, 5, 5, 849–853, 2012.
107. Wang, K., Wang, M., Yu, J., Liao, D., Shi, H., Wang, X., Yu, H., $BiVO_4$ microparticles decorated with Cu@Au core-shell nanostructures for photocatalytic H_2O_2 production. *ACS Appl. Nano Mater.*, 4, 12, 13158–13166, 2021.
108. Tsuji, E., Nanbu, R., Degami, Y., Hirao, K., Watanabe, T., Matsumoto, N., Suganuma, S., Katada, N., Brownmillerite-type crystalline Ca_2FeCoO_5 ultrasmall particles with single-nanometer dimensions as an active cocatalyst for oxygen photoevolution reaction. *Part. Part. Syst. Charact.*, 37, 5, 2000053, 2020.
109. Feng, J., Liu, J., Cheng, X., Liu, J., Xu, M., Zhang, J., Hydrothermal cation exchange enabled gradual evolution of Au@ZnS-AgAuS yolk-shell

nanocrystals and their visible light photocatalytic applications. *Adv. Sci.*, 5, 1, 1700376, 2018.
110. Sun, J., Duan, L., Wu, Q., Yao, W., Synthesis of MoS_2 quantum dots cocatalysts and their efficient photocatalytic performance for hydrogen evolution. *Chem. Eng. J.*, 332, 449–455, 2018.
111. Cheng, L., Zhang, P., Wen, Q., Fan, J., Xiang, Q., Copper and platinum dual-single-atoms supported on crystalline graphitic carbon nitride for enhanced photocatalytic CO_2 reduction. *Chin. J. Catal.*, 43, 2, 451–460, 2022.
112. Li, Y., Li, B., Zhang, D., Cheng, L., Xiang, Q., Crystalline carbon nitride supported copper single atoms for photocatalytic CO_2 reduction with nearly 100% CO selectivity. *ACS Nano*, 14, 8, 10552–10561, 2020.
113. Zhao, Q., Sun, J., Li, S., Huang, C., Yao, W., Chen, W., Zeng, T., Wu, Q., Xu, Q., Single nickel atoms anchored on nitrogen-doped graphene as a highly active cocatalyst for photocatalytic H_2 evolution. *ACS Catal.*, 8, 12, 11863–11874, 2018.
114. Wang, M., Xu, S., Zhou, Z., Dong, C., Guo, X., Chen, J., Huang, Y., Shen, S., Chen, Y., Guo, L., Burda, C., Atomically dispersed janus nickel sites on red phosphorus for photocatalytic overall water splitting. *Angew. Chem. Int. Ed.*, 61, 29, e202204711, 2022.
115. Yao, J., Zheng, Y., Jia, X., Duan, L., Wu, Q., Huang, C., An, W., Xu, Q., Yao, W., Highly active $Pt_3Sn\{110\}$-excavated nanocube cocatalysts for photocatalytic hydrogen production. *ACS Appl. Mater. Interfaces*, 11, 29, 25844–25853, 2019.
116. Ma, B., Zhang, J., Lin, K., Li, D., Liu, Y., Yang, X., Improved photocatalytic hydrogen evolution of CdS using earth-abundant cocatalyst Mo_2N with rod shape and large capacitance. *ACS Sustain. Chem. Eng.*, 7, 15, 13569–13575, 2019.
117. Shao, Z., Meng, X., Lai, H., Zhang, D., Pu, X., Su, C., Li, H., Ren, X., Geng, Y., Coralline-like Ni_2P decorated novel tetrapod-bundle $Cd_{0.9}Zn_{0.1}S$ ZB/WZ homojunctions for highly efficient visible-light photocatalytic hydrogen evolution. *Chin. J. Catal.*, 42, 3, 439–449, 2021.
118. Su, L., Luo, L., Wang, J., Song, T., Tu, W., Wang, Z., Lamellar flower-like porous MoS_2 as an efficient cocatalyst to boost photocatalytic hydrogen evolution of CdS. *Catal. Sci. Technol.*, 11, 4, 1292–1297, 2021.
119. Ye, F., Wang, F., Meng, C., Bai, L., Li, J., Xing, P., Teng, B., Zhao, L., Bai, S., Crystalline phase engineering on cocatalysts: A promising approach to enhancement on photocatalytic conversion of carbon dioxide to fuels. *Appl. Catal. B Environ.*, 230, 145–153, 2018.
120. Jia, X., Zhao, J., Lv, Y., Fu, X., Jian, Y., Zhang, W., Wang, Y., Sun, H., Wang, X., Long, J., Yang, P., Gu, Q., Gao, Z., Low-crystalline PdCu alloy on large-area ultrathin 2D carbon nitride nanosheets for efficient photocatalytic Suzuki coupling. *Appl. Catal. B Environ.*, 300, 120756, 2022.

121. Niu, P., Pan, Z., Wang, S., Wang, X., Tuning crystallinity and surface hydrophobicity of a cobalt phosphide cocatalyst to boost CO_2 photoreduction performance. *ChemSusChem*, 14, 5, 1302–1307, 2021.
122. Peng, Z., Jianping, L., Yonghua, T., Yuguang, C., Fei, L., Shaojun, G., Amorphous FeCoPO$_x$ nanowires coupled to g-C$_3$N$_4$ nanosheets with enhanced interfacial electronic transfer for boosting photocatalytic hydrogen production. *Appl. Catal. B Environ.*, 238, 161–167, 2018.
123. Liu, W., Wang, X., Yu, H., Yu, J., Direct photoinduced synthesis of amorphous CoMoS$_x$ cocatalyst and its improved photocatalytic H$_2$-evolution activity of CdS. *ACS Sustain. Chem. Eng.*, 6, 9, 12436–12445, 2018.
124. Yu, H., Yuan, R., Gao, D., Xu, Y., Yu, J., Ethyl acetate-induced formation of amorphous MoS$_x$ nanoclusters for improved H$_2$-evolution activity of TiO$_2$ photocatalyst. *Chem. Eng. J.*, 375, 121934, 2019.
125. Du, F., Lu, H., Lu, S., Wang, J., Xiao, Y., Xue, W., Cao, S., Photodeposition of amorphous MoS$_x$ cocatalyst on TiO$_2$ nanosheets with {001} facets exposed for highly efficient photocatalytic hydrogen evolution. *Int. J. Hydrogen Energy*, 43, 6, 3223–3234, 2018.
126. Xu, J., Zhong, W., Yu, H., Hong, X., Fan, J., Yu, J., Triethanolamine-assisted photodeposition of non-crystalline Cu$_x$P nanodots for boosting photocatalytic H$_2$ evolution of TiO$_2$. *J. Mater. Chem. C*, 8, 44, 15816–15822, 2020.
127. Liu, Y., Wang, B., Zhang, Q., Yang, S., Li, Y., Zuo, J., Wang, H., Peng, F., A novel bicomponent Co$_3$S$_4$/Co@C cocatalyst on CdS, accelerating charge separation for highly efficient photocatalytic hydrogen evolution. *Green Chem.*, 22, 1, 238–247, 2020.
128. Wang, Z., Li, L., Liu, M., Miao, T., Ye, X., Meng, S., Chen, S., Fu, X., A new phosphidation route for the synthesis of NiP and their cocatalytic performances for photocatalytic hydrogen evolution over g-C$_3$N$_4$. *J. Energy Chem.*, 48, 241–249, 2020.
129. Bi, L., Gao, X., Zhang, L., Wang, D., Zou, X., Xie, T., Enhanced photocatalytic hydrogen evolution of NiCoP/g-C$_3$N$_4$ with improved separation efficiency and charge transfer efficiency. *ChemSusChem*, 11, 1, 276–284, 2018.
130. He, J., Zhong, W., Xu, Y., Yu, H., Fan, J., Yu, J., Few-layered Mo$_x$W$_{1-x}$S$_2$-modified CdS photocatalyst: One-step synthesis with bifunctional precursors and improved H$_2$-evolution activity. *Sol. RRL*, 5, 10, 2100387, 2021.
131. Zhao, Y., Lu, Y., Chen, L., Wei, X., Zhu, J., Zheng, Y., Redox dual-cocatalyst-modified CdS double-heterojunction photocatalysts for efficient hydrogen production. *ACS Appl. Mater. Interfaces*, 12, 41, 46073–46083, 2020.
132. Song, L. and Zhang, S., A novel cocatalyst of NiCoP significantly enhances visible-light photocatalytic hydrogen evolution over cadmium sulfide. *J. Ind. Eng. Chem.*, 61, 197–205, 2018.
133. Chao, Y., Zhou, P., Lai, J., Zhang, W., Yang, H., Lu, S., Chen, H., Yin, K., Li, M., Tao, L., Shang, C., Tong, M., Guo, S., Ni$_{1-x}$Co$_x$Se$_2$C/ZnIn$_2$S$_4$ hybrid nanocages with strong 2D/2D hetero-interface interaction enable efficient H$_2$-releasing photocatalysis. *Adv. Funct. Mater.*, 31, 24, 2100923, 2021.

134. Saleh, M.R. and El-Bery, H.M., Unraveling novel Cu/Cu$_x$P@N-doped C composite as effective cocatalyst for photocatalytic hydrogen production under UV and visible irradiation. *Appl. Surf. Sci.*, 580, 152280, 2022.
135. Cheng, L., Li, Y., Chen, A., Zhu, Y., Li, C., Subnano-sized Pt-Au alloyed clusters as enhanced cocatalyst for photocatalytic hydrogen evolution. *Chem. Asian J.*, 14, 12, 2112–2115, 2019.
136. Zhong, S., Xi, Y., Wu, S., Liu, Q., Zhao, L., Bai, S., Hybrid cocatalysts in semiconductor-based photocatalysis and photoelectrocatalysis. *J. Mater. Chem. A*, 8, 30, 14863–14894, 2020.
137. Shen, R., Jiang, C., Xiang, Q., Xie, J., Li, X., Surface and interface engineering of hierarchical photocatalysts. *Appl. Surf. Sci.*, 471, 43–87, 2019.
138. Qin, Z., Chen, Y., Wang, X., Guo, X., Guo, L., Intergrowth of cocatalysts with host photocatalysts for improved solar-to-hydrogen conversion. *ACS Appl. Mater. Interfaces*, 8, 2, 1264–1272, 2016.
139. Liu, Y., Yang, S., Zhang, S., Wang, H., Yu, H., Cao, Y., Peng, F., Design of cocatalyst loading position for photocatalytic water splitting into hydrogen in electrolyte solutions. *Int. J. Hydrogen Energy*, 43, 11, 5551–5560, 2018.
140. Zuo, G., Liu, S., Wang, L., Song, H., Zong, P., Hou, W., Li, B., Guo, Z., Meng, X., Du, Y., Wang, T., Roy, V.A.L., Finely dispersed Au nanoparticles on graphitic carbon nitride as highly active photocatalyst for hydrogen peroxide production. *Catal. Commun.*, 123, 69–72, 2019.
141. Ding, J., Li, X., Chen, L., Zhang, X., Yin, H., Tian, X., Site-selective deposition of reductive and oxidative dual cocatalysts to improve the photocatalytic hydrogen production activity of CaIn$_2$S$_4$ with a surface nanostep structure. *ACS Appl. Mater. Interfaces*, 11, 1, 835–845, 2019.
142. Qi, M., Li, Y., Zhang, F., Tang, Z., Xiong, Y., Xu, Y., Switching light for site-directed spatial loading of cocatalysts onto heterojunction photocatalysts with boosted redox catalysis. *ACS Catal.*, 10, 5, 3194–3202, 2020.
143. Chen, S., Shen, S., Liu, G., Qi, Y., Zhang, F., Li, C., Interface engineering of a CoO$_x$/Ta$_3$N$_5$ photocatalyst for unprecedented water oxidation performance under visible-light-irradiation. *Angew. Chem. Int. Ed.*, 54, 10, 3047–3051, 2015.
144. Qiu, B., Cai, L., Zhang, N., Tao, X., Chai, Y., A ternary dumbbell structure with spatially separated catalytic sites for photocatalytic overall water splitting. *Adv. Sci.*, 7, 17, 1903568, 2020.
145. Wei, S., Chang, S., Qian, J., Xu, X., Selective cocatalyst deposition on ZnTiO$_{3-x}$N$_y$ hollow nanospheres with efficient charge separation for solar-driven overall water splitting. *Small*, 17, 11, 2100084, 2021.
146. Chen, Y., Zhao, S., Wang, X., Peng, Q., Lin, R., Wang, Y., Shen, R., Cao, X., Zhang, L., Zhou, G., Li, J., Xia, A., Li, Y., Synergetic integration of Cu$_{1.94}$S-Zn$_x$Cd$_{1-x}$S heteronanorods for enhanced visible-light-driven photocatalytic hydrogen production. *J. Am. Chem. Soc.*, 138, 13, 4286–4289, 2016.
147. Zhu, Q., Xu, Z., Qiu, B., Xing, M., Zhang, J., Emerging cocatalysts on g-C$_3$N$_4$ for photocatalytic hydrogen evolution. *Small*, 17, 40, e2101070, 2021.

148. Wang, X., Tian, X., Sun, Y., Zhu, J., Li, F., Mu, H., Zhao, J., Enhanced schottky effect of a 2D-2D CoP/g-C_3N_4 interface for boosting photocatalytic H_2 evolution. *Nanoscale*, 10, 26, 12315–12321, 2018.
149. Yi, S., Wulan, B., Yan, J., Jiang, Q., Highly efficient photoelectrochemical water splitting: Surface modification of cobalt-phosphate-loaded Co_3O_4/Fe_2O_3 p-n heterojunction nanorod arrays. *Adv. Funct. Mater.*, 29, 11, 1801902, 2019.
150. Chen, Y., Feng, X., Guo, X., Zheng, W., Toward a fundamental understanding of factors affecting the function of cocatalysts in photocatalytic water splitting. *Curr. Opin. Green Sustain. Chem.*, 17, 21–28, 2019.
151. Xu, H., Yi, J., She, X., Liu, Q., Song, L., Chen, S., Yang, Y., Song, Y., Vajtai, R., Lou, J., Li, H., Yuan, S., Wu, J., Ajayan, P.M., 2D heterostructure comprised of metallic 1T-MoS_2/Monolayer O-g-C_3N_4 towards efficient photocatalytic hydrogen evolution. *Appl. Catal. B Environ.*, 220, 379–385, 2018.
152. Qin, Z., Wang, M., Li, R., Chen, Y., Novel Cu_3P/g-C_3N_4 p-n heterojunction photocatalysts for solar hydrogen generation. *Sci. China Mater.*, 61, 6, 861–868, 2018.
153. Tian, H., Liu, M., Zheng, W., Constructing 2D graphitic carbon nitride nanosheets/layered MoS_2/graphene ternary nanojunction with enhanced photocatalytic activity. *Appl. Catal. B Environ.*, 225, 468–476, 2018.
154. Bie, C., Cheng, B., Fan, J., Ho, W., Yu, J., Enhanced solar-to-chemical energy conversion of graphitic carbon nitride by two-dimensional cocatalysts. *EnergyChem*, 3, 2, 100051, 2021.
155. Wu, J., Sun, B., Wang, H., Li, Y., Zuo, Y., Wang, W., Lin, H., Li, S., Wang, L., Efficient spatial charge separation in unique 2D tandem heterojunction $Cd_xZn_{1-x}In_2S_4$–CdS–MoS_2 rendering highly-promoted visible-light-induced H_2 generation. *J. Mater. Chem. A*, 9, 1, 482–491, 2021.
156. Cheng, C., Mao, L., Shi, J., Xue, F., Zong, S., Zheng, B., Guo, L., $NiCo_2O_4$ nanosheets as a novel oxygen-evolution-reaction cocatalyst *in situ* bonded on the g-C_3N_4 photocatalyst for excellent overall water splitting. *J. Mater. Chem. A*, 9, 20, 12299–12306, 2021.
157. Okamoto, H., Kodera, M., Hisatomi, T., Katayama, M., Minegishi, T., Domen, K., Effects of annealing conditions on the oxygen evolution activity of a $BaTaO_2N$ photocatalyst loaded with cobalt species. *Catal. Today*, 354, 204–210, 2020.
158. Liu, J., Li, Y., Zhou, X., Jiang, H., Yang, H., Li, C., Positively charged Pt-based cocatalysts: an orientation for achieving efficient photocatalytic water splitting. *J. Mater. Chem. A*, 8, 1, 17–26, 2020.
159. Cheng, C., Shi, J., Du, F., Zong, S., Guan, X., Zhang, Y., Liu, M., Guo, L., Simply blending Ni nanoparticles with typical photocatalysts for efficient photocatalytic H_2 production. *Catal. Sci. Technol.*, 9, 24, 7016–7022, 2019.
160. Yan, J., Liu, J., Ji, Y., Batmunkh, M., Li, D., Liu, X., Cao, X., Li, Y., Liu, S., Ma, T., Surface engineering to reduce the interfacial resistance for enhanced photocatalytic water oxidation. *ACS Catal.*, 10, 15, 8742–8750, 2020.

161. Chang, K., Mei, Z., Wang, T., Kang, Q., Ouyang, S., Ye, J., MoS_2/graphene cocatalyst for efficient photocatalytic H_2 evolution under visible light irradiation. *ACS Nano*, 8, 7, 7078–7087, 2014.
162. Li, J., Zhang, L., Li, J., An, P., Hou, Y., Zhang, J., Nanoconfined growth of carbon-encapsulated cobalts as cocatalysts for photocatalytic hydrogen evolution. *ACS Sustain. Chem. Eng.*, 7, 16, 14023–14030, 2019.
163. Liu, M., Chen, Y., Su, J., Shi, J., Wang, X., Guo, L., Photocatalytic hydrogen production using twinned nanocrystals and an unanchored NiS_x co-catalyst. *Nat. Energy.*, 1, 16151, 2016.
164. Wu, K. and Lian, T., Quantum confined colloidal nanorod heterostructures for solar-to-fuel conversion. *Chem. Soc. Rev.*, 45, 14, 3781–3810, 2016.
165. Wang, P., Mao, Y., Li, L., Shen, Z., Luo, X., Wu, K., An, P., Wang, H., Su, L., Li, Y., Zhan, S., Unraveling the interfacial charge migration pathway at the atomic level in a highly efficient Z-scheme photocatalyst. *Angew. Chem. Int. Ed.*, 58, 33, 11329–11334, 2019.
166. Wu, K., Chen, Z., Lv, H., Zhu, H., Hill, C.L., Lian, T., Hole removal rate limits photodriven H_2 generation efficiency in CdS-Pt and CdSe/CdS-Pt semiconductor nanorod–metal Tip heterostructures. *J. Am. Chem. Soc.*, 136, 21, 7708–7716, 2014.
167. Liu, G., Kolodziej, C., Jin, R., Qi, S., Lou, Y., Chen, J., Jiang, D., Zhao, Y., Burda, C., MoS_2-stratified CdS-Cu_{2-x}S core-shell nanorods for highly efficient photocatalytic hydrogen production. *ACS Nano*, 14, 5, 5468–5479, 2020.
168. Indra, A., Menezes, P.W., Kailasam, K., Hollmann, D., Schroder, M., Thomas, A., Bruckner, A., Driess, M., Nickel as a co-catalyst for photocatalytic hydrogen evolution on graphitic-carbon nitride (sg-CN): What is the nature of the active species? *Chem. Commun.*, 52, 1, 104–107, 2016.
169. Ida, S., Sato, K., Nagata, T., Hagiwara, H., Watanabe, M., Kim, N., Shiota, Y., Koinuma, M., Takenaka, S., Sakai, T., Ertekin, E., Ishihara, T., A cocatalyst that stabilizes a hydride intermediate during photocatalytic hydrogen evolution over a rhodium-doped TiO_2 nanosheet. *Angew. Chem. Int. Ed.*, 57, 29, 9073–9077, 2018.
170. Benisti, I., Shaik, F., Xing, Z., Ben-Refael, A., Amirav, L., Paz, Y., The effect of Pt cocatalyst on the performance and transient IR spectrum of photocatalytic g-C_3N_4 nanospheres. *Appl. Surf. Sci.*, 542, 148432, 2021.
171. Sivasankaran, R.P., Rockstroh, N., Kreyenschulte, C.R., Bartling, S., Lund, H., Acharjya, A., Junge, H., Thomas, A., Brückner, A., Influence of MoS_2 on activity and stability of carbon nitride in photocatalytic hydrogen production. *Catalysts*, 9, 8, 695, 2019.
172. Gao, C., Zhuang, C., Li, Y., Qi, H., Chen, G., Sun, Z., Zou, J., Han, X., *In situ* liquid cell transmission electron microscopy guiding the design of large-sized cocatalysts coupled with ultra-small photocatalysts for highly efficient energy harvesting. *J. Mater. Chem. A*, 9, 22, 13056–13064, 2021.

Index

AAO, 304–305
ABPE, 325–326
Absorbance, 308, 315, 317, 320, 325
Activation energy, 227, 233
Active sites, 376, 378, 381, 384–386, 388–391
Agglomeration, 378
Aluminum, 69, 77, 88
Amorphous, 306–307, 309, 315
Amorphous cocatalyst, 386–387
Anabaena, 179
Anatase, 299
Annealing, 300–307, 326
Anode, 341, 345–348, 350, 351, 353–361
Auto-combustion technique, 303

Band gap, 343, 348, 356, 358
Basic modeling results, 121
BCM (BaCe$_{0.25}$Mn$_{0.75}$O$_3$), 31, 32, 39
Biomass conversion, 166
Bond order conservation, 28, 29
Born-Haber cycle, 35–37
Bradyrhizobium, 181
Built-in electric field, 386, 390–391
Built-in potential, 291, 308, 314

Calcination, 297, 300–304
Caldicellulosiruptor, 180
Carbon neutral, 341, 342, 361
Catalysts, 342, 352, 353, 362, 365, 371, 372
Cathode, 341, 343–345, 347, 348, 350, 351, 354–361, 363

C-C functionalization, 192
CCTM [(Ca,Ce)(Ti,Mn)O$_3$], 32, 39
Ceria (CeO$_2$), 11, 12, 16, 17, 26–31, 33, 37–42, 44, 46
C-H functionalization, 186, 189
Chalcogenide PV cells, 356
Charge carrier separation and transfer, 376
Chlamydomonas, 179
Chlorella, 181
Chlorococcum, 178
Climate change, 4, 6, 42
Clostridium, 180
Cluster expansion, 17, 23, 24
CO$_2$ reduction, 192, 194
Collector, 5, 6
Collision contact, 393
Compound energy formalism, 22
Concentrated solar energy, 137, 155
Concentrated solar technologies (CSTs), 4–8, 13
Conduction band, 291, 293, 307, 312, 317, 320, 326, 343
Convex hull, 17, 24, 35, 36
Cost comparison, 130
Cost-effective, 168
Counter electrode, 291, 293
Cross coupling, 186, 199
Crystal bond dissociation (E$_b$), 28, 30, 35–37
Crystal reduction potential (V$_r$), 35–37
Cu-Cl cycles, 103, 104, 108–110

411

Density functional theory, 18, 19, 21, 24, 25, 36, 314
Depletion layer, 329
Diatoms, microbial, 167
Diffusion length, 313, 324
Dish, 5–7
Dopant, 10, 27–29, 31
Doping, 311, 313–314, 317–318, 320, 324, 327
Drop-casting, 300
Dye degradation, 311, 314–315, 318, 320–322, 326

Effective interaction, 219
Efficiency, 5–7, 9, 10, 13, 16, 26, 28
  Carnot, 5–7
  receiver, 5–7
  solar-to-fuel, 5–7, 31
Efficient oxygen evolution (OER), 212
Electric dipole, 289
Electrodeposition, 312
Electrolyzer, 341, 342, 344–361, 370, 372
Electrophoretic deposition, 301
Enterobacter, 179
Entropy, 14, 22–25, 33, 37–41
  configurational (ΔS_c), 17, 22–24, 32, 33, 39–41
  electronic (ΔS_e), 22–24, 38–40
  vibrational (ΔS_p), 22, 24, 25, 38, 40
Escherichia, 180

Ferroelectric perovskite-type oxide semiconductors,
  barium titanate, 290
  bismuth ferrite, 290
  bismuth manganite, 290
  lanthanum ferrite, 290
  lead titanate, 290
  lithium niobate, 290
  potassium niobate, 290
  sodium niobate, 290
  strontium titanate, 290
  yttrium ferrite, 290
  zinc stannate, 290
  zirconate titanate, 290
Ferroelectricity, 289, 314, 323
Ferrotronics, 288
Few-layer graphene (FLG), 223, 232
Flat band potential, 312
Fossil fuels, 341, 342, 348
Fresh and marine water algae, 169

Global warming CO_2, 342
Group III-V, 345, 352, 354

H2, 171
Heliostat, 5, 6
History of Cu-Cl cycle, 103
History of thermochemical cycles, 102, 103
Holographic memory, 323
Hot electrons, 311, 316
Hubbard U, 17, 19–21
Hydrocarbon fuels, 220, 223, 228, 235
Hydrogen absorption-free energy, 378
Hydrogen applications, 100
Hydrogen atom transfer, 191
Hydrogen evolution reaction (HER), 208, 308–309, 311, 319, 342–344, 348, 350
Hydrogen production, 341–344, 348, 361–365, 367, 369–371
Hydrogen production pathways, 100
Hydrothermal, 295–300, 328
Hydroxyl radical, 292

In situ electron paramagnetic resonance (EPR) spectroscopy, 395
In situ liquid cell transmission electron microscopy, 395
Indirect contact, 392
Integrated system description, 111–113
Interfacial photoelectrochemistry, 197

INDEX   413

Intimate contact, 382, 385–386, 388, 392
Inversion symmetry, 295

Janus structure, 384–385

Klebsiella, 180

Light absorption, 376–380, 382, 384, 389
LSPR, 308, 311, 315–316, 320, 322

Magnesium, 69, 76, 85
Magnetron sputtering,
  DC, 305
  RF, 305
Matching factor, 345–347, 349–352, 354–357, 359–361, 372
Mesoporous, 304, 310
Metal complexes, 383–384
Metal fuels,
  aerospace rockets, 68
  direct combustion, 69, 71–72
  engines, 73–74
  metal-air flame, 70
Metal oxides and hydroxides, 380–381
Metal phosphides and carbides, 382–383
Metal sulfides, 381–382
Modeling input data, 118, 119
Modulator, 323
Molten salt and thermal energy storage, 111, 115, 116
Monophasic perovskite, 304

Nano heterostructures (NHS), 223
Nanobelts, 294
Nanocrystalline, 302, 304, 319
Nanocrystallites, 317
Nanocubes, 296–298, 315, 320
Nanofibers, 294, 310, 323
Nanofilms, 294
Nanoflakes, 294
Nanoheterostructure, 296, 298–300, 302, 310, 313, 315–316, 321, 326–327

Nanoparticles, 294, 297–300, 311–312
Nanopillars, 317
Nanosheets, 295, 297, 320, 326
Nanotubes, 294, 317, 321
Nanowebs, 310
Nanowires, 294
Nitrogen-doped graphene, 228, 234, 235
Noble metal-free cocatalysts, 379–380
Noncentrosymmetric structure, 325
Non-governmental, 167
Non-photochemical, 171
Non-renewable, 164, 167
Nonstoichiometric oxide, 137–138
N-type semiconductor, 187
Nucleation, 295–296

Onset potential, 307–308, 312–314, 319–320
Open-circuit voltage (VOC), 344, 349, 351, 358, 363
Operating voltage, 347, 350
Optimization, 120, 121, 128–130
Organic heterojunction PV cells, 359
Organic pollutants, 321
Orthorhombic, 296–297, 301, 319
Overpotential, 344, 355, 363, 364, 376
Oxides reduction,
  carbothermal reduction, 77, 83, 86
  catalysts, 78
  reoxidation, 80
  thermodynamics and kinetics, 75–77
Oxygen evolving reaction (OER), 310, 312–313, 342–344, 348, 350, 355
Oxygen exchange, 143–144, 154
Oxygen vacancy (V_O), 11, 17, 21, 22, 26, 28, 32–41, 44–46
  formation energy (E_v), 17, 18, 21, 24, 25, 28, 30–37, 45, 46
  formation enthalpy (ΔH_v), 27, 33, 37
  formation entropy (ΔS_v), 22–25, 32, 33, 37–40
Oxygen-containing functional groups, 209, 219

Parametric study, 123–128
PEC cell,
    heterojunction photoelectrode, 187
    PEC tandem cell, 187
    PV-electrolyzer, 187
    PV-PEC tandem, 187
    single light absorber, 187
Perovskites, 12, 30–38, 40–42
Perovskite PV cells, 358, 359
Photo(electro)catalysis, 288–290, 292, 294, 327–328
Photoanode, 290–291, 293–294, 297, 300, 302–304, 307, 309–310, 313–315, 317–325, 343, 344, 362, 363, 366
Photocanode, 187, 189
Photocatalysts, 342, 343, 365
Photocatalytic (PC), 342, 344, 364, 365
Photocatalytic systems (PC),
    fixed bed photoreactor, 254–256
    microreactor, 259–260
    monolith photoreactor, 255–256
    optical-fiber photoreactor, 255–256
    optofluidic microreactor, 260
    photoreduction of CO_2, 249–251
    slurry photoreactor, 252–254
    twin (membrane) photoreactor, 256–259
Photocathode, 187, 189, 343, 344
Photochemical dimethoxylation of furan, 190
Photoelectrochemical (PEC), 342, 344, 348, 353, 363–366, 372, 373
Photoelectrochemical (PEC) systems,
    cathode–photoanode system, 269–271
    gas diffusion electrodes (GDE), 268–269
    photocathode–anode system, 269–271
    photocathode-photoanode system, 269–271
    photoelectrochemical (PEC) cell, 249, 267–269

Photoelectrochemical cell, 187, 291
Photoreactor, 189
Photoredox catalysis, 190, 194, 196
Photostability, 376–377, 393
Photosynthetic system (PS), 249–251
Photovoltaic cell, 187
Photovoltaic electrochemical systems (PV+EC),
    anion exchange membrane (AEM), 266
    bipolar membrane (BPM), 266
    cation exchange membrane (CEM), 266
    electroreduction of CO_2, 249, 260–263
    H-cell, 263–264
    membrane flow cells, 265–266
    microfluidic cell, 266–267
Photovoltaic-electrolyzer (PVE) systems, 341–346, 348–350, 352–361
Physico-chemical, 165, 177
Piezoelectricity, 289–290, 326
Piezopotential, 288, 293, 316
Piezotronics, 288
Polarization, 288–290, 292–295, 307–308, 311, 314–316, 318, 320, 322–326, 328
Poling,
    negative, 320–322
    positive, 309, 311, 321
Polycrystalline material, 289
P-type semiconductor, 187
Pulsed laser ablation, 310
Pulsed laser deposition, 305, 307
PV cells, 344, 345, 347–361
Pyridine nitrogen sites, 224
Pyroelectricity, 289–290

Quantum,
    confinement, 289, 294–295, 328
    shape, 289, 294–295, 328
    size, 289, 294–295, 328

Quantum efficiency, 253
Quantum mechanics simulations, 17, 25, 41, 42

Receiver, 5–7, 10
Redox exfoliation, 209, 236
Redox potential, 291, 314, 319
Redox-active material, 5, 8, 10–14, 16–19, 24, 26, 30–33, 38–40, 42
  multi-component, 10, 11, 23
  non-volatile, 9, 10, 12, 13
  off-stoichiometric, 10–13, 15, 17, 26, 27, 29, 31, 33, 37–39, 41, 42
  single-component, 10–12
  stoichiometric, 10–13, 18, 33, 38
  volatile, 9, 10, 12, 13
Review of solar Cu-Cl studies, 105–107
Rhodobacter, 179
Rutile, 299

Scanning ion conductance microscope, 395
Scenedesmus, 181
Screen printing, 312
Selectively deposition, 389
Shape control, 385–386
Shomate equations, 114
Short circuit current density, 349, 351
Si PV cells, 353, 254
Single-crystalline, 296–298, 310
Solar cells, 342, 360, 363, 365–369, 371, 373
Solar harnessing systems, 100, 101, 114, 115, 121–123
Solar reactors, 82–84, 87
Solar thermochemical cycles, 6, 8, 12, 13, 38
  multi-step, 12, 13
  two-step, 8–10, 12–14, 17, 42
Solar thermolysis, 6–8, 10, 43
Solar to hydrogen (STH), 342–350, 352, 354–361, 371

Solar-driven water splitting, 341–344, 348, 350, 352, 353, 360, 361, 372
Sol-gel route, 300–302
Solvothermal, 295–297, 299
Space charge region, 316, 317, 329
Special quasi-random structures (SQSs), 17, 25
Spin-coating, 304–305
Spray pyrolysis, 303
Stacked nanostructure, 382
Storage of solar energy, 137–138, 141, 151–155
Structural defects, 209
Sublattice models, 22, 23
Sustainable strategies,
  circular economy, 67
  climate change, 66
  energy transition, 67
  solar recycling, 68, 75, 81, 83, 86
  transportation, 66
Synergistic effect, 227, 231
Synthetic organic electrochemistry, 194
System performance criteria, 116, 117

Tandem cell, 310–311
Template-assisted, 304
Tetraselmis, 181
Thermal reduction, 9–14, 16, 18, 25–29, 31, 33, 38, 39, 44
Thermodynamic modeling, 113, 114
Thermodynamic potential, 344
Time-resolved spectra technology, 393
Total revenue requirement (TRR) method, 115, 116
Tower, 5, 6, 16
Transition metal, 301, 313, 324
Transition metal dichalcogenides (TMDs), 382
Transition-metal-based semiconductors, 387
Transport of solar energy, 137, 155
Tunneling, 309

Ultrasonication, 301

Valence band, 291, 312, 320, 343
Vapor phase deposition, 305

Water splitting, 5, 7–11, 13–16, 26, 32, 33, 42, 101, 102, 137–140, 143–144, 149–151, 156, 288–291, 294, 297, 305, 307, 310–311, 313–315, 317, 319–320, 322–325, 327–328, 341–344, 348, 350, 352–354, 356, 358–362, 364–366, 371–373

Well-crystalized cocatalysts, 386
Wet chemical method, 328
Working electrode, 291

XC functional, 17, 19, 20
X-ray absorption fine structure (XAFS) spectroscopy, 395

Printed and bound by CPI Group (UK) Ltd, Croydon, CR0 4YY
30/05/2023
03222980-0005